PENGUIN CLASSICS

Fellow Citizens

ROBERT V. REMINI is a professor emeritus of history and research professor emeritus of humanities at the University of Illinois at Chicago. He is the winner of the National Book Award for the third volume of his study of Andrew Jackson and is now the historian for the U.S. House of Representatives.

TERRY GOLWAY teaches U.S. history at Kean University in Union, N.J. His previous books include *Washington's General, So Others Might Live, The Irish in America, For the Cause of Liberty,* and *Irish Rebel.* He coauthored with Robert Dallek *Let Every Nation Know: John F. Kennedy in His Own Words.*

Fellow Citizens

THE PENGUIN BOOK OF U.S. PRESIDENTIAL ADDRESSES

Edited with an Introduction
and Commentaries
by

ROBERT V. REMINI *and* TERRY GOLWAY

PENGUIN BOOKS

PENGUIN BOOKS

Published by the Penguin Group

Penguin Group (USA) Inc., 375 Hudson Street,
New York, New York 10014, U.S.A.

Penguin Group (Canada), 90 Eglinton Avenue East, Suite 700, Toronto,
Ontario, Canada M4P 2Y3 (a division of Pearson Penguin Canada Inc.)

Penguin Books Ltd, 80 Strand, London WC2R 0RL, England

Penguin Ireland, 25 St Stephen's Green, Dublin 2,
Ireland (a division of Penguin Books Ltd)

Penguin Group (Australia), 250 Camberwell Road, Camberwell,
Victoria 3124, Australia (a division of Pearson Australia Group Pty Ltd)

Penguin Books India Pvt Ltd, 11 Community Centre,
Panchsheel Park, New Delhi–110 017, India

Penguin Group (NZ), 67 Apollo Drive, Rosedale, North Shore 0632,
New Zealand (a division of Pearson New Zealand Ltd)

Penguin Books (South Africa) (Pty) Ltd, 24 Sturdee Avenue,
Rosebank, Johannesburg 2196, South Africa

Penguin Ltd, Registered Offices:
80 Strand, London WC2R 0RL, England

First published in Penguin Books 2008

1   3   5   7   9   10   8   6   4   2

LIBRARY OF CONGRESS CATALOGING-IN-PUBLICATION DATA

Fellow citizens : the Penguin book of U.S. presidential inaugural
addresses / edited with an introduction and commentaries
by Robert V. Remini and Terry Golway.
p.   cm.
Includes bibliographical references.
ISBN 978-0-14-311453-6
1. Presidents—United States—Inaugural addresses.   2. Speeches,
addresses, etc., American.   3. United States—Politics and government.
4. Presidents—United States—Election—History.   I. Remini, Robert
Vincent, 1921–   II. Golway, Terry, 1955–
J81.4.F45 2008
352.23'860973—dc22
2008019970

Printed in the United States of America
Set in Minion with Excelsior Display
Designed by Elke Sigal

Contents

Introduction

The practice of addressing the American people at the start of a President's term in office began with the first chief executive, George Washington. There was nothing in the Constitution that mandated such an address, but it did say that he was to give to Congress, from time to time, "Information of the State of the Union."

A presidential inaugural address is quite different from a speech on the State of the Union. Washington decided that as part of the celebrations involved in his assumption of office he would take advantage of the opportunity to speak to members of Congress, and through them to the nation, about what the event meant to him. A man of enormous dignity and self-importance, and a gentleman of the first rank, he felt he had a duty to express his appreciation for the honor conferred upon him by his unanimous election as the first President of the United States. He did not believe it proper to simply stand mute and say nothing except to recite the oath of office as prescribed in the Constitution. So he asked a fellow Virginian, James Madison, the man who rightfully became known as the Father of the Constitution, to help him in the preparation of this address. Madison may have accepted the invitation in order to encourage Washington to mention that he thought the Constitution should be immediately amended to provide a bill of rights and thereby answer some of the objections of those who opposed ratification of the Constitution.

The idea of a bill of rights had been discussed in the Constitutional Convention and set aside as inappropriate in a document concerned with the mechanics of government. Initially, Madison himself opposed the idea of such a bill but soon found that his constituents decidedly favored it. So he changed his mind and agreed that the Constitution should be altered to include a statement about basic civil rights. He succeeded in convincing Washington that it should be mentioned in his inaugural address. Such support from the Great Man himself helped Madison win passage of the Bill of Rights in Congress, of which ten amendments won ratification on December 15, 1791, by all but two of the states in the Union. In any event, the first inaugural address delivered by Washington provided the first request to Congress for a significant change in the Constitution. It was a momentous event.

On the day of the inauguration, April 30, 1789, Washington appeared in public elegantly dressed and was driven to Federal Hall in New York City where the ceremony would take place. New York City had been the seat of the American government since 1785. On arrival, he alighted from his carriage, walked to the Senate chamber on the second floor, greeted those in attendance, and was escorted by Vice President John Adams to the outer gallery, which overlooked the street at the second-story level. Once the huge crowd below spotted him, they saluted him with a thunderous burst of applause. As they quieted, Washington then took the oath of office, which was administered by Chancellor Robert R. Livingston, the highest legal officer in New York.

After the oath taking, Washington returned to the Senate chamber and addressed both houses of Congress in a speech that would provide a splendid guide to all future inaugural addresses. As he read from his manuscript, the President spoke in a deep voice, "his aspect grave," almost to the point of "sadness." Ever conscious of his rank and importance, still "this great Man was agitated and embarrassed," recalled Senator William Maclay of Pennsylvania, ". . . he trembled and several times could scarce make out to read."[1]

After one long opening sentence, Washington added this memorable statement: To this office "I was summoned by my country, whose voice I can never hear but with veneration and love." He then went on to comment in the broadest terms on such things as virtue, freedom, duty, and

---

1. Fisher Ames to George Richards Minot, May 3, 1789, in *Works of Fisher Ames*, vol. 1, ed. Seth Ames (Boston, 1854), 34; *Documentary History of the First Federal Congress: The Diary of William Maclay*, vol. 9, eds. Kenneth R. Bowling and Helen Viet (Baltimore: Johns Hopkins University Press, 1988), 13.

the need for guidance by "that Almighty Being who rules over the universe." He said that he recognized his responsibility to recommend legislation but felt that this was not the time or place for specifics, except to mention Madison's desire to amend the Constitution to answer the growing demand for a bill of rights. Still, he cautioned against going too far in changing their republican system. He was certain that "you [will] carefully avoid every alteration which might endanger the benefits of a united and effective government."

Since George Washington had initiated the practice of delivering an inaugural address, all the succeeding Presidents who were elected to the office followed suit. (Those who assumed the presidency on the death of a predecessor did not usually feel obliged to provide an inaugural address.) In creating what became a tradition, John Adams deserves a share of the credit—as does Thomas Jefferson. Considering Washington's apathetic effort in his second inaugural, a speech of only 135 words, the practice of delivering an inaugural address as an important statement by the chief executive at the outset of his term in office might have died if Adams, the second President of the United States, had not revived the idea. Jefferson followed his lead, and all future Presidents now feel bound to continue the tradition.

Of course in the early years of the Republic there was no guarantee that any tradition would survive. For example, it should be remembered that the practice of the President appearing in person to deliver his State of the Union address to Congress each year, started by Washington and continued by Adams, ended when Jefferson chose to send his message to Congress rather than read it in person. The absence of the chief executive continued for the rest of the nineteenth century, and not until Woodrow Wilson became President was the tradition of a personal appearance revived and is continued to this day.

Many Presidents have tried to capture the tone and dignity of the first President. By and large, most of them include a tribute to the greatness of the country and its values, and conclude with an invocation asking God to continue to pour forth His blessings on the American people and to help them maintain their independence and democratic institutions.

But what is an inaugural address exactly? And why is it important? Is it a statement of governing principles? Is it a rehash of campaign promises? Should it offer clues as to how the President will govern, or should it be a list of goals the President will pursue? Should it be an occasion for ceremonial eloquence? These are some of the questions Presidents (and their speechwriters) ask themselves when they sit down to compose the speech. And it is clear from reading these addresses that many Presidents have had a different set of answers. Herbert Hoover and Lyndon B. Johnson, for example, composed very detail-oriented inaugurals, even dividing them into subsections. But the speeches that are remembered and frequently quoted by historians are those that are broad and sweeping in their overview of the conditions in the country and are either eloquent, and qualify as literature, or capture the current national mood and inspire listeners because they contain important truths that make them unforgettable. Among these notable examples are Jefferson's first inaugural, Lincoln's first and second, Franklin Roosevelt's first, Kennedy's, and Reagan's first.

Any analyses of these outstanding orations will find that on the whole they seek to redefine the greatness of the nation and its mission. For the most part, they are idea driven; that is, they are tinged with a philosophical tone about the Presidents' fundamental beliefs in the American system of government and the country's destiny.

It helps when these beliefs are spoken with eloquence. As might be expected, Thomas Jefferson was the first to achieve that distinction. He called upon his fellow citizens to "unite with one heart and with one mind. Let us restore to social intercourse, that harmony and affection without which liberty and even life itself are but dreary things." There will be differences among us, he recognized, "But every difference of opinion is not a difference of principle. We have called by different names brethren of the same principle. We are all Republicans, we are all Federalists. If there be any among us who would wish to dissolve this Union or to change its republican form, let them stand undisturbed as monuments of the safety with which error of opinion may be tolerated where reason is left free to combat it."

The fact that Jefferson continued the tradition of the inaugural address but not the State of the Union speech to Congress, underscored the importance of the former tradition. As a member of the Republican party, following Washington and Adams who belonged to the Federalist party, and by imitating their example by giving the address, he skillfully eased the transition from one political party to another. And it was done without inciting discord or conflict. It also helped him validate his right to serve as President of the United States. He had no such need when it came to delivering the State of the Union address, and being an essentially shy man he decided to skip the ordeal of presenting it to Congress in person.

So today the inaugural address completes the ritual by which a new administration begins its term in office, and that is why it is important. It has become an essential part of the ceremonies marking the start of a new administration. The President-elect first takes the oath of office then follows it with his inaugural speech. Interestingly, in the early part of the nineteenth century, the newly elected chief executive first gave his address and then took the oath of office.

Obviously, an inaugural address provides the President with a splendid opportunity to speak to the American people about his vision for the immediate future. But how seriously do Presidents take their inaugural efforts? Some, like Kennedy and Nixon, considered it very important. In fact, Kennedy asked his speechwriter, Ted Sorensen, to study the Jefferson and Lincoln speeches. He was determined to deliver an address that would be as memorable as those of these two predecessors. On the other hand, Theodore Roosevelt gave the opportunity short shrift, regarding it as little more than a ceremonial ritual. He may have been right. Today, the inaugural address is essentially ceremonial in character, and that may explain why so few of them are remembered or read. Roosevelt's attitude is particularly interesting when one recalls that he understood the importance of the President's "bully pulpit." Is the inaugural ceremony not a bully pulpit?

Also remarkable is the fact that these speeches usually offer no clue to future events that ultimately wind up defining an administration. For example, there is no mention of territorial expansion in Jefferson's inaugural, and yet his administration is remembered chiefly for the Louisiana Purchase of 1803. There is no hint of an impending economic calamity in Hoover's inaugural, no indication of Watergate in Nixon's, little mention of Vietnam in Johnson's, and no suggestion of the challenge of global terrorism in George W. Bush's first inaugural. History can always upend a President's goals and purposes. Events can and do set aside a President's attempts to control his administration. But there are exceptions, of course. Both James Knox Polk's inaugural and Franklin Roosevelt's first serve as fairly substantial outlines of what they wound up achieving.

Those Presidents who take office during difficult times face an imposing task. They must find a way of acknowledging the crisis without sounding too frightening and without finding fault with their predecessor or with Congress. The way in which Lincoln spoke about the crises at the start of both his terms in office is an excellent example of how to handle the problem skillfully. And he is not alone. John Q. Adams, Franklin D. Roosevelt, Richard Nixon, and Gerald Ford also managed to acknowledge the existence of crises that they faced without placing blame or causing undue alarm.

Today, virtually all Presidents-elect seek help in the composition of their inaugural addresses. No doubt they all have a hand in its composition, but to what degree? It is hard to evaluate. It probably depends on how much the President-elect cares about the address and whether he feels he has something to say that is important. We know that John F. Kennedy edited multiple drafts of his speech and worked laboriously with Ted Sorensen in perfecting it. Employing a speechwriter, even an unpaid one, goes back to the man who began the tradition of the inaugural address. George Washington had help from James Madison, and the latter succeeded in slipping in a few ideas of his own about what Washington should say. Jefferson, like Lincoln, needed no editorial assistance, and Andrew Jackson wrote his own inaugural, having had long experience in composing addresses to his soldiers during the years of his military career. Unfortunately, after Jackson completed the work he showed it to some friends in Nashville and Washington, and they decided it would not do. So they prepared a second and then a third draft. On the whole, the third draft—which is the one Jackson read at his inaugural—is the weakest of the three and comes nowhere near the sweep and vitality of Jackson's original version. The revisions were aimed at winning approval from all classes and sections of the country, and so Jackson's eloquent efforts were discarded.

There is a lesson here. No one has a greater stake in the composition of an inaugural address than the President-elect. The final version of the speech, the one to be read at the ceremony and published later, must reflect the thoughts and character of the President, and it is his obligation to see to it that the ideas as well as the style and spirit of the piece are his.

Occasionally, we learn something about a President in discovering which inaugurals of the past he admires and attempts to emulate. It is not surprising to find that Jimmy Carter's favorite address was Woodrow Wilson's first inaugural, a document that was very moralistic and that broke with the tradition of celebrating the nation's strengths. Instead, Wilson focused on the ways the nation had been led astray.

It is also instructive to note how certain themes appear and then disappear as the nation matures and grows stronger and more self-confident. One example is the theme of Manifest Destiny, or the notion of continental expansion. It appeared early in the nineteenth century and slowly petered out as the frontier came to a close—only to be revived again by President Kennedy in proclaiming a New Frontier. Other important themes include the situation of the American Indians and, later, the plight of free blacks, both of which disappear by the beginning of the twentieth century. And Grover Cleveland was the first President to talk about the "working man" and the issues that affected him. Farmers were a special part of inaugural rhetoric, particularly in the early nineteenth century, but no more. And surprisingly little was said about the suffrage movement or, indeed, any of the larger demands of women during the Progressive Era.

In discussing the greatness of the nation, many of these early speeches note how the country transformed the lives of its citizens and those who came to its shores in search of freedom and a better life, and how it grew from a handful of states huddled along the coastline and moved steadily westward until it encompassed a continent.

Today, there is no need to spell out the transformation that takes place when immigrants arrive in this country. Indeed, there is even a fear that too many immigrants are gaining entrance illegally. With the rise of global terrorism, a new issue has arrived that will command presidential attention. The world is changing, and as it does it provides new challenges and new dangers for Presidents to address as they begin their administrations. It will be interesting to hear what, in the future, he or she will have to say about the role the country will play in world events and how the United States will find partners in its efforts to preserve freedom and prevent another world war. It is an ongoing story.

Fellow Citizens

George Washington

· FIRST INAUGURAL ADDRESS ·

Thursday, April 30, 1789

On the afternoon of April 30, 1789, a fine spring day in New York City, President-elect George Washington climbed into a yellow carriage hitched to six white horses and attended by four footmen, and rode several blocks from his official residence to New York's former City Hall on Wall Street. The building was newly renovated to accommodate both houses of Congress, and the city's leaders hoped the new national government would be impressed enough to keep the nation's capital in New York. The building even bore a new name: Federal Hall.

Washington arrived on Wall Street at twelve thirty, wearing white silk stockings and a plain brown cloth suit with silver buttons he procured from his friend Henry Knox. A ceremonial sword hung from his side, glistening in the midday sun. Awaiting him at Federal Hall were members of Congress, cabinet members, and the new vice president, John Adams. If the dignitaries seemed uncertain about their roles in the forthcoming event, it was understandable. There was no precedent for the ceremony that was about to unfold. In fact, for quite a time nobody knew for sure how to address the new President. Should he be called His Most Benign Highness, as Adams proposed? Washington, who was called His Excellency during the Revolution, rather liked the title His High Mightiness. Some of the founders believed a grand title was necessary if the new republic were to be taken seriously in Europe. In the end, however, Congress decided to skip grandeur in favor of authentic republicanism. George Washington and his successors would be called, simply, the President of the United States.

In honor of the occasion, New Yorkers gathered in the city's narrow streets, hoping to catch a glimpse of the war hero who was about to become head of their new government. It was a moment without precedent in the new nation's life, and perhaps for that reason members of Congress had debated at great length about how this day should proceed. A special congressional committee studied the logistical issues, issued a report, and, on April 25, Congress fixed April 30 for the swearing-in ceremony. Washington already was in town, having been summoned from Mount Vernon on April 14, the day he received news that electors had chosen him to be President. He arrived on April 23 and waited while Congress polished its plans for the swearing-in.

The heart of the inaugural ceremony required no congressional tinkering, for it was outlined plainly in the new Constitution. According to the document, George Washington was to utter an oath of office consisting of thirty-five words. Nothing more was required of this day or this ceremony. With a simple few words, the nation would have the chief executive it so sorely lacked during the long War of Independence and the chaotic years following the signing of the Treaty of Paris in 1783.

After Washington climbed the steps leading into Federal Hall, he reemerged on a balcony so he could take the oath of office in front of the crowds gathered on Wall Street. As he did so, he placed his right hand on an open Bible, a gesture made possible only after the man who administered the oath, Robert Livingston, chancellor of the state of New York, helped find one after none turned up in Federal Hall. Washington's hand covered a page open to the book of Genesis, 49:13. The selection had no particular significance. For all the careful planning and thought which went into the ceremony, the Bible's central role in the proceedings was executed with unseemly haste. After a search party found a Bible on short notice, nobody had time to select a particularly relevant passage to serve as a resting place for Washington's left hand.

As prescribed in article 2, section 1 of the Constitution, the oath of office ended with the president-elect promising to "preserve, protect and defend the Constitution of the United States." With those words, George Washington became President of the United States. But Washington was not through: he kissed the Bible and ad-libbed an additional phrase, "so help me, God." Those four words have been included in the oath recited by new Presidents ever since, a tribute to Washington's solemn gesture.

Livingston administered the oath because he was the highest judicial figure in the land. There was not yet a Supreme Court, so there was no chief justice to swear in Washington. When the new President completed the oath, Livingston added an impromptu gesture of his own, calling out a regal-sounding cheer: "Long live George Washington, President of the United States." Cannons fired a salute, and the crowd roared in approval.

In a sense, the business of the day was over. But Washington believed the occasion called for more. So, after accepting the cheers of onlookers, he returned to the Senate chamber to deliver a speech to members of the House and Senate. It was the first inaugural address in United States history. And so another precedent was established.

Although Washington was not known for his oratory, the spoken word had served him well during the Revolution. In late December 1776, after the shocking American victory in Trenton, New Jersey, Washington persuaded thousands of troops to remain in the army even though their enlistments were up. His eloquence saved the army and allowed him to march to Princeton to follow up his victory in Trenton. Years later, on March 15, 1783, Washington again used the power of words to aid his infant nation when he persuaded mutinous officers in Newburgh, New York, to abandon talk of demanding their back pay at the point of a gun.

As the new President prepared to address members of Congress, circumstances were not as dire as they were in late 1776 or as anxious as they were in Newburgh in 1783. But it was, all the same, a critical moment in the life of the young American republic. The weak government enshrined in the Articles of Confederation had proved disastrous. In desperation, the founders junked the articles in 1787 and devised a new system based on three branches of government, including a strong executive. With Washington's inauguration, the transition to a new form of government was nearly complete. Nobody knew how, or if, this new government, with its competing centers of power, would work. Indeed, two states, North Carolina and Rhode Island, had yet to ratify the document.

The choice of Washington to preside over this experiment was hardly surprising. He had not only been commander in chief of the Continental army but had been a delegate to the Constitutional Convention and to the Continental Congress before being appointed general. He knew his fellow founders, and they knew—and for the most part revered—him.

As he addressed members of the House and Senate, gathered in the Senate chamber at Federal Hall, Washington employed few rhetorical flourishes or meaningful gestures. He never took his eyes off the sheaf of paper in front of him, and witnesses said it was difficult to hear him. The man who calmed troops under fire suddenly seemed nervous; some said his hands trembled as he spoke.

His sentences, crafted by James Madison, were long and ornate, requiring listeners to follow closely as assertions were followed by caveats, or protestations of unworthiness were piled one upon another. Early on in his speech, he sought to explain his "conflict of emotions" upon hearing news that he had been elected President. "On the one hand," he said, "I was summoned by my country, whose voice I can never hear but with veneration and love, from a retreat which I had chosen with the fondest predilection, and, in my flattering hopes, with an immutable decision, as the asylum of my declining years—a retreat which was rendered every day more necessary as well as more dear to me by the addition of habit to inclination, and of frequent interruptions in my health to the gradual waste committed on it by time. On the other hand, the magnitude and difficulty of the trust to which the voice of my country called me, being sufficient to awaken in the wisest and most experienced of her citizens a distrustful scrutiny into his qualifications, could not but overwhelm with despondence one who (inheriting inferior endowments from nature and unpracticed in the duties of civil administration) ought to be peculiarly conscious of his own deficiencies."

In other words, the new President was a modest fellow who would have been content to spend the rest of his years at Mount Vernon. But duty called, and he had no choice but to respond.

The speech contained few policy goals, no real vision of where the new President might wish to lead the nation. He did, however, invoke the blessings of "that Almighty Being who rules over the universe," and he pledged his support for a bill of rights to be added to the Constitution.

When he finished his speech, Washington shook hands with his fellow founders and then left Federal Hall for a short walk up Wall Street to St. Paul's Chapel, where an Episcopal bishop conducted a special service and prayed for blessings on the new administration.

FELLOW CITIZENS OF THE SENATE AND OF
THE HOUSE OF REPRESENTATIVES:

Among the vicissitudes incident to life no event could have filled me with greater anxieties than that of which the notification was transmitted by your order, and received on the 14th day of the present month. On the one hand, I was summoned by my country, whose voice I can never hear but with veneration and love, from a retreat which I had chosen with the fondest predilection, and, in my flattering hopes, with an immutable decision, as the asylum of my declining years—a retreat which was rendered every day more necessary as well as more dear to me by the addition of habit to inclination, and of frequent interruptions in my health to the gradual waste committed on it by time. On the other hand, the magnitude and difficulty of the trust to which the voice of my country called me, being sufficient to awaken in the wisest and most experienced of her citizens a distrustful scrutiny into his qualifications, could not but overwhelm with despondence one who (inheriting inferior endowments from nature and unpracticed in the duties of civil administration) ought to be peculiarly conscious of his own deficiencies. In this conflict of emotions all I dare aver is that it has been my faithful study to collect my duty from a just appreciation of every circumstance by which it might be affected. All I dare hope is that if, in executing this task, I have been too much swayed by a grateful remembrance of former instances, or by an affectionate sensibility to this transcendent proof of the confidence of my fellow citizens, and have thence too little consulted my incapacity as well as disinclination for the weighty and untried cares before me, my error will be palliated by the motives which mislead me, and its consequences be judged by my country with some share of the partiality in which they originated.

Such being the impressions under which I have, in obedience to the public summons, repaired to the present station, it would be peculiarly improper to omit in this first official act my fervent supplications to that Almighty Being who rules over the universe, who presides in the councils of nations, and whose providential aids can supply every human defect, that His benediction may consecrate to the liberties and happiness of the people of the United States a government instituted by themselves for these essential purposes, and may enable every instrument employed in its administration to execute with success the functions allotted to his charge. In tendering this homage to the Great Author of every public and private good, I assure myself that it expresses your sentiments not less than my own, nor those of my fellow citizens at large less than either. No people can be bound to acknowledge and adore the Invisible Hand which conducts the affairs of men more than those of the United States. Every step by which they have advanced to the character of an independent nation seems to have been distinguished by some token of providential agency; and in the important revolution just accomplished in the system of their united government the tranquil deliberations and voluntary consent of so many distinct communities from which the event has resulted cannot be compared with the means by which most governments have been established without some return of pious gratitude, along with an humble anticipation of the future blessings which the past seems to presage. These reflections, arising out of the present crisis, have forced themselves too strongly on my mind to be suppressed. You will join with me, I trust, in thinking that there are none under the influence of which the proceedings of a new and free government can more auspiciously commence.

By the article establishing the executive department it is made the duty of the President "to recommend to your consideration such measures as he shall judge necessary and expedient." The circumstances under which I now meet you will acquit me from entering into that subject further than to refer to the great constitutional charter under which you are assembled, and which, in defining your powers, designates the objects to which your attention is to be given. It will be more consistent with those circumstances, and far more congenial with the feelings which actuate me, to substitute, in place of a recommendation of particular measures, the tribute that is due to the talents, the rectitude, and the patriotism which adorn the characters selected to devise and adopt them. In these honorable qualifications I behold the surest pledges that as on one side no local prejudices or attachments, no separate views nor party animosities, will misdirect the comprehensive and equal eye which ought to watch over this great assemblage of communities and interests, so, on another, that the foundation of our national policy will be laid in the pure and immutable principles of private morality, and the preeminence of free government be exemplified by all the attributes which can win the affections of its citizens and command the respect of the world. I dwell on this prospect with every satisfaction which an ardent love for my country can inspire, since there is no truth more thoroughly established than that there exists in the economy and course of nature an indissoluble union between virtue and happiness; between duty and advantage; between the genuine maxims of an honest and magnanimous policy and the solid rewards of public prosperity and felicity; since we ought to be no less persuaded that the propitious smiles of heaven can never be expected on a nation that disregards the eternal rules of order and right which heaven itself has ordained; and since the preservation of the sacred fire of liberty and the destiny of the republican model of government are justly considered, perhaps, as deeply, as finally, staked on the experiment entrusted to the hands of the American people.

Besides the ordinary objects submitted to your care, it will remain with your judgment to decide how far an exercise of the occasional power delegated by the fifth article of the Constitution is rendered expedient at the present juncture by the nature of objections which have been urged against the system, or by the degree of inquietude which has given birth to them. Instead of undertaking particular recommendations on this subject, in which I could be guided by no lights derived from official opportunities, I shall again give way to my entire confidence in your discernment and pursuit of the public good; for I assure myself that whilst you carefully avoid every alteration which might endanger the benefits of a united and effective government, or which ought to await the future lessons of experience, a reverence for the characteristic rights of freemen and a regard for the public harmony will sufficiently influence your deliberations on the question of how far the former can be impregnably fortified or the latter be safely and advantageously promoted.

To the foregoing observations I have one to add, which will be most properly addressed to the House of Representatives. It concerns myself, and will therefore be as brief as possible. When I was first honored with a call into the service of my country, then on the eve of an arduous struggle for its liberties, the light in which I contemplated my duty required that I should renounce every pecuniary compensation. From this resolution I have in no instance departed; and being still under the impressions which produced it, I must decline as inapplicable to myself any share in the personal emoluments which may be indispensably included in a permanent provision for the executive department, and must accordingly pray that the pecuniary estimates for the station in which I am placed may during my continuance in it be limited to such actual expenditures as the public good may be thought to require.

Having thus imparted to you my sentiments as they have been awakened by the occasion which brings us together, I shall take my present leave; but not without resorting once more to the benign Parent of the Human Race in humble supplication that, since He has been pleased to favor the American people with opportunities for deliberating in perfect tranquillity, and dispositions for deciding with unparalleled unanimity on a form of government for the security of their union and the advancement of their happiness, so His divine blessing may be equally conspicuous in the enlarged views, the temperate consultations, and the wise measures on which the success of this government must depend.

George Washington

Monday, March 4, 1793

George Washington won a second four-year term as President in 1792 and was sworn in on March 4, 1793, in Philadelphia. In contrast to his first inauguration, when he spoke after taking the oath of office, Washington delivered his remarks in Philadelphia beforehand. At 135 words, Washington's second inaugural address is the shortest on record.

Brief as it is, it offers a remarkable insight into Washington's character and his times. He noted that if he were to violate the oath of office, not only would he incur the Constitution's punishment, but he would be "subject to the upbraidings of all who are now witnesses of the present solemn ceremony." George Washington and his colleagues put a high premium on honor. To be upbraided in public was to die a thousand deaths and was, perhaps, a worse fate than the Constitution's formal remedy of impeachment. This was an age that understood and feared the power of shame.

As he entered his second term as President, however, Washington had little reason to feel ashamed. The constitutional experiment appeared to be working, the executive department was running well and attracting some of the nation's brightest people, foreign relations were established, the Bill of Rights had passed, and the nation's finances were stabilized. Washington mediated growing factional disputes focused on two of his cabinet members, Secretary of State Thomas Jefferson and Treasury Secretary Alexander Hamilton. He established the presidency not as a figurehead or as a mere administrator but as a powerful force in running the nation. As President, he believed he represented the entire nation, not just a single state or district, and he conducted himself accordingly.

While Washington's second inaugural address contained no memorable phrases and captured no great historical moment, he was not quite finished with speech making. Near the end of his second and last term—he chose not to run for a third term, establishing yet another precedent, one that would remain intact until Franklin Roosevelt won a third term in 1940—he delivered a farewell address that remains his most memorable speech. In it, he reminded the nation that it was "our true policy to steer clear of permanent alliances with any portion of the foreign world." The speech did not include a warning against "entangling alliances"—that was Thomas Jefferson's phrase, not Washington's—but it did encourage the United States to be wary of the Old World's ways and wiles.

Washington's Farewell Address is the only one of his presidential speeches regularly cited in collections of great speeches. His inaugural speeches are not milestones of oratory and argument. Nevertheless, they established the means by which a new President could speak directly to American citizens.

FELLOW CITIZENS:

I am again called upon by the voice of my country to execute the functions of its chief magistrate. When the occasion proper for it shall arrive, I shall endeavor to express the high sense I entertain of this distinguished honor, and of the confidence which has been reposed in me by the people of united America.

Previous to the execution of any official act of the President the Constitution requires an oath of office. This oath I am now about to take, and in your presence: that if it shall be found during my administration of the government I have in any instance violated willingly or knowingly the injunctions thereof, I may (besides incurring constitutional punishment) be subject to the upbraidings of all who are now witnesses of the present solemn ceremony.

John Adams

Saturday, March 4, 1797

As the nation's first vice president, John Adams experienced frustrations that would sound familiar to any one of his forty-six successors. Adams quickly realized that his job held little power, and he learned that his opinions on matters of state mattered a good deal less than those of George Washington's cabinet officers. His main task was to preside over the United States Senate and, when the occasion called for it, to cast a vote to break a tie. That he did thirty-one times.

Adams did appreciate that while his office was rather powerless, it gave him an advantage over those who sought to succeed Washington. As the election year of 1796 approached, Adams seemed like a natural successor to Washington—and he knew it. "I am heir apparent, you know," he wrote to his wife, Abigail, in early 1796, several months before the first President announced his retirement.[1] Washington, of course, was eligible to run for a third term. But he loathed the increasingly bitter tenor of debate in the country and especially resented criticism of a treaty with Great Britain that some politicians and editors believed was too favorable to the British. Washington longed for Mount Vernon anyway, and political faction fighting made his decision to retire that much easier.

Not everybody agreed on the inevitability of an Adams presidency. He was not nearly as popular as Washington—who was?—and with the great leader leaving the scene, other political leaders saw no reason to disguise their ambitions. What's more, divisions between the nation's leaders had coalesced into a party system, leading to the first election between two contending political factions. The Federalists, allied with Washington, rallied around Adams, one of their own, while the Republicans supported

---

1. David McCullough, *John Adams* (New York: Simon & Schuster, 2001), 458.

Thomas Jefferson, who retired from Washington's cabinet in 1793. The Federalists were an urban party that supported a strong central government and envisioned a powerful manufacturing nation presided over by elite decision makers. They abhorred the French Revolution. The Republicans, or Democratic-Republicans, believed in small, localized government and a more grass-roots democracy. Party members tended to sympathize with the revolutionaries in France. The two parties established the framework for the two-party politics of twenty-first-century America.

When votes from the Electoral College were counted in Congress on February 8, 1797, Adams barely had enough support to claim victory: seventy-one votes to Jefferson's sixty-eight. As the second-place finisher in a multicandidate field, Jefferson became the nation's second vice president, even though he and Adams represented competing political parties.

On March 4, 1797, a Saturday, the formal transfer of executive power from Washington to Adams took place in Congress Hall in Philadelphia. No troops were required to enforce this transfer. No citizens stormed public buildings to protest or prevent the change in leadership. At a time of terror and counterterror in revolutionary France, the peaceful and efficient change of administrations in Philadelphia seemed proof that the Constitution worked and that the American experiment could serve as an example of orderly democracy.

But Inauguration Day 1797 was more than an exercise in civic pride. It was, too, an occasion for nostalgia and sentiment, for it marked the final appearance of George Washington as a public figure. The members of Congress, Supreme Court justices, diplomats, and other officials in attendance were well aware of the day's significance. When Washington appeared in the hall, before the arrival of the new President and vice president, the audience burst into applause. Jefferson soon followed, and then came Adams. As Adams's biographer David McCullough has noted, the new President was a short, stout man, qualities that must have been exaggerated as he stood in the company of the two towering men from Virginia.[2]

Adams was a nervous wreck. He had hardly slept the night before—he could not banish a fear that he would collapse onstage, and so he tossed and turned through the night and into early morning. His ailing wife, Abigail, and their family did not join him for the ceremony in Philadelphia,

2. Ibid., 467.

so he felt very much alone at this solemn hour, as he prepared to succeed a living icon, George Washington.

The first President's eight years in office stabilized the new constitutional government, presided over the entrance of three new states into the Union, formalized relations with other nations, and settled questions about the young republic's finances. These accomplishments were part of Washington's legacy to Adams. But that inheritance was not without problems. As Jefferson noted of Washington, "The President is fortunate to get off just as the bubble is bursting, leaving others to hold the bag."[3]

The bag contained a series of problems overseas. Relations with Great Britain over trade and its continued military presence in the Great Lakes region had nearly reached the breaking point in 1796, until John Jay negotiated a treaty with the British that prevented war. Critics feared that the treaty drew the United States too close to Britain, so recently seen as an oppressive enemy. For a moment, when there was talk about the House refusing to fund some of the treaty's provisions, Adams feared the treaty's critics would provoke a constitutional crisis that would lead to war with Britain. The treaty passed, but not the sense of crisis. The bag Adams was about to grasp was heavy, indeed.

When the time came for him to deliver the nation's third inaugural address, Adams rose from his chair on a dais and faced his audience. Seated in front of him, with members of the Supreme Court, was his predecessor-to-be, George Washington. As Washington did in his second inaugural, Adams spoke before taking the oath of office, so, technically, he was still president-elect when he delivered the address. (Jefferson already had been sworn in as the new vice president.)

It was a remarkably humble speech, with little reference to the role he played in the nation's founding and an outright acknowledgement that he was out of the country, as U.S. ambassador to Britain, when the Constitution was written. Seated in front of him as he spoke was the icon he was about to replace. Adams acknowledged George Washington without ever mentioning his name. He did not have to do so. When he spoke of "a citizen who, by a long course of great actions . . . has merited the gratitude of his fellow-citizens," everyone knew to whom he was referring. Eyes overflowed as he spoke, although the new President observed that Washington himself seemed as "serene and unclouded as the day." But Adams thought

---

3. Ralph Adams Brown, *The Presidency of John Adams* (Lawrence, KS.: University Press of Kansas, 1975), 22.

he saw something beneath that serenity. "Methought I heard him say, 'Ay! I am fairly out and you fairly in. See which of us will be the happiest.'"[4]

Clues about his priorities as President were buried in a gargantuan sentence of more than seven hundred words near the speech's conclusion. Amid a riot of semicolons and dependent clauses were several remarkable promises and observations. Adams pledged too seek "equity and humanity" toward Native Americans, to remain neutral in the war "among the belligerent powers of Europe," and to "preserve . . . friendship" with France. The mention of France is noteworthy because it is one of very few specifics in the speech, evidence of just how grave relations were between the two republics.

Following Washington's tradition, Adams asked the Almighty for blessings before concluding his speech. He then took the oath of office from the chief justice of the Supreme Court, Oliver Ellsworth, the first chief justice to play such a role at an inauguration. Washington left his place of honor to congratulate his successor. He "wished my administration might be happy, successful, and honorable," Adams told Abigail.[5] The first presidential transition in American history was complete.

WHEN IT WAS FIRST PERCEIVED, in early times, that no middle course for America remained between unlimited submission to a foreign legislature and a total independence of its claims, men of reflection were less apprehensive of danger from the formidable power of fleets and armies they must determine to resist than from those contests and dissensions which would certainly arise concerning the forms of government to be instituted over the whole and over the parts of this extensive country. Relying, however, on the purity of their intentions, the justice of their cause, and the integrity and intelligence of the people, under an overruling Providence which had so signally protected this country from the first, the representatives of this nation, then consisting of little more than half its present number, not only broke to pieces the chains which were forging and the rod of iron that was lifted up, but frankly cut asunder the ties which had bound them, and launched into an ocean of uncertainty.

4. Ibid., 4–5.

5. Ibid., 5.

The zeal and ardor of the people during the Revolutionary War, supplying the place of government, commanded a degree of order sufficient at least for the temporary preservation of society. The confederation which was early felt to be necessary was prepared from the models of the Batavian and Helvetic confederacies, the only examples which remain with any detail and precision in history, and certainly the only ones which the people at large had ever considered. But reflecting on the striking difference in so many particulars between this country and those where a courier may go from the seat of government to the frontier in a single day, it was then certainly foreseen by some who assisted in Congress at the formation of it that it could not be durable.

Negligence of its regulations, inattention to its recommendations, if not disobedience to its authority, not only in individuals but in states, soon appeared with their melancholy consequences—universal languor, jealousies and rivalries of states, decline of navigation and commerce, discouragement of necessary manufactures, universal fall in the value of lands and their produce, contempt of public and private faith, loss of consideration and credit with foreign nations, and at length in discontents, animosities, combinations, partial conventions, and insurrection, threatening some great national calamity.

In this dangerous crisis the people of America were not abandoned by their usual good sense, presence of mind, resolution, or integrity. Measures were pursued to concert a plan to form a more perfect union, establish justice, insure domestic tranquillity, provide for the common defense, promote the general welfare, and secure the blessings of liberty. The public disquisitions, discussions, and deliberations issued in the present happy Constitution of government.

Employed in the service of my country abroad during the whole course of these transactions, I first saw the Constitution of the United States in a foreign country. Irritated by no literary altercation, animated by no public debate, heated by no party animosity, I read it with great satisfaction, as the result of good heads prompted by good hearts, as an experiment better adapted to the genius, character, situation, and relations of this nation and country than any which had ever been proposed or suggested. In its general principles and great outlines it was conformable to such a system of government as I had ever most esteemed, and in some states, my own native state in particular, had contributed to establish. Claiming a right of suffrage, in common with my fellow citizens, in the adoption or rejection of a constitution which was to rule me and my posterity, as well as them and theirs, I did not hesitate to express my approbation of it on all occasions, in public and in private. It was not then, nor has been since, any objection to it in my mind that the executive and senate were not more permanent. Nor have I ever entertained a thought of promoting any alteration in it but such as the people themselves, in the course of their experience, should see and feel to be necessary or expedient, and by their representatives in Congress and the state legislatures, according to the Constitution itself, adopt and ordain.

Returning to the bosom of my country after a painful separation from it for ten years, I had the honor to be elected to a station under the new order of things, and I have repeatedly laid myself under the most serious obligations to support the Constitution. The operation of it has equaled the most sanguine expectations of its friends, and from an habitual attention to it, satisfaction in its administration, and delight in its effects upon the peace, order, prosperity, and happiness of the nation I have acquired an habitual attachment to it and veneration for it.

What other form of government, indeed, can so well deserve our esteem and love?

There may be little solidity in an ancient idea that congregations of men into cities and nations are the most pleasing objects in the sight of superior intelligences, but this is very certain, that to a benevolent human mind there can be no spectacle presented by any nation more pleasing, more noble, majestic, or august, than an assembly like that which has so often been seen in this and the other chamber of Congress, of a government in which the executive authority, as well as that of all the branches of the legislature, are exercised by citizens selected at regular periods by their neighbors to make and execute laws for the general good. Can anything essential, anything more than mere ornament and decoration, be added to this by robes and diamonds? Can authority be more amiable and respectable when it descends from accidents or institutions established in remote antiquity than when it springs fresh from the hearts and judgments of an honest and enlightened people? For it is the people only that are represented. It is their power and majesty that is reflected, and only for their good, in every legitimate government, under whatever form it may appear. The existence of such a government as ours for any length of time is a full proof of a general dissemination of knowledge and virtue throughout the whole body of the people. And what object or consideration more pleasing than this can be presented to the human mind? If national pride is ever justifiable or excusable it is when it springs, not from power or riches, grandeur or glory, but from conviction of national innocence, information, and benevolence.

In the midst of these pleasing ideas we should be unfaithful to ourselves if we should ever lose sight of the danger to our liberties if anything partial or extraneous should infect the purity of our free, fair, virtuous, and independent elections. If an election is to be determined by a majority of a single vote, and that can be procured by a party through artifice or corruption, the government may be the choice of a party for its own ends, not of the nation for the national good. If that solitary suffrage can be obtained by foreign nations by flattery or menaces, by fraud or violence, by terror, intrigue, or venality, the government may not be the choice of the American people, but of foreign nations. It may be foreign nations who govern us, and not we, the people, who govern ourselves; and candid men will acknowledge that in such cases choice would have little advantage to boast of over lot or chance.

Such is the amiable and interesting system of government (and such are some of the abuses to which it may be exposed) which the people of America have exhibited to the admiration and anxiety of the wise and virtuous of all nations for eight years under the administration of a citizen who, by a long course of great actions, regulated by prudence, justice, temperance, and fortitude, conducting a people inspired with the same virtues and animated with the same ardent patriotism and love of liberty to independence and peace, to increasing wealth and unexampled prosperity, has merited the gratitude of his fellow citizens, commanded the highest praises of foreign nations, and secured immortal glory with posterity.

In that retirement which is his voluntary choice may he long live to enjoy the delicious recollection of his services, the gratitude of mankind, the happy fruits of them to himself and the world, which are daily increasing, and that splendid prospect of the future fortunes of this country which is opening from year to year. His name may be still a rampart, and the knowledge that he lives a bulwark, against all open or secret enemies of his country's peace. This example has been recommended to the imitation of his successors by both houses of Congress and by the voice of the legislatures and the people throughout the nation.

On this subject it might become me better to be silent or to speak with diffidence; but as something may be expected, the occasion, I hope, will be admitted as an apology if I venture to say that if a preference, upon principle, of a free republican government, formed upon long and serious reflection, after a diligent and impartial inquiry after truth; if an attach-

ment to the Constitution of the United States, and a conscientious determination to support it until it shall be altered by the judgments and wishes
of the people, expressed in the mode prescribed in it; if a respectful attention to the constitutions of the individual states and a constant caution
and delicacy toward the state governments; if an equal and impartial regard to the rights, interest, honor, and happiness of all the states in the
Union, without preference or regard to a northern or southern, an eastern
or western, position, their various political opinions on unessential points
or their personal attachments; if a love of virtuous men of all parties and
denominations; if a love of science and letters and a wish to patronize
every rational effort to encourage schools, colleges, universities, academies, and every institution for propagating knowledge, virtue, and religion among all classes of the people, not only for their benign influence on
the happiness of life in all its stages and classes, and of society in all its
forms, but as the only means of preserving our Constitution from its natural enemies, the spirit of sophistry, the spirit of party, the spirit of intrigue, the profligacy of corruption, and the pestilence of foreign influence,
which is the angel of destruction to elective governments; if a love of equal
laws, of justice, and humanity in the interior administration; if an inclination to improve agriculture, commerce, and manufacturers for necessity,
convenience, and defense; if a spirit of equity and humanity toward the
aboriginal nations of America, and a disposition to meliorate their condition by inclining them to be more friendly to us, and our citizens to be
more friendly to them; if an inflexible determination to maintain peace
and inviolable faith with all nations, and that system of neutrality and impartiality among the belligerent powers of Europe which has been adopted
by this government and so solemnly sanctioned by both houses of Congress and applauded by the legislatures of the states and the public opinion, until it shall be otherwise ordained by Congress; if a personal esteem
for the French nation, formed in a residence of seven years chiefly among
them, and a sincere desire to preserve the friendship which has been so
much for the honor and interest of both nations; if, while the conscious
honor and integrity of the people of America and the internal sentiment of
their own power and energies must be preserved, an earnest endeavor to
investigate every just cause and remove every colorable pretense of com-
plaint; if an intention to pursue by amicable negotiation a reparation for
the injuries that have been committed on the commerce of our fellow citizens by whatever nation, and if success cannot be obtained, to lay the facts
before the legislature, that they may consider what further measures the

honor and interest of the government and its constituents demand; if a resolution to do justice as far as may depend upon me, at all times and to all nations, and maintain peace, friendship, and benevolence with all the world; if an unshaken confidence in the honor, spirit, and resources of the American people, on which I have so often hazarded my all and never been deceived; if elevated ideas of the high destinies of this country and of my own duties toward it, founded on a knowledge of the moral principles and intellectual improvements of the people deeply engraven on my mind in early life, and not obscured but exalted by experience and age; and, with humble reverence, I feel it to be my duty to add, if a veneration for the religion of a people who profess and call themselves Christians, and a fixed resolution to consider a decent respect for Christianity among the best recommendations for the public service, can enable me in any degree to comply with your wishes, it shall be my strenuous endeavor that this sagacious injunction of the two houses shall not be without effect.

With this great example before me, with the sense and spirit, the faith and honor, the duty and interest, of the same American people pledged to support the Constitution of the United States, I entertain no doubt of its continuance in all its energy, and my mind is prepared without hesitation to lay myself under the most solemn obligations to support it to the utmost of my power.

And may that Being who is supreme over all, the Patron of Order, the Fountain of Justice, and the Protector in all ages of the world of virtuous liberty, continue His blessing upon this nation and its government and give it all possible success and duration consistent with the ends of His providence.

Thomas Jefferson

Wednesday, March 4, 1801

The election of 1800 was yet another milestone for the fledgling American republic. While the election four years earlier was significant because it transferred power from one person, George Washington, to another, John Adams, the results in 1800 led a governing political party, the Federalists, to surrender power to its partisan enemies, the Republicans of Thomas Jefferson.

Americans in the twenty-first century are familiar with sharp-edged rhetoric and personal attacks between competing political figures. Journalists and voters alike often seem to yearn for an imagined past when politicians disagreed without being disagreeable, when political campaigns featured high-minded debates about public policy. While some presidential elections no doubt featured graceful orations and intellectual give-and-take, the election of 1800 was not one of them.

Against the pleas of no less a wise man than George Washington, who warned about the evils of political factions, the country became increasingly divided between Federalists and Republicans during the Adams administration. Adams's bid for a second term was challenged by his vice president, Jefferson, and two other candidates, Aaron Burr of New York and Charles Pinckney of South Carolina. Members of Adams's own party abandoned him because he refused all-out war after three French officials demanded a bribe from American diplomats in return for allowing the Americans access to France's foreign minister, Charles-Maurice de Talleyrand-Périgord. The Americans refused to pay and returned home.

Adams, under criticism from Jeffersonians who blamed him for the diplomatic impasse, released documents about the scandal, identifying the French officials only as X, Y, and Z. The XYZ affair led to what historians have called a "quasi-war" with France, with warships from both nations engaged in combat on the high seas. While Federalists were demanding full-scale hostilities, Adams pursued negotiation.

In an age when partisan newspapers thrived on scurrilous rumors, the campaign featured no end of innuendo and gossip. Pro-Jefferson editors questioned President Adams's sanity and denounced him as a hopeless Anglophile who preferred monarchy to American republicanism. The Adams faction did not turn the other cheek. Its broadsheets suggested that Jefferson would import the violence and revolution associated with France's radicals. There were whispers, proved true many years later, about Jefferson's relationship with women he enslaved. Most damning of all, Jefferson was accused of being an atheist.

The charges and countercharges left the country badly divided, as the returns showed. For the first and only time in U.S. history, two candidates received the same number of votes in the Electoral College. But neither of them was the incumbent, John Adams. Jefferson and Burr both received seventy-three votes, with Adams, badly hurt when fellow Federalists like Hamilton abandoned and attacked him, finishing with sixty-five. The fourth candidate, Pinckney, had sixty-three. At the time, members of the Electoral College cast two votes for President. The first-place finisher became President; the second-place finisher became vice president. This was the last time that formula was used.

The vote presented another test for the Constitution, which provided that such deadlocks should be decided in the House of Representatives. Each of the sixteen states had one vote, so the winning candidate needed nine votes. The House voted thirty-five times without a clear-cut winner, because the Vermont and Maryland delegations were evenly split and so cast blank ballots. Jefferson finally prevailed on the thirty-sixth ballot when Burr supporters in Vermont and Maryland gave up. Burr, as the second-place finisher, became vice president.

Jefferson took the oath of office on March 4, 1801, after walking from his rooms to the construction site that was the U.S. Capitol, a building which, like the nation itself, was very much a work in progress. Jefferson was the first President to be inaugurated in the new federal district rising along the banks of the Potomac River. John Adams, the first President to lose his bid for reelection, was not among the dignitaries gathered in the Senate chamber of the rising Capitol buildings. The outgoing President left Washington before dawn, perhaps out of pique, perhaps out of some sense of propriety. After all, there was no precedent then for a defeated

President to witness the inauguration of his successor. Adams was, at the time, the only President to have suffered such a fate.

And so the former President did not hear the words for which Jefferson's first inaugural is best known. "We are all Republicans, we are all Federalists," Jefferson said in a low voice.

While Jefferson's call for unity may be standard fare in ceremonial speeches today, in 1801 the phrase served as an important reminder that whatever divisions may have split the founders during the election, they shared basic principles and were partners in developing a new democratic nation. The election of 1800 had been a divisive affair from beginning to end. In his inaugural, Jefferson sought to ease the tensions among members of the founding generation. Comrades in the crucible of war, they had grown apart during the 1790s, leading to the all-out political war of 1800. It was Jefferson's task in his first speech as President to assure the outgoing party and its supporters that "the minority possess their equal rights, which equal law must protect, and to violate would be oppression."

But Jefferson sought more than political reconciliation. What made this speech exceptional was the new President's argument on behalf of democratic government, his sense of mission—Americans were "high-minded"; they lived in a "chosen country" where rights were based "not from birth" but "from our actions"—and his confidence in the will of the people. "Sometimes it is said that man cannot be trusted with the government of himself," he said. "Can he, then, be trusted with the government of others? Or have we found angels in the forms of kinds to govern him? Let history answer this question."

History has done just that. Historians agree that Jefferson's first inaugural is among the finest ever given, in part because of its statement of principles, its conciliatory approach, and its moral certainty. Here, for the first time, a new President addressed himself not simply to his colleagues in government, but to all his fellow Americans. He addressed himself to his "fellow citizens," and used the phrase at several points in the speech, putting on an equal plane his influential audience and those with less access to power. All were invited to join in this call for national purpose: "Let us, then . . . pursue our own Federal and Republican principles." And again, "let us hasten to retrace our steps" if "we wander" from "the creed of our political faith."

Here is the essence of Jeffersonian republicanism, a flawed but noteworthy attempt to include not just the people's representatives but all "fellow citizens" as part of the national dialogue. Of course, Jefferson's idea of "fellow citizens" did not include African Americans, Native Americans, or women. Nevertheless, the rhetoric of inclusion, the humble usage of "let us," with its reliance on persuasion rather than dictation, suggest a more democratic tone of presidential address.

Included in the speech is a phrase often attributed to George Washington but actually coined here by Jefferson: the promise to avoid "entangling alliances." While Washington, like many other fervent American republicans, certainly did not wish to see the United States become embroiled in the quarrels of the Old World—and said so in his last speech as President—it was Jefferson who formulated the phrase that would be cited in the centuries to come.

FRIENDS AND FELLOW CITIZENS:

Called upon to undertake the duties of the first executive office of our country, I avail myself of the presence of that portion of my fellow citizens which is here assembled to express my grateful thanks for the favor with which they have been pleased to look toward me, to declare a sincere consciousness that the task is above my talents, and that I approach it with those anxious and awful presentiments which the greatness of the charge and the weakness of my powers so justly inspire. A rising nation, spread over a wide and fruitful land, traversing all the seas with the rich productions of their industry, engaged in commerce with nations who feel power and forget right, advancing rapidly to destinies beyond the reach of mortal eye—when I contemplate these transcendent objects, and see the honor, the happiness, and the hopes of this beloved country committed to the issue, and the auspices of this day, I shrink from the contemplation, and humble myself before the magnitude of the undertaking. Utterly, indeed, should I despair did not the presence of many whom I here see remind me that in the other high authorities provided by our Constitution I shall find resources of wisdom, of virtue, and of zeal on which to rely under all difficulties. To you, then, gentlemen, who are charged with the sovereign functions of legislation, and to those associated with you, I look with encouragement for that guidance and support which may enable us to steer with safety the vessel in which we are all embarked amidst the conflicting elements of a troubled world.

During the contest of opinion through which we have passed the animation of discussions and of exertions has sometimes worn an aspect which might impose on strangers unused to think freely and to speak and to write what they think; but this being now decided by the voice of the nation, announced according to the rules of the Constitution, all will, of course, arrange themselves under the will of the law, and unite in common efforts for the common good. All, too, will bear in mind this sacred principle, that though the will of the majority is in all cases to prevail, that will to be rightful must be reasonable; that the minority possess their equal rights, which equal law must protect, and to violate would be oppression. Let us, then, fellow citizens, unite with one heart and one mind. Let us restore to social intercourse that harmony and affection without which liberty and even life itself are but dreary things. And let us reflect that, having banished from our land that religious intolerance under which mankind so long bled and suffered, we have yet gained little if we countenance a political intolerance as despotic, as wicked, and capable of as bitter and bloody persecutions. During the throes and convulsions of the ancient world, during the agonizing spasms of infuriated man, seeking through blood and slaughter his long-lost liberty, it was not wonderful that the agitation of the billows should reach even this distant and peaceful shore; that this should be more felt and feared by some and less by others, and should divide opinions as to measures of safety. But every difference of opinion is not a difference of principle. We have called by different names brethren of the same principle. We are all Republicans, we are all Federalists. If there be any among us who would wish to dissolve this Union or to change its republican form, let them stand undisturbed as monuments of the safety with which error of opinion may be tolerated where reason is left free to combat it. I know, indeed, that some honest men fear that a republican government can not be strong, that this government is not strong enough; but would the honest patriot, in the full tide of successful experiment, abandon a government which has so far kept us free and firm on the theoretic and visionary fear that this government, the world's best hope, may by possibility want energy to preserve itself? I trust not. I believe this, on the contrary, the strongest government on earth. I believe it the only one where every man, at the call of the law, would fly to the standard of the law, and would meet invasions of the public order as his own personal concern. Sometimes it is said that man cannot be trusted with the government of himself. Can he, then, be trusted with the government of others? Or have we found angels in the forms of kings to govern him? Let history answer this question.

Let us, then, with courage and confidence pursue our own federal and republican principles, our attachment to union and representative government. Kindly separated by nature and a wide ocean from the exterminating havoc of one quarter of the globe; too high-minded to endure the degradations of the others; possessing a chosen country, with room enough for our descendants to the thousandth and thousandth generation; entertaining a due sense of our equal right to the use of our own faculties, to the acquisitions of our own industry, to honor and confidence from our fellow citizens, resulting not from birth, but from our actions and their sense of them; enlightened by a benign religion, professed, indeed, and practiced in various forms, yet all of them inculcating honesty, truth, temperance, gratitude, and the love of man; acknowledging and adoring an overruling Providence, which by all its dispensations proves that it delights in the happiness of man here and his greater happiness hereafter—with all these blessings, what more is necessary to make us a happy and a prosperous people? Still one thing more, fellow citizens— a wise and frugal government, which shall restrain men from injuring one another, shall leave them otherwise free to regulate their own pursuits of industry and improvement, and shall not take from the mouth of labor the bread it has earned. This is the sum of good government, and this is necessary to close the circle of our felicities.

About to enter, fellow citizens, on the exercise of duties which comprehend everything dear and valuable to you, it is proper you should understand what I deem the essential principles of our government, and consequently those which ought to shape its administration. I will compress them within the narrowest compass they will bear, stating the general principle, but not all its limitations. Equal and exact justice to all men, of whatever state or persuasion, religious or political; peace, commerce, and honest friendship with all nations, entangling alliances with none; the support of the state governments in all their rights, as the most competent administrations for our domestic concerns and the surest bulwarks against anti-republican tendencies; the preservation of the general government in its whole constitutional vigor, as the sheet anchor of our peace at home and safety abroad; a jealous care of the right of election by the people—a mild and safe corrective of abuses which are lopped by the sword of revolution where peaceable remedies are unprovided; absolute acquiescence in the decisions of the majority, the vital principle of republics, from which is no appeal but to force, the vital principle and immediate parent of despotism; a well disciplined militia, our best reliance in peace and for the first moments of war, till regulars may relieve them; the supremacy of the civil over the military authority; economy in the public expense, that labor may be lightly burthened; the honest payment of our debts and sacred preservation of the public faith; encouragement of agriculture, and of commerce as its handmaid; the diffusion of information and arraignment of all abuses at the bar of the public reason; freedom of religion; freedom of the press, and freedom of person under the protection of the habeas corpus, and trial by juries impartially selected. These principles form the bright constellation which has gone before us and guided our steps through an age of revolution and reformation. The wisdom of our sages and blood of our heroes have been devoted to their attainment. They should be the creed of our political faith, the text of civic instruction, the touchstone by which to try the services of those we trust; and should we wander from them in moments of error or of alarm, let us hasten to retrace our steps and to regain the road which alone leads to peace, liberty, and safety.

I repair, then, fellow citizens, to the post you have assigned me. With experience enough in subordinate offices to have seen the difficulties of this the greatest of all, I have learnt to expect that it will rarely fall to the lot of imperfect man to retire from this station with the reputation and the favor which bring him into it. Without pretensions to that high confidence you reposed in our first and greatest revolutionary character, whose preeminent services had entitled him to the first place in his country's love and destined for him the fairest page in the volume of faithful history, I ask so much confidence only as may give firmness and effect to the legal administration of your affairs. I shall often go wrong through defect of judgment. When right, I shall often be thought wrong by those whose positions will not command a view of the whole ground. I ask your indulgence for my own errors, which will never be intentional, and your support against the errors of others, who may condemn what they would not if seen in all its parts. The approbation implied by your suffrage is a great consolation to me for the past, and my future solicitude will be to retain the good opinion of those who have bestowed it in advance, to conciliate that of others by doing them all the good in my power, and to be instrumental to the happiness and freedom of all.

Relying, then, on the patronage of your good will, I advance with obedience to the work, ready to retire from it whenever you become sensible how much better choice it is in your power to make. And may that Infinite Power which rules the destinies of the universe lead our councils to what is best, and give them a favorable issue for your peace and prosperity.

Thomas Jefferson

· SECOND INAUGURAL ADDRESS ·

Monday, March 4, 1805

*T*he United States doubled in size during Thomas Jefferson's first term in office, an astonishing expansion made possible by an equally enlarged view of the powers of the presidency. Thomas Jefferson believed in small government, decentralized power, and a strict interpretation of the U.S. Constitution. Unlike Alexander Hamilton, he did not envision the new nation as a great power with an expansionist appetite. And yet, when the French emperor Napoléon offered to sell the Louisiana Territory to the United States for a scant fifteen million dollars, Jefferson seized the opportunity, putting aside doubts about his power to purchase land from a foreign government.

The Louisiana Purchase stands as one of the great achievements of any President. Yet when Jefferson addressed the issue in his second inaugural, he made a point of noting that the acquisition had "been disapproved by some" because they—Federalists, mostly—believed such a huge expansion "would endanger its union." History would justify their fears, for the question of slavery in new territories carved out of the purchase truly did endanger the Union and helped inspire a movement to dissolve it. At the time, however, Jefferson argued that a larger country would help dilute "local passions."

Thomas Jefferson's second inaugural lacked the eloquent framework and soaring statement of principles that made his first one of the finest. The new President, who certainly took great pride in his writing, noted that his first speech was a general statement of principles, appropriate for an incoming President. His second, he said, was a summation of the ways in which he carried out the vision of the first speech. Jefferson's second inaugural address, then, is more a self-evaluation than a philosophical exposition.

The fifth inauguration in the young nation's history provided yet another political milestone, for it marked the first time an incumbent President began a second term with a new vice president. Aaron Burr, whom the House of Representatives chose as vice president in 1801, quickly fell out of Jefferson's graces and was not renominated for vice president by the Republican Party in 1804. Instead, the party chose New York governor George Clinton as Jefferson's running mate. (The election of 1804 was the first in which electors voted separately for President and vice president, thanks to changes in the voting procedure outlined in the Twelfth Amendment.)

So Aaron Burr became the first vice president to be unceremoniously dumped from the ticket. Seeking a measure of political vindication, he promptly ran for governor of New York in the spring of 1804 but lost after Hamilton campaigned aggressively against him. The two men famously decided to settle their bitter differences with pistols, leading to Hamilton's death in a duel with Burr on July 11, 1804. Burr was finished as a political figure, although he did continue to serve as vice president until Inauguration Day 1805, when Clinton officially succeeded him.

The Burr-Hamilton duel left the Federalist Party without its strongest figure and helped hasten the party's demise. The party barely put up a fight in the 1804 election, nominating Charles C. Pinckney of South Carolina and Rufus King from New York. They won just three states and 14 electoral votes, to 162 votes and fifteen states for the Jefferson-Clinton ticket.

Jefferson's second inauguration took place in the shadow of the shocking duel, but, of course, the President made no mention of his former vice president's role in the affair. He had more positive developments to hail, among the first being the elimination of internal taxes. He noted with pride that he had presided over the "suppression of unnecessary offices, of useless establishments and expenses," all of which allowed him to eliminate taxes. Jefferson noted that "it may be the pleasure and the pride of an American to ask, What farmer, what mechanic, what laborer ever sees a taxgatherer of the United States?" But one group absent from that list did, in fact, see a taxgatherer from time to time: the rich. Jefferson continued to tax foreign-made goods, arguing that the tax was paid "chiefly by those who can afford to add foreign luxuries to domestic comforts."

The Louisiana Purchase added more than just territory to U.S. control. It also added people, the Native Americans who lived along the Mississippi Valley, in the plains of the central Midwest, and in the rugged lands of the

Pacific Northwest. In his speech, Jefferson directly confronted the existence of Native Americans with a mix of admiration and paternal impatience. At a time when many of his fellow citizens regarded the Indians as uncivilized barbarians, Jefferson insisted that they were "endowed with the faculties and rights of men, breathing an ardent love of liberty and independence, and occupying a country which left them no desire but to be undisturbed." But Jefferson believed that the white man's attempts to civilize the Indians was doomed, because of the Indians' "ignorance" and "pride."

While Jefferson enjoyed a measure of popularity that eluded his predecessor, he could not let the moment pass without venting about his critics in the press. He complained, "During this course of administration, and in order to disturb it, the artillery of the press has been leveled against us." He made it clear that he wished somebody had done something about these meddlesome journalists. The abuses of the press, he said, "might, indeed, have been corrected by the wholesome punishments reserved to and provided by the laws of the several states against falsehood and defamation."

Some of Jefferson's critics focused their attention on the President's religious beliefs or, in their view, lack of the same. Jefferson's refusal to designate days of prayer and thanksgiving seemed to confirm suspicions that the man had suspect beliefs. Early in his administration, he wrote to members of the Danbury Baptist Association in Connecticut, asserting his belief that there should be a wall of separation between church and state. For some critics, the President's attitude bordered on outright atheism.

Jefferson sought to explain himself in his second inaugural, insisting that he believed the "free exercise" of religion was beyond government's control, and so he refused to "prescribe . . . religious exercises."

Just in case his listeners thought they detected a whiff of heresy in these sentiments, Jefferson concluded his address with a plea for "the favor of that Being in whose hands we are . . . who has covered our infancy with His providence and our riper years with His wisdom and power." While not a declaration of membership in any particular denomination, Jefferson's prayer certainly showed a spiritual side of one of the nation's foremost intellects and one of its most complicated Presidents.

PROCEEDING, FELLOW CITIZENS, to that qualification which the Constitution requires before my entrance on the charge again conferred on me, it is my duty to express the deep sense I entertain of this new proof of confidence from my fellow citizens at large, and the zeal with which it inspires me so to conduct myself as may best satisfy their just expectations.

On taking this station on a former occasion I declared the principles on which I believed it my duty to administer the affairs of our commonwealth. My conscience tells me I have on every occasion acted up to that declaration according to its obvious import and to the understanding of every candid mind.

In the transaction of your foreign affairs we have endeavored to cultivate the friendship of all nations, and especially of those with which we have the most important relations. We have done them justice on all occasions, favored where favor was lawful, and cherished mutual interests and intercourse on fair and equal terms. We are firmly convinced, and we act on that conviction, that with nations as with individuals our interests soundly calculated will ever be found inseparable from our moral duties, and history bears witness to the fact that a just nation is trusted on its word when recourse is had to armaments and wars to bridle others.

At home, fellow citizens, you best know whether we have done well or ill. The suppression of unnecessary offices, of useless establishments and expenses, enabled us to discontinue our internal taxes. These, covering our land with officers and opening our doors to their intrusions, had already begun that process of domiciliary vexation which once entered is scarcely to be restrained from reaching successively every article of property and produce. If among these taxes some minor ones fell which had not been inconvenient, it was because their amount would not have paid the officers who collected them, and because, if they had any merit, the state authorities might adopt them instead of others less approved.

The remaining revenue on the consumption of foreign articles is paid chiefly by those who can afford to add foreign luxuries to domestic comforts, being collected on our seaboard and frontiers only, and incorporated with the transactions of our mercantile citizens, it may be the pleasure and the pride of an American to ask, What farmer, what mechanic, what laborer ever sees a taxgatherer of the United States? These contributions enable us to support the current expenses of the government, to fulfill contracts with foreign nations, to extinguish the native right of soil within our limits, to extend those limits, and to apply such a surplus to our public debts as places at a short day their final redemption, and that redemption once effected the revenue thereby liberated may, by a just repartition of it among the states and a corresponding amendment of the Constitution, be applied in time of peace to rivers, canals, roads, arts, manufactures, education, and other great objects within each state. In time of war, if injustice by ourselves or others must sometimes produce war, increased as the same revenue will be by increased population and consumption, and aided by other resources reserved for that crisis, it may meet within the year all the expenses of the year without encroaching on the rights of future generations by burthening them with the debts of the past. War will then be but a suspension of useful works, and a return to a state of peace, a return to the progress of improvement.

I have said, fellow citizens, that the income reserved had enabled us to extend our limits, but that extension may possibly pay for itself before we are called on, and in the meantime may keep down the accruing interest; in all events, it will replace the advances we shall have made. I know that the acquisition of Louisiana had been disapproved by some from a candid apprehension that the enlargement of our territory would endanger its union. But who can limit the extent to which the federative principle may operate effectively? The larger our association the less will it be shaken by local passions; and in any view is it not better that the opposite bank of the Mississippi should be settled by our own brethren and children than by strangers of another family? With which should we be most likely to live in harmony and friendly intercourse?

In matters of religion I have considered that its free exercise is placed by the Constitution independent of the powers of the general government. I have therefore undertaken on no occasion to prescribe the religious exercises suited to it, but have left them, as the Constitution found them, under the direction and discipline of the church or state authorities acknowledged by the several religious societies.

The aboriginal inhabitants of these countries I have regarded with the commiseration their history inspires. Endowed with the faculties and the rights of men, breathing an ardent love of liberty and independence, and occupying a country which left them no desire but to be undisturbed, the stream of overflowing population from other regions directed itself on these shores; without power to divert or habits to contend against it, they have been overwhelmed by the current or driven before it; now reduced within limits too narrow for the hunter's state, humanity enjoins us to teach them agriculture and the domestic arts; to encourage them to that industry which alone can enable them to maintain their place in existence and to prepare them in time for that state of society which to bodily comforts adds the improvement of the mind and morals. We have therefore liberally furnished them with the implements of husbandry and household use; we have placed among them instructors in the arts of first necessity, and they are covered with the aegis of the law against aggressors from among ourselves.

But the endeavors to enlighten them on the fate which awaits their present course of life, to induce them to exercise their reason, follow its dictates, and change their pursuits with the change of circumstances have powerful obstacles to encounter; they are combated by the habits of their bodies, prejudices of their minds, ignorance, pride, and the influence of interested and crafty individuals among them who feel themselves something in the present order of things and fear to become nothing in any other. These persons inculcate a sanctimonious reverence for the customs of their ancestors; that whatsoever they did must be done through all time; that reason is a false guide, and to advance under its counsel in their physical, moral, or political condition is perilous innovation; that their duty is to remain as their Creator made them, ignorance being safety and knowledge full of danger; in short, my friends, among them also is seen the action and counteraction of good sense and of bigotry; they too have their anti-philosophists who find an interest in keeping things in their present state, who dread reformation, and exert all their faculties to maintain the ascendancy of habit over the duty of improving our reason and obeying its mandates.

In giving these outlines I do not mean, fellow citizens, to arrogate to myself the merit of the measures. That is due, in the first place, to the reflecting character of our citizens at large, who, by the weight of public opinion, influence and strengthen the public measures. It is due to the sound discretion with which they select from among themselves those to whom they confide the legislative duties. It is due to the zeal and wisdom of the characters thus selected, who lay the foundations of public happiness in wholesome laws, the execution of which alone remains for others, and it is due to the able and faithful auxiliaries, whose patriotism has associated them with me in the executive functions.

During this course of administration, and in order to disturb it, the artillery of the press has been leveled against us, charged with whatsoever its licentiousness could devise or dare. These abuses of an institution so important to freedom and science are deeply to be regretted, inasmuch as they tend to lessen its usefulness and to sap its safety. They might, indeed, have been corrected by the wholesome punishments reserved to and provided by the laws of the several states against falsehood and defamation, but public duties more urgent press on the time of public servants, and the offenders have therefore been left to find their punishment in the public indignation.

Nor was it uninteresting to the world that an experiment should be fairly and fully made, whether freedom of discussion, unaided by power, is not sufficient for the propagation and protection of truth—whether a government conducting itself in the true spirit of its constitution, with zeal and purity, and doing no act which it would be unwilling the whole world should witness, can be written down by falsehood and defamation. The experiment has been tried; you have witnessed the scene; our fellow citizens looked on, cool and collected; they saw the latent source from which these outrages proceeded; they gathered around their public functionaries, and when the Constitution called them to the decision by suffrage, they pronounced their verdict, honorable to those who had served them and consolatory to the friend of man who believes that he may be trusted with the control of his own affairs.

No inference is here intended that the laws provided by the states against false and defamatory publications should not be enforced; he who has time renders a service to public morals and public tranquillity in reforming these abuses by the salutary coercions of the law; but the experiment is noted to prove that, since truth and reason have maintained their ground against false opinions in league with false facts, the press, confined to truth, needs no other legal restraint; the public judgment will correct false reasoning and opinions on a full hearing of all parties; and no other definite line can be drawn between the inestimable liberty of the press and its demoralizing licentiousness. If there be still improprieties which this rule would not restrain, its supplement must be sought in the censorship of public opinion.

Contemplating the union of sentiment now manifested so generally as auguring harmony and happiness to our future course, I offer to our country sincere congratulations. With those, too, not yet rallied to the same point the disposition to do so is gaining strength; facts are piercing through the veil drawn over them, and our doubting brethren will at length see that the mass of their fellow citizens with whom they can not yet resolve to act as to principles and measures, think as they think and desire what they desire; that our wish as well as theirs is that the public efforts may be directed honestly to the public good, that peace be cultivated, civil and religious liberty unassailed, law and order preserved, equality of rights maintained, and that state of property, equal or unequal, which results to every man from his own industry or that of his father's. When satisfied of these views it is not in human nature that they should not approve and support them. In the meantime let us cherish them with patient affection, let us do them justice, and more than justice, in all competitions of interest; and we need not doubt that truth, reason, and their own interests will at length prevail, will gather them into the fold of their country, and will complete that entire union of opinion which gives to a nation the blessing of harmony and the benefit of all its strength.

I shall now enter on the duties to which my fellow citizens have again called me, and shall proceed in the spirit of those principles which they have approved. I fear not that any motives of interest may lead me astray; I am sensible of no passion which could seduce me knowingly from the path of justice, but the weaknesses of human nature and the limits of my own understanding will produce errors of judgment sometimes injurious to your interests. I shall need, therefore, all the indulgence which I have heretofore experienced from my constituents; the want of it will certainly not lessen with increasing years. I shall need, too, the favor of that Being in whose hands we are, who led our fathers, as Israel of old, from their native land and planted them in a country flowing with all the necessaries and comforts of life; who has covered our infancy with His providence and our riper years with His wisdom and power, and to whose goodness I ask you to join in supplications with me that He will so enlighten the minds of your servants, guide their councils, and prosper their measures that whatsoever they do shall result in your good, and shall secure to you the peace, friendship, and approbation of all nations.

James Madison

Saturday, March 4, 1809

$\mathscr{B}$efore turning over the presidency to his secretary of state, Thomas Jefferson told a friend: "Never did a prisoner, released from his chains, feel such relief as I shall on shaking off the shackles of power."[1] The headaches of Jefferson's last few years in office were numerous. Like his predecessor, Jefferson tried to keep the young nation out of the Old World's quarrels, and, like John Adams, he paid a political price for doing so, although for different reasons.

In late 1807, at the urging of the Jefferson administration, Congress passed the Embargo Act, which banned American ships from sailing from the United States to the ports of Europe. It was an astounding blunder. Jefferson, infuriated by British interference with American commerce, intended to declare a pox on all the warring houses of Europe, believing they would change their hostility toward neutral shipping. But the Embargo Act simply served to devastate American maritime commerce. The U.S. merchant fleet lay at anchor in the ports of the Northeast, businesses collapsed, and, inevitably, smugglers responded to unabated market demand. It did not pass without notice that a southern planter put in place restrictions on trade based in the industrializing North. Jefferson eventually reversed himself by dropping the blanket ban but substituting an embargo against the two principal antagonists in Europe, England and France.

The Embargo Act was just one of the crises, embarrassments, and frustrations that Jefferson faced in his second term. Aaron Burr's conspiracy to seize Louisiana from the United States and Mexico from Spain unraveled late in his term, a diplomatic embarrassment and a domestic scandal.

---

1. Paul Johnson, *A History of the American People* (New York: Harper Perennial, 1999), 257.

Tensions with Great Britain, embroiled in the horrific Napoleonic wars, were deteriorating to outright hostility. At home, Jefferson complained that he could not get anything done because Federalists still were in charge of the nation's court system. His private affairs were in similar disarray as he sunk further into debt. Like many of his two-term successors, Thomas Jefferson was grouchy and tired after eight years, certainly in no mood to run for a third term.

In stepping aside voluntarily after eight years, Jefferson wittingly or unwittingly solidified the unofficial two-term limit that remained in place until 1940, when Franklin D. Roosevelt successfully sought a third term. While Jefferson said he simply was following Washington's example, there is no evidence that Washington sought to impose such a limit on his successors. There's an argument to be made, then, that Jefferson invented the notion that incumbent Presidents ought to follow the first President's precedent by retiring after eight years.

If Jefferson's ordeal was coming to an end, Madison's was just beginning, for the world was no less placid, no less dangerous, on March 4, 1809, than it had been during Jefferson's final few years. But the volatility of international relations played to Madison's strength: he was Jefferson's only secretary of state, an experienced diplomat in addition to being a framer of the Constitution and one of the writers of *The Federalist*. In the election of 1808, Madison easily defeated perennial Federalist candidate Charles Pinckney as well as Jefferson's vice president, George Clinton, in the general election. But, in another constitutional quirk, Vice President Clinton remained in office despite his loss in the presidential race, making him the first person, and only one of two in U.S. history, to serve as vice president under two Presidents.

Inauguration Day 1805 was a Saturday, and it began with the concussions of cannon fire, an early morning salute to the new President. Thousands waved as Madison made his way to the Capitol along Pennsylvania Avenue. The outgoing President was among the dignitaries waiting on his arrival. Thomas Jefferson did not duck out of town, as John Adams did, before witnessing the swearing-in of his successor.

According to Madison's biographer Irving Brant, the new President looked pale and nervous as he began his inaugural address. An accomplished politician and diplomat, Madison was hardly a novice as a public speaker. Perhaps the moment unnerved him, or perhaps it was the content of his speech. In the second paragraph of his inaugural address, Madison observed that "The present situation of the world is indeed without parallel, and that of our own country full of difficulties." Despite the country's material prosperity and maturation, it remained vulnerable to the rivalries and blood feuds of the Old World. That vulnerability might well have made even an experienced leader like Madison nervous.

Most of his speech is taken up with the crisis in Europe, with its implications for American commerce caught up in the Old World's embargoes and blockades. If Madison saw war on the horizon, he preferred to direct the nation's gaze to the more pleasant vista of benevolent neutrality. The United States, he said, was at peace because it observed its "neutral obligations with the most scrupulous impartiality."

As the fourth person to become President, Madison had the luxury of turning to precedents as he prepared for his swearing-in. It certainly seems possible that as he contemplated what he might say in his first speech, he may have taken a glance at Adams's address. For stuck in the middle of Madison's speech is an ungainly sentence of nearly five hundred words, ponderous nouns and verbs held together by fragile semicolons. The gist of the sentence, if there is gist to be had, seems to be that under a Madison administration the United States would stand aside from Europe's quarrels, obey the Constitution (not a surprise, unless Madison chose to disown his own work), pay down the national debt, advance commerce and science, and continue the "conversion of our aboriginal neighbors" from their "savage life" to a "civilized state."

Joining Madison in a postinaugural reception was his wife, Dorothea, commonly remembered as Dolley. She was the first First Lady since Abigail Adams, for Jefferson was a widower during his presidency, and she resembled neither the cerebral Mrs. Adams nor the matronly Martha Washington. She turned forty-one in the spring of 1809, seventeen years her husband's junior and much younger than her predecessors. By force of personality and character, she turned the new White House into a salon where Washington power brokers gathered to talk politics.

It was not long before those conversations turned to talk of war.

Bibliographic Note
Irving Brant, *The Fourth President: A Life of James Madison*
(New York: The Bobbs-Merrill Company, 1970).

UNWILLING TO DEPART FROM EXAMPLES of the most revered authority, I avail myself of the occasion now presented to express the profound impression made on me by the call of my country to the station to the duties of which I am about to pledge myself by the most solemn of sanctions. So distinguished a mark of confidence, proceeding from the deliberate and tranquil suffrage of a free and virtuous nation, would under any circumstances have commanded my gratitude and devotion, as well as filled me with an awful sense of the trust to be assumed. Under the various circumstances which give peculiar solemnity to the existing period, I feel that both the honor and the responsibility allotted to me are inexpressibly enhanced.

The present situation of the world is indeed without a parallel and that of our own country full of difficulties. The pressure of these, too, is the more severely felt because they have fallen upon us at a moment when the national prosperity being at a height not before attained, the contrast resulting from the change has been rendered the more striking. Under the benign influence of our republican institutions, and the maintenance of peace with all nations whilst so many of them were engaged in bloody and wasteful wars, the fruits of a just policy were enjoyed in an unrivaled growth of our faculties and resources. Proofs of this were seen in the improvements of agriculture, in the successful enterprises of commerce, in the progress of manufacturers and useful arts, in the increase of the public revenue and the use made of it in reducing the public debt, and in the valuable works and establishments everywhere multiplying over the face of our land.

It is a precious reflection that the transition from this prosperous condition of our country to the scene which has for some time been distressing us is not chargeable on any unwarrantable views, nor, as I trust, on any involuntary errors in the public councils. Indulging no passions which trespass on the rights or the repose of other nations, it has been the true glory of the United States to cultivate peace by observing justice, and to entitle themselves to the respect of the nations at war by fulfilling their neutral obligations with the most scrupulous impartiality. If there be candor in the world, the truth of these assertions will not be questioned; posterity at least will do justice to them.

This unexceptionable course could not avail against the injustice and violence of the belligerent powers. In their rage against each other, or impelled by more direct motives, principles of retaliation have been introduced equally contrary to universal reason and acknowledged law. How long their arbitrary edicts will be continued in spite of the demonstrations that not even a pretext for them has been given by the United States, and of the fair and liberal attempt to induce a revocation of them, cannot be anticipated. Assuring myself that under every vicissitude the determined spirit and united councils of the nation will be safeguards to its honor and its essential interests, I repair to the post assigned me with no other discouragement than what springs from my own inadequacy to its high duties. If I do not sink under the weight of this deep conviction it is because I find some support in a consciousness of the purposes and a confidence in the principles which I bring with me into this arduous service.

To cherish peace and friendly intercourse with all nations having correspondent dispositions; to maintain sincere neutrality toward belligerent nations; to prefer in all cases amicable discussion and reasonable accommodation of differences to a decision of them by an appeal to arms; to exclude foreign intrigues and foreign partialities, so degrading to all countries and so baneful to free ones; to foster a spirit of independence too just to invade the rights of others, too proud to surrender our own, too liberal to indulge unworthy prejudices ourselves and too elevated not to look down upon them in others; to hold the union of the states as the basis of their peace and happiness; to support the Constitution which is the cement of the Union, as well in its limitations as in its authorities; to respect the rights and authorities reserved to the states and to the people as equally incorporated with and essential to the success of the general system; to avoid the slightest interference with the right of conscience or the functions of religion, so wisely exempted from civil jurisdiction; to preserve in their full energy the other salutary provisions in behalf of private and personal rights, and of the freedom of the press; to observe economy in public expenditures; to liberate the public resources by an honorable discharge of the public debts; to keep within the requisite limits a standing military force, always remembering that an armed and trained militia is the firmest bulwark of republics—that without standing armies their liberty can never be in danger, nor with large ones safe; to promote by authorized means improvements friendly to agriculture, to manufactures, and to external as well as internal commerce; to favor in like manner the advancement of science and the diffusion of information as the best aliment to true liberty; to carry on the benevolent plans which have been so meritoriously applied to the conversion of our aboriginal neighbors from the degradation and wretchedness of savage life to a participation of the improvements of which the human mind and manners are susceptible in a civilized state—as far as sentiments and intentions such as these can aid the fulfillment of my duty, they will be a resource which cannot fail me.

It is my good fortune, moreover, to have the path in which I am to tread lighted by examples of illustrious services successfully rendered in the most trying difficulties by those who have marched before me. Of those of my immediate predecessor it might least become me here to speak. I may, however, be pardoned for not suppressing the sympathy with which my heart is full in the rich reward he enjoys in the benedictions of a beloved country, gratefully bestowed on exalted talents zealously devoted through a long career to the advancement of its highest interest and happiness.

But the source to which I look or the aids which alone can supply my deficiencies is in the well-tried intelligence and virtue of my fellow citizens, and in the counsels of those representing them in the other departments associated in the care of the national interests. In these my confidence will under every difficulty be best placed, next to that which we have all been encouraged to feel in the guardianship and guidance of that Almighty Being whose power regulates the destiny of nations, whose blessings have been so conspicuously dispensed to this rising Republic, and to whom we are bound to address our devout gratitude for the past, as well as our fervent supplications and best hopes for the future.

James Madison

· SECOND INAUGURAL ADDRESS ·

Thursday, March 4, 1813

*N*ever before in the short history of the United States had an election taken place in the midst of war; never before had there been a formal inauguration even as soldiers and sailors were in harm's way. On March 4, 1813, a wartime President, James Madison, addressed the nation with just one mission in mind: to justify the war raging along the Canadian border, in today's Midwest, and in the Great Lakes and the Atlantic Ocean.

Madison, the onetime diplomat and architect of Thomas Jefferson's foreign policy, had failed to avoid open conflict with Great Britain, even though the British were far more concerned with beating back the challenge of Napoléon in Europe. The President seemed content to wait on events, hoping that the conflict across the Atlantic would end and with it the threat to American merchant shipping. The British continued to abduct American merchant sailors on the high seas, tried to block U.S. ships from trading with continental ports, and maintained forts along the northern border of the United States despite agreements to dismantle them. France, which had sought to stop maritime commerce with Britain, changed its policy in 1810. It was a shrewd move, because the British continued to enforce its blockade of the Continent, worsening tensions between the British and the Americans. Still, Madison made no move to strengthen an army that had just over eleven thousand troops.

In Congress, however, a group of pro-war members of his own party, the Democratic-Republicans, beat the drums for war, convinced that Britain was so overextended in Europe that it would be unable to resist a quick American march on Canada. For the so-called war hawks, a splendid little war would chase the British and their ally, Spain, out of North America, and would offer an opportunity to punish another British ally, the Native American tribes in the Midwest, for their continued resistance to American settlement.

The Federalists, now little more than a regional faction based in New England, opposed any move to war. New England was the maritime capital of the country, and while its fleet and sailors bore the brunt of British harassment on the high seas, its political and business leaders believed a war with the world's best navy would only make matters much worse.

Despite these partisan and regional differences, by the middle of 1812, Madison decided that war with Britain was the nation's only option. But as the principal author of the Constitution, he had helped put in place safeguards designed to make sure that questions of war were properly debated. Accordingly, on June 1, 1812, he sent a message to the Senate and the House of Representatives outlining British offenses against the United States, adding that the question of war or peace "is a solemn question which the Constitution wisely confides to the legislative department of the government." So the constitutional procedure for declaring war was tested for the first time. Democratic-Republicans such as Henry Clay argued the case for war, while Federalists were almost unanimously against a formal declaration of hostilities. The debate raged until June 18, when, for the first time, the two houses of Congress voted on a war declaration. It passed in the House, seventy-nine to forty-nine, and in the Senate, nineteen to thirteen. Some members of Madison's party either abstained or, in the case of John Randolph, who hailed from the President's home state of Virginia, voted against the declaration. Ironically, the war's critics took to calling the war "Mr. Madison's war," although the President was a reluctant warrior. The war would have been better labeled "Mr. Clay's war," after one of its more enthusiastic advocates.

About two weeks after war was declared, President Madison was renominated by the Democratic-Republican congressional caucus that had supported the war resolution. But a rump group of Democratic-Republicans got behind the candidacy of New York's De Witt Clinton, who was, ironically enough, the nephew of Madison's vice president, George Clinton, who had died in office. The antiwar Federalists agreed to join the Democratic-Republican dissidents in supporting Clinton.

The wartime election of 1812 reflected the country's bitter divisions. Clinton carried his home state of New York and New England, except for Vermont, while Madison swept the South, except for Maryland, and West.

As Madison became the first President to deliver a wartime inaugural address, it was clear that the war hawks' dream of marching on Canada and expelling the British was just that—a dream, or perhaps more of a nightmare. From late summer into the fall of 1812, American forces were humiliated along the Canadian border. The worst fiasco took place in Detroit, where General William Hull was supposed to launch an invasion of Canada. Instead, he ordered a hasty surrender to a small force of British troops and Indian allies.

Madison's address was an attempt to rally the country around the flag and to stir up resentment against the British, especially because of their dangerous alliance with Native American tribes. Madison raised the issue of Britain's allies in blunt language, asserting that they "have let loose the savages armed with these cruel instruments; have allured them into their service, and carried them to battle by their sides, eager to glut their savage thirst with the blood of the vanquished and to finish the work of torture and death on maimed and defenseless captives." This was the Indian card, and Madison clearly believed that playing it would help isolate critics who believed the war was unnecessary.

The speech did not have the rallying effect Madison hoped it would have. Indeed, more disasters followed, culminating in the burning of Washington in 1814. A group of New England leaders gathered in Hartford, Connecticut, in late 1814 and early 1815 to discuss secession. Even as they did, the governor of Massachusetts sent secret messages to Great Britain, offering a separate peace between the state and the enemy.

Once the fortunes of war turned, and especially after Andrew Jackson's spectacular victory at the Battle of New Orleans on January 8, 1815, the Hartford Convention was almost instantly discredited and the Federalist Party mortally wounded. Madison's war, so terribly unpopular for more than two years, ended with an explosion of patriotism and nationalism.

Bibliographic Note
Irving Brant, *The Fourth President: A Life of James Madison*
(New York: The Bobbs-Merrill Company, 1970).

ABOUT TO ADD THE SOLEMNITY OF AN OATH to the obligations imposed by a second call to the station in which my country heretofore placed me, I find in the presence of this respectable assembly an opportunity of publicly repeating my profound sense of so distinguished a confidence and of the responsibility united with it. The impressions on me are strengthened by such an evidence that my faithful endeavors to discharge my arduous duties have been favorably estimated, and by a consideration of the momentous period at which the trust has been renewed. From the weight and magnitude now belonging to it I should be compelled to shrink if I had less reliance on the support of an enlightened and generous people, and felt less deeply a conviction that the war with a powerful nation, which forms so prominent a feature in our situation, is stamped with that justice which invites the smiles of heaven on the means of conducting it to a successful termination.

May we not cherish this sentiment without presumption when we reflect on the characters by which this war is distinguished?

It was not declared on the part of the United States until it had been long made on them, in reality though not in name; until arguments and postulations had been exhausted; until a positive declaration had been received that the wrongs provoking it would not be discontinued; nor until this last appeal could no longer be delayed without breaking down the spirit of the nation, destroying all confidence in itself and in its political institutions, and either perpetuating a state of disgraceful suffering or regaining by more costly sacrifices and more severe struggles our lost rank and respect among independent powers.

On the issue of the war are staked our national sovereignty on the high seas and the security of an important class of citizens whose occupations give the proper value to those of every other class. Not to contend for such a stake is to surrender our equality with other powers on the element common to all and to violate the sacred title which every member of the society has to its protection. I need not call into view the unlawfulness of the practice by which our mariners are forced at the will of every cruising officer from their own vessels into foreign ones, nor paint the outrages inseparable from it. The proofs are in the records of each successive administration of our government, and the cruel sufferings of that portion of the American people have found their way to every bosom not dead to the sympathies of human nature.

As the war was just in its origin and necessary and noble in its objects, we can reflect with a proud satisfaction that in carrying it on no principle of justice or honor, no usage of civilized nations, no precept of courtesy or humanity, have been infringed. The war has been waged on our part with scrupulous regard to all these obligations, and in a spirit of liberality which was never surpassed.

How little has been the effect of this example on the conduct of the enemy!

They have retained as prisoners of war citizens of the United States not liable to be so considered under the usages of war.

They have refused to consider as prisoners of war, and threatened to punish as traitors and deserters, persons emigrating without restraint to the United States, incorporated by naturalization into our political family, and fighting under the authority of their adopted country in open and honorable war for the maintenance of its rights and safety. Such is the avowed purpose of a government which is in the practice of naturalizing by thousands citizens of other countries, and not only of permitting but compelling them to fight its battles against their native country.

They have not, it is true, taken into their own hands the hatchet and the knife, devoted to indiscriminate massacre, but they have let loose the savages armed with these cruel instruments; have allured them into their service, and carried them to battle by their sides, eager to glut their savage thirst with the blood of the vanquished and to finish the work of torture and death on maimed and defenseless captives. And, what was never before seen, British commanders have extorted victory over the unconquerable valor of our troops by presenting to the sympathy of their chief captives awaiting massacre from their savage associates. And now we find them, in further contempt of the modes of honorable warfare, supplying the place of a conquering force by attempts to disorganize our political society, to dismember our confederated Republic. Happily, like others, these will recoil on the authors; but they mark the degenerate counsels from which they emanate, and if they did not belong to a sense of unexampled inconsistencies might excite the greater wonder as proceeding from a government which founded the very war in which it has been so long engaged on a charge against the disorganizing and insurrectional policy of its adversary.

To render the justice of the war on our part the more conspicuous, the reluctance to commence it was followed by the earliest and strongest manifestations of a disposition to arrest its progress. The sword was scarcely out of the scabbard before the enemy was apprised of the reasonable terms on which it would be resheathed. Still more precise advances were repeated, and have been received in a spirit forbidding every reliance not placed on the military resources of the nation.

These resources are amply sufficient to bring the war to an honorable issue. Our nation is in number more than half that of the British Isles. It is composed of a brave, a free, a virtuous, and an intelligent people. Our country abounds in the necessaries, the arts, and the comforts of life. A general prosperity is visible in the public countenance. The means employed by the British cabinet to undermine it have recoiled on themselves; have given to our national faculties a more rapid development, and, draining or diverting the precious metals from British circulation and British vaults, have poured them into those of the United States. It is a propitious consideration that an unavoidable war should have found this seasonable facility for the contributions required to support it. When the public voice called for war, all knew, and still know, that without them it could not be carried on through the period which it might last, and the patriotism, the good sense, and the manly spirit of our fellow citizens are pledges for the cheerfulness with which they will bear each his share of the common burden. To render the war short and its success sure, animated and systematic exertions alone are necessary, and the success of our arms now may long preserve our country from the necessity of another resort to them. Already have the gallant exploits of our naval heroes proved to the world our inherent capacity to maintain our rights on one element. If the reputation of our arms has been thrown under clouds on the other, presaging flashes of heroic enterprise assure us that nothing is wanting to correspondent triumphs there also but the discipline and habits which are in daily progress.

James Monroe

Tuesday, March 4, 1817

*I*t was an unseasonably warm late-winter's day when James Monroe became the fourth Virginian to take the oath of office as President. Thousands turned out for the inaugural ceremonies, eager to catch a glimpse of the new President and to celebrate their liberation from winter's cold grasp.

The city of Washington, D.C., bore scars from a brief British occupation during the War of 1812. The Capitol building and the White House were burned when British forces entered the city in August 1814, seeking to avenge the American burning of the Canadian parliament building earlier in the war. Bad weather forced the British to withdraw from the capital after a little more than a day, so reconstruction of the government buildings began almost immediately.

The renovations on Capitol Hill were still under way on Inauguration Day, so the new President took the oath on a platform constructed in front of a building that housed Congress until members could return to the Capitol. The ceremony itself was not exactly an occasion for nonpartisan collegiality and good cheer. Members of the House of Representatives and the Senate feuded bitterly over the inaugural ceremonies. The representatives barred the senators from bringing their armchairs into the temporary House chamber, where the ceremony was scheduled to take place. The senators refused to back down, and so the ceremony was moved to an open-air space in front of the temporary houses of Congress.

The petty feud between the House and Senate did not bode well for one of the new President's avowed goals: to realize, finally, George Washington's dream of a government devoid of party and faction. The opportunity seemed to be within reach, because the Federalists were, in fact, a party on the verge of extinction. They hadn't elected a President since John

Adams in 1796. The Madison administration had co-opted some signature Federalist issues, including support for the Second Bank of the United States and protective tariffs. Finally, in the aftermath of the War of 1812, they were seen, not as principled critics of a dubious war, but as carping traitors who seemed ready to separate from the Union during a time of war. Although the War of 1812 was decidedly unpopular for a time, Andrew Jackson's victory at the Battle of New Orleans persuaded many Americans that they had beaten the British again. The wave of patriotism was hardly mitigated by the fact that the Battle of New Orleans actually was fought after the Americans and British signed the Treaty of Ghent on Christmas Eve 1814 but before the United States ratified the document. Technically, then, the battle took place while both sides were at war. The battle's outcome, which left two thousand British dead compared to a dozen Americans, affected the way in which American opinion came to view the war.

The Federalists were crushed in the presidential election of 1816. James Monroe, candidate of the Democratic-Republicans, won 183 electoral votes, while the Federalist candidate, Rufus King, won just 34. King would be the last serious Federalist contender for the presidency.

Coming off such a smashing win, Monroe entertained the hope that he might develop a consensus government that would include Federalists as well as members of his own party. Andrew Jackson urged the new President to name William Drayton, a prominent Federalist, to be secretary of war. Others urged Monroe to appoint Daniel Webster, the great orator and New England Federalist, as secretary of state. In the end, however, Monroe chose convention over innovation, choosing only fellow Democratic-Republicans for his cabinet and other top positions. He had no lack of talent from which to choose: He named John Quincy Adams as secretary of state and John C. Calhoun (all of thirty-five years old) as secretary of war. Both men were immensely practical, the sort of politicians who could craft the kind of consensus-driven government Monroe wished to pursue. Before long, it was hard to find an ambitious politician in Washington, D.C., who did not describe himself as a Democratic-Republican. But that renewed sense of common purpose could no more hide reality than the Capitol renovations could erase memories of the British invasion. Slavery, sectional differences, and other issues promised divisions and bitterness to come.

About eight thousand people gathered in the open air to hear Monroe's inaugural address. As was his custom, the new President was dressed in knee breeches, which seemed old-fashioned even then. Although a Virginian, he was not manor-born. His speech was devoid of the long, baroque sentences of some of his predecessors. He was not an accomplished speaker, but his first inaugural address is easier to read than his predecessors' speeches. Hidden in its relative simplicity, however, are some revealing comments and startling assertions.

In his recitation of the nation's growing glories, Monroe offered a glimpse of the ways in which the founders saw themselves and their government. Citing the glories of the Constitution, Monroe asked: "On whom has oppression fallen in any quarter of our Union? Who has been deprived of any right of person or property?" The question today seems absurd: oppression was an everyday burden in Monroe's native Virginia, where thousands were enslaved.

Befitting a postwar inaugural address, Monroe spent a good portion of his speech on national defense, the first inaugural to do so. The War of 1812 demonstrated that the United States might not be able to stand aloof from the Old World's conflicts. Monroe demanded that his listeners acknowledge the possibility that "the United States may be again involved in war." He said the nation ought to have a professional army and navy sufficient to defend it against potential invaders. In addition, the nation's militias ought to be better organized and trained "as to be prepared for any emergency." This sounded more like the Federalist program than that of a committed Jeffersonian, but Monroe wasn't finished. He also insisted that the nation's manufacturers would require "the systematic and fostering care of the government." In other words, he was calling for protective tariffs. This, too, was a break from Jeffersonian-style republicanism, with its emphasis on small government and yeomen farmers.

It was Jefferson who so famously insisted that "We are all Republicans, we are all Federalists." It was Monroe, nearly two decades later, who sought to go beyond platitudes by moving the nation's debate to the center. Historians note that the ensuing "era of good feelings" was more illusion than reality, but Monroe's conciliatory, inclusive approach certainly had one tangible effect: when he ran for reelection in 1820, he faced no opposition.

I SHOULD BE DESTITUTE OF FEELING if I was not deeply affected by the strong proof which my fellow citizens have given me of their confidence in calling me to the high office whose functions I am about to assume. As the expression of their good opinion of my conduct in the public service, I derive from it a gratification which those who are conscious of having done all that they could to merit it can alone feel. My sensibility is increased by a just estimate of the importance of the trust and of the nature and extent of its duties, with the proper discharge of which the highest interests of a great and free people are intimately connected. Conscious of my own deficiency, I cannot enter on these duties without great anxiety for the result. From a just responsibility I will never shrink, calculating with confidence that in my best efforts to promote the public welfare my motives will always be duly appreciated and my conduct be viewed with that candor and indulgence which I have experienced in other stations.

In commencing the duties of the chief executive office it has been the practice of the distinguished men who have gone before me to explain the principles which would govern them in their respective administrations. In following their venerated example my attention is naturally drawn to the great causes which have contributed in a principal degree to produce the present happy condition of the United States. They will best explain the nature of our duties and shed much light on the policy which ought to be pursued in future.

From the commencement of our Revolution to the present day almost forty years have elapsed, and from the establishment of this Constitution twenty-eight. Through this whole term the government has been what may emphatically be called self-government. And what has been the effect? To whatever object we turn our attention, whether it relates to our foreign or domestic concerns, we find abundant cause to felicitate ourselves in the excellence of our institutions. During a period fraught with difficulties and marked by very extraordinary events the United States have flourished beyond example. Their citizens individually have been happy and the nation prosperous.

Under this Constitution our commerce has been wisely regulated with foreign nations and between the states; new states have been admitted into our Union; our territory has been enlarged by fair and honorable treaty, and with great advantage to the original states; the states, respectively protected by the national government under a mild, parental system against foreign dangers, and enjoying within their separate spheres, by a wise partition of power, a just proportion of the sovereignty, have improved their police, extended their settlements, and attained a strength and maturity which are the best proofs of wholesome laws well administered. And if we look to the condition of individuals what a proud spectacle does it exhibit! On whom has oppression fallen in any quarter of our Union? Who has been deprived of any right of person or property? Who restrained from offering his vows in the mode which he prefers to the Divine Author of his being? It is well known that all these blessings have been enjoyed in their fullest extent; and I add with peculiar satisfaction that there has been no example of a capital punishment being inflicted on anyone for the crime of high treason.

Some who might admit the competency of our government to these beneficent duties might doubt it in trials which put to the test its strength and efficiency as a member of the great community of nations. Here too experience has afforded us the most satisfactory proof in its favor. Just as this Constitution was put into action several of the principal states of Europe had become much agitated and some of them seriously convulsed. Destructive wars ensued, which have of late only been terminated. In the course of these conflicts the United States received great injury from several of the parties. It was their interest to stand aloof from the contest, to demand justice from the party committing the injury, and to cultivate by a fair and honorable conduct the friendship of all. War became at length inevitable, and the result has shown that our government is equal to that, the greatest of trials, under the most unfavorable circumstances. Of the virtue of the people and of the heroic exploits of the army, the navy, and the militia I need not speak.

Such, then, is the happy government under which we live—a government adequate to every purpose for which the social compact is formed; a government elective in all its branches, under which every citizen may by his merit obtain the highest trust recognized by the Constitution; which contains within it no cause of discord, none to put at variance one portion of the community with another; a government which protects every citizen in the full enjoyment of his rights, and is able to protect the nation against injustice from foreign powers.

Other considerations of the highest importance admonish us to cherish our Union and to cling to the government which supports it. Fortunate as we are in our political institutions, we have not been less so in other circumstances on which our prosperity and happiness essentially depend. Situated within the temperate zone, and extending through many degrees of latitude along the Atlantic, the United States enjoy all the varieties of climate, and every production incident to that portion of the globe. Penetrating internally to the Great Lakes and beyond the sources of the great rivers which communicate through our whole interior, no country was ever happier with respect to its domain. Blessed, too, with a fertile soil, our produce has always been very abundant, leaving, even in years the least favorable, a surplus for the wants of our fellow men in other countries. Such is our peculiar felicity that there is not a part of our Union that is not particularly interested in preserving it. The great agricultural interest of the nation prospers under its protection. Local interests are not less fostered by it. Our fellow citizens of the North engaged in navigation find great encouragement in being made the favored carriers of the vast productions of the other portions of the United States, while the inhabitants of these are amply recompensed, in their turn, by the nursery for seamen and naval force thus formed and reared up for the support of our common rights. Our manufactures find a generous encouragement by the policy which patronizes domestic industry, and the surplus of our produce a steady and profitable market by local wants in less-favored parts at home.

Such, then, being the highly favored condition of our country, it is the interest of every citizen to maintain it. What are the dangers which menace us? If any exist they ought to be ascertained and guarded against.

In explaining my sentiments on this subject it may be asked, What raised us to the present happy state? How did we accomplish the Revolution? How remedy the defects of the first instrument of our Union, by infusing into the national government sufficient power for national purposes, without impairing the just rights of the states or affecting those of individuals? How sustain and pass with glory through the late war? The government has been in the hands of the people. To the people, therefore, and to the faithful and able depositaries of their trust is the credit due. Had the people of the United States been educated in different principles, had they been less intelligent, less independent, or less virtuous can it be believed that we should have maintained the same steady and consistent career or been blessed with the same success? While, then, the constituent body retains its present sound and healthful state everything will be safe. They will choose competent and faithful representatives for every department. It is only when the people become ignorant and corrupt, when they degenerate into a populace, that they are incapable of exercising the sovereignty. Usurpation is then an easy attainment, and an usurper soon found. The people themselves become the willing instruments of their own debasement and ruin. Let us, then, look to the great cause, and endeavor to preserve it in full force. Let us by all wise and constitutional measures promote intelligence among the people as the best means of preserving our liberties.

Dangers from abroad are not less deserving of attention. Experiencing the fortune of other nations, the United States may be again involved in war, and it may in that event be the object of the adverse party to overset our government, to break our Union, and demolish us as a nation. Our distance from Europe and the just, moderate, and pacific policy of our government may form some security against these dangers, but they ought to be anticipated and guarded against. Many of our citizens are engaged in commerce and navigation, and all of them are in a certain degree dependent on their prosperous state. Many are engaged in the fisheries. These interests are exposed to invasion in the wars between other powers, and we should disregard the faithful admonition of experience if we did not expect it. We must support our rights or lose our character, and with it, perhaps, our liberties. A people who fail to do it can scarcely be said to hold a place among independent nations. National honor is national property of the highest value. The sentiment in the mind of every citizen is national strength. It ought therefore to be cherished.

To secure us against these dangers our coast and inland frontiers should be fortified, our army and navy, regulated upon just principles as to the force of each, be kept in perfect order, and our militia be placed on the best practicable footing. To put our extensive coast in such a state of defense as to secure our cities and interior from invasion will be attended with expense, but the work when finished will be permanent, and it is fair to presume that a single campaign of invasion by a naval force superior to our own, aided by a few thousand land troops, would expose us to greater expense, without taking into the estimate the loss of property and distress of our citizens, than would be sufficient for this great work. Our land and naval forces should be moderate, but adequate to the necessary purposes—the former to garrison and preserve our fortifications and to meet the first invasions of a foreign foe, and, while constituting the elements of a greater force, to preserve the science as well as all the necessary implements of war in a state to be brought into activity in the event of war; the latter, retained within the limits proper in a state of peace, might aid in maintaining the neutrality of the United States with dignity in the wars of other powers and in saving the property of their citizens from spoliation. In time of war, with the enlargement of which the great naval resources of the country render it susceptible, and which should be duly fostered in time of peace, it would contribute essentially, both as an auxiliary of defense and as a powerful engine of annoyance, to diminish the calamities of war and to bring the war to a speedy and honorable termination.

But it ought always to be held prominently in view that the safety of these states and of everything dear to a free people must depend in an eminent degree on the militia. Invasions may be made too formidable to be resisted by any land and naval force which it would comport either with the principles of our government or the circumstances of the United States to maintain. In such cases recourse must be had to the great body of the people, and in a manner to produce the best effect. It is of the highest importance, therefore, that they be so organized and trained as to be prepared for any emergency. The arrangement should be such as to put at the command of the government the ardent patriotism and youthful vigor of the country. If formed on equal and just principles, it can not be oppressive. It is the crisis which makes the pressure, and not the laws which provide a remedy for it. This arrangement should be formed, too, in time of peace, to be the better prepared for war. With such an organization of such a people the United States have nothing to dread from foreign invasion. At its approach an overwhelming force of gallant men might always be put in motion.

Other interests of high importance will claim attention, among which the improvement of our country by roads and canals, proceeding always with a constitutional sanction, holds a distinguished place. By thus facilitating the intercourse between the states we shall add much to the convenience and comfort of our fellow citizens, much to the ornament of the country, and, what is of greater importance, we shall shorten distances, and, by making each part more accessible to and dependent on the other, we shall bind the Union more closely together. Nature has done so much for us by intersecting the country with so many great rivers, bays, and lakes, approaching from distant points so near to each other, that the inducement to complete the work seems to be peculiarly strong. A more interesting spectacle was perhaps never seen than is exhibited within the limits of the United States—a territory so vast and advantageously situated, containing objects so grand, so useful, so happily connected in all their parts!

Our manufacturers will likewise require the systematic and fostering care of the government. Possessing as we do all the raw materials, the fruit of our own soil and industry, we ought not to depend in the degree we have done on supplies from other countries. While we are thus dependent the sudden event of war, unsought and unexpected, cannot fail to plunge us into the most serious difficulties. It is important, too, that the capital which nourishes our manufacturers should be domestic, as its influence in that case instead of exhausting, as it may do in foreign hands, would be felt advantageously on agriculture and every other branch of industry. Equally important is it to provide at home a market for our raw materials, as by extending the competition it will enhance the price and protect the cultivator against the casualties incident to foreign markets.

With the Indian tribes it is our duty to cultivate friendly relations and to act with kindness and liberality in all our transactions. Equally proper is it to persevere in our efforts to extend to them the advantages of civilization.

The great amount of our revenue and the flourishing state of the Treasury are a full proof of the competency of the national resources for any emergency, as they are of the willingness of our fellow citizens to bear the burdens which the public necessities require. The vast amount of vacant lands, the value of which daily augments, forms an additional resource of great extent and duration. These resources, besides accomplishing every other necessary purpose, put it completely in the power of the United States to discharge the national debt at an early period. Peace is the best time for improvement and preparation of every kind; it is in peace that our commerce flourishes most, that taxes are most easily paid, and that the revenue is most productive.

The executive is charged officially in the departments under it with the disbursement of the public money, and is responsible for the faithful application of it to the purposes for which it is raised. The legislature is the watchful guardian over the public purse. It is its duty to see that the disbursement has been honestly made. To meet the requisite responsibility every facility should be afforded to the executive to enable it to bring the public agents intrusted with the public money strictly and promptly to account. Nothing should be presumed against them; but if, with the requisite facilities, the public money is suffered to lie long and uselessly in their hands, they will not be the only defaulters, nor will the demoralizing effect be confined to them. It will evince a relaxation and want of tone in the administration which will be felt by the whole community. I shall do all I can to secure economy and fidelity in this important branch of the administration, and I doubt not that the legislature will perform its duty with equal zeal. A thorough examination should be regularly made, and I will promote it.

It is particularly gratifying to me to enter on the discharge of these duties at a time when the United States are blessed with peace. It is a state most consistent with their prosperity and happiness. It will be my sincere desire to preserve it, so far as depends on the executive, on just principles with all nations, claiming nothing unreasonable of any and rendering to each what is due.

Equally gratifying is it to witness the increased harmony of opinion which pervades our Union. Discord does not belong to our system. Union is recommended as well by the free and benign principles of our government, extending its blessings to every individual, as by the other eminent advantages attending it. The American people have encountered together great dangers and sustained severe trials with success. They constitute one great family with a common interest. Experience has enlightened us on some questions of essential importance to the country. The progress has been slow, dictated by a just reflection and a faithful regard to every interest connected with it. To promote this harmony in accord with the principles of our republican government and in a manner to give them the most complete effect, and to advance in all other respects the best interests of our Union, will be the object of my constant and zealous exertions.

Never did a government commence under auspices so favorable, nor ever was success so complete. If we look to the history of other nations, ancient or modern, we find no example of a growth so rapid, so gigantic, of a people so prosperous and happy. In contemplating what we have still to perform, the heart of every citizen must expand with joy when he reflects how near our government has approached to perfection; that in respect to it we have no essential improvement to make; that the great object is to preserve it in the essential principles and features which characterize it, and that is to be done by preserving the virtue and enlightening the minds of the people; and as a security against foreign dangers to adopt such arrangements as are indispensable to the support of our independence, our rights, and liberties. If we persevere in the career in which we have advanced so far and in the path already traced, we cannot fail, under the favor of a gracious Providence, to attain the high destiny which seems to await us.

In the administrations of the illustrious men who have preceded me in this high station, with some of whom I have been connected by the closest ties from early life, examples are presented which will always be found highly instructive and useful to their successors. From these I shall endeavor to derive all the advantages which they may afford. Of my immediate predecessor, under whom so important a portion of this great and successful experiment has been made, I shall be pardoned for earnest wishes that he may long enjoy in his retirement the affections of a grateful country, the best reward of exalted talents, and the most faithful and meritorious service. Relying on the aid to be derived from the other departments of the government, I enter on the trust to which I have been called by the suffrages of my fellow citizens with my fervent prayers to the Almighty that He will be graciously pleased to continue to us that protection which He has already so conspicuously displayed in our favor.

James Monroe

Monday, March 5, 1821

*J*ames Monroe faced no opposition to his bid for a second term in 1820. Only George Washington has ever won so resounding a victory, although Monroe's victory was not unanimous. One elector, Governor William Plumer of New Hampshire, cast a solitary ballot for Monroe's secretary of state, John Quincy Adams.

Monroe's victory suggested that the politics of consensus was at hand. But, in fact, the magnitude of the incumbent's victory masked increasingly bitter divisions in the country at large and even with his own cabinet. Secretary of State Adams, Secretary of War Calhoun, and Treasury Secretary William Crawford all entertained visions of one day sitting in Monroe's place. Conflict among these brilliant, ambitious politicians was inevitable, as time would tell.

What's more, Monroe's unopposed reelection bid said more about the state of the opposition party, the Federalists, than it did about the popularity of Monroe's policies. The Federalist Party was about to reach the painful end of its death spiral, incapable of providing serious opposition to the Democratic-Republican juggernaut.

One measure of the opposition's weakness was its inability to seize on the hard times that followed a recession in 1819. Monroe, in his annual State of the Union message in 1820, blamed the recession on "convulsions" in Europe, where the Napoleonic wars at last had come to an end. The "sudden transition to a state of peace," Monroe said, had upset a market accustomed to wartime trade.

The recession would have political and economic effects long after Monroe left office. Popular unrest focused on the Second Bank of the United States, that cherished Federalist institution adopted by Monroe's Democratic-Republican predecessor, Madison. Critics charged that the bank's move to tighten credit in 1818 helped bring on the recession.

Adding to the tensions during the supposed "era of good feelings" was the fight over Missouri's petition to join the Union, which was a surrogate war over the future of slavery. Monroe stayed aloof from the Missouri debate, allowing it to be thrashed out among the great debaters and orators who walked the halls of Congress. The Missouri debate revealed fissures in the Democratic-Republican Party along sectional lines, with Southerners resisting attempts to regulate slavery and Northerners attempting to prevent its spread. Under the compromise, Missouri was admitted with the understanding that it would allow slavery, while Maine was admitted as a free state. Slavery was prohibited in new territories north of the parallel 36°30', except for Missouri.

The debate over Missouri foreshadowed the divisions over slavery and section that would dominate politics for the postrevolutionary generation. Monroe himself believed the compromise saved the Union, according to biographer Harry Ammon. But another Virginian, Thomas Jefferson, was horrified. A dividing line between slave states and free had implications far beyond the political necessities of the moment, the former President said. He called the compromise "a fire bell in the night . . . the knell of the Union." One of the last members of the revolutionary generation, Jefferson believed the nation he helped found was destined for further, and perhaps fatal, conflict. "A geographical line, coinciding with a marked principle, moral and political, once conceived and held up to the angry passions of men, will never be obliterated; and every new irritation will mark it deeper and deeper," he wrote.[1]

The designation "era of good feelings" now appears more ironic than truly representative of reality during the Monroe years.

Monroe took the oath of office for the second time on March 5, rather than March 4, a Sunday. It was the first time since Washington's first inaugural that the President was sworn in on a date other than March 4. The ceremony took place in the new House chamber, which was filled to overflowing with visitors who made little effort to keep quiet while the President read his second inaugural.

The speech itself was lengthy, more than forty-four hundred words, and Monroe believed he had an urgent message to deliver. The War of 1812 had shown the nation's vulnerability to invasion and to the wars and feuds of the Old World. Monroe used the occasion to argue for better coastal de-

---

1. Thomas Jefferson to John Holmes, 22 April 1820, Jefferson Papers, Library of Congress.

fenses against seaborne invaders and the harassment of warring naval powers. Without increased fortifications, he said, an enemy fleet "might go where he pleased." The message was pointed, for Congress had cut money for defensive fortifications in the previous budget.

The threat from the Old World was hypothetical in 1821, but dangers from elsewhere in the New World were very much real. During Monroe's first term, the Spanish empire in the Americas was going the way of the Federalist Party. Revolutionaries, inspired by the U.S. example, were agitating against Spanish rule in South and Central America. The United States worried that other European powers would intervene to crush the independence movements south of the border. But it had to tread carefully, because it hoped to purchase Florida from Spain, and that required good relations between the two nations. Meanwhile, the Seminole and Creek Indians in the Southeast were resisting the spread of white settlement in the area, leading to Andrew Jackson's campaign against them in 1817–18.

Beginning with John Adams, Presidents regularly addressed the plight of Native Americans in their inaugural speeches, but in his second inaugural Monroe devoted more time and words to the issue than his predecessors had. He acknowledged that the "care of Indian tribes within our limits" has "not been executed in a manner to accomplish all the objects intended by it." He noted, with some evident sympathy, that the "progress of our settlements westward, supported as they are by a dense population, has constantly driven them [Native Americans] back, with almost the total sacrifice of the lands which they have been compelled to abandon." Monroe insisted that the tribes had "claims on the magnanimity and, I may add, on the justice of this nation . . ."

He proposed that the United States act as a "Great Father" to the Indians by ending their claims to sovereignty over "vast territories." In return for ceded lands, he insisted that the tribes be compensated with other land and that government funds assist in educating Indian children.

This would not be the policy of the man Monroe sent to the Southeast, Andrew Jackson.

Bibliographic Note
Harry Ammon, *James Monroe: The Quest for National Identity*
(New York: McGraw-Hill, 1971).

I shall not attempt to describe the grateful emotions which the new and very distinguished proof of the confidence of my fellow citizens, evinced by my reelection to this high trust, has excited in my bosom. The approbation which it announces of my conduct in the preceding term affords me a consolation which I shall profoundly feel through life. The general accord with which it has been expressed adds to the great and never-ceasing obligations which it imposes. To merit the continuance of this good opinion, and to carry it with me into my retirement as the solace of advancing years, will be the object of my most zealous and unceasing efforts.

Having no pretensions to the high and commanding claims of my predecessors, whose names are so much more conspicuously identified with our Revolution, and who contributed so preeminently to promote its success, I consider myself rather as the instrument than the cause of the union which has prevailed in the late election. In surmounting, in favor of my humble pretensions, the difficulties which so often produce division in like occurrences, it is obvious that other powerful causes, indicating the great strength and stability of our Union, have essentially contributed to draw you together. That these powerful causes exist, and that they are permanent, is my fixed opinion; that they may produce a like accord in all questions touching, however remotely, the liberty, prosperity, and happiness of our country will always be the object of my most fervent prayers to the Supreme Author of All Good.

In a government which is founded by the people, who possess exclusively the sovereignty, it seems proper that the person who may be placed by their suffrages in this high trust should declare on commencing its duties the principles on which he intends to conduct the administration. If the person thus elected has served the preceding term, an opportunity is afforded him to review its principal occurrences and to give such further explanation respecting them as in his judgment may be useful to his constituents. The events of one year have influence on those of another, and, in like manner, of a preceding on the succeeding administration. The movements of a great nation are connected in all their parts. If errors have been committed they ought to be corrected; if the policy is sound it ought to be supported. It is by a thorough knowledge of the whole subject that our fellow citizens are enabled to judge correctly of the past and to give a proper direction to the future.

Just before the commencement of the last term the United States had concluded a war with a very powerful nation on conditions equal and honorable to both parties. The events of that war are too recent and too deeply impressed on the memory of all to require a development from me. Our commerce had been in a great measure driven from the sea, our Atlantic and inland frontiers were invaded in almost every part; the waste of life along our coast and on some parts of our inland frontiers, to the defense of which our gallant and patriotic citizens were called, was immense, in addition to which not less than $120 million were added at its end to the public debt.

As soon as the war had terminated, the nation, admonished by its events, resolved to place itself in a situation which should be better calculated to prevent the recurrence of a like evil, and, in case it should recur, to mitigate its calamities. With this view, after reducing our land force to the basis of a peace establishment, which has been further modified since, provision was made for the construction of fortifications at proper points through the whole extent of our coast and such an augmentation of our naval force as should be well adapted to both purposes. The laws making this provision were passed in 1815 and 1816, and it has been since the constant effort of the executive to carry them into effect.

The advantage of these fortifications and of an augmented naval force in the extent contemplated, in a point of economy, has been fully illustrated by a report of the Board of Engineers and Naval Commissioners lately communicated to Congress, by which it appears that in an invasion by twenty thousand men, with a correspondent naval force, in a campaign of six months only, the whole expense of the construction of the works would be defrayed by the difference in the sum necessary to maintain the force which would be adequate to our defense with the aid of those works and that which would be incurred without them. The reason of this difference is obvious. If fortifications are judiciously placed on our great inlets, as distant from our cities as circumstances will permit, they will form the only points of attack, and the enemy will be detained there by a small regular force a sufficient time to enable our militia to collect and repair to that on which the attack is made. A force adequate to the enemy, collected at that single point, with suitable preparation for such others as might be menaced, is all that would be requisite. But if there were no fortifications, then the enemy might go where he pleased, and, changing his position and sailing from place to place, our force must be called out and spread in vast numbers along the whole coast and on both sides of every bay and river as high up in each as it might be navigable for ships of war. By these fortifications, supported by our Navy, to which they would afford like support, we should present to other powers an armed front from St. Croix to the Sabine, which would protect in the event of war our whole coast and interior from invasion; and even in the wars of other powers, in which we were neutral, they would be found eminently useful, as, by keeping their public ships at a distance from our cities, peace and order in them would be preserved and the government be protected from insult.

It need scarcely be remarked that these measures have not been resorted to in a spirit of hostility to other powers. Such a disposition does not exist toward any power. Peace and good will have been, and will hereafter be, cultivated with all, and by the most faithful regard to justice. They have been dictated by a love of peace, of economy, and an earnest desire to save the lives of our fellow citizens from that destruction and our country from that devastation which are inseparable from war when it finds us unprepared for it. It is believed, and experience has shown, that such a preparation is the best expedient that can be resorted to prevent war. I add with much pleasure that considerable progress has already been made in these measures of defense, and that they will be completed in a few years, considering the great extent and importance of the object, if the plan be zealously and steadily persevered in.

The conduct of the government in what relates to foreign powers is always an object of the highest importance to the nation. Its agriculture, commerce, manufactures, fisheries, revenue, in short, its peace, may all be affected by it. Attention is therefore due to this subject.

At the period adverted to the powers of Europe, after having been engaged in long and destructive wars with each other, had concluded a peace, which happily still exists. Our peace with the power with whom we had been engaged had also been concluded. The war between Spain and the colonies in South America, which had commenced many years before, was then the only conflict that remained unsettled. This being a contest between different parts of the same community, in which other powers had not interfered, was not affected by their accommodations.

This contest was considered at an early stage by my predecessor a civil war in which the parties were entitled to equal rights in our ports. This decision, the first made by any power, being formed on great consideration of the comparative strength and resources of the parties, the length of time, and successful opposition made by the colonies, and of all other circumstances on which it ought to depend, was in strict accord with the law of nations. Congress has invariably acted on this principle, having made no change in our relations with either party. Our attitude has therefore been that of neutrality between them, which has been maintained by the government with the strictest impartiality. No aid has been afforded to either, nor has any privilege been enjoyed by the one which has not been equally open to the other party, and every exertion has been made in its power to enforce the execution of the laws prohibiting illegal equipments with equal rigor against both.

By this equality between the parties their public vessels have been received in our ports on the same footing; they have enjoyed an equal right to purchase and export arms, munitions of war, and every other supply, the exportation of all articles whatever being permitted under laws which were passed long before the commencement of the contest; our citizens have traded equally with both, and their commerce with each has been alike protected by the government.

Respecting the attitude which it may be proper for the United States to maintain hereafter between the parties, I have no hesitation in stating it as my opinion that the neutrality heretofore observed should still be adhered to. From the change in the government of Spain and the negotiation now depending, invited by the Cortes and accepted by the colonies, it may be presumed, that their differences will be settled on the terms proposed by the colonies. Should the war be continued, the United States, regarding its occurrences, will always have it in their power to adopt such measures respecting it as their honor and interest may require.

Shortly after the general peace a band of adventurers took advantage of this conflict and of the facility which it afforded to establish a system of buccaneering in the neighboring seas, to the great annoyance of the commerce of the United States, and, as was represented, of that of other powers. Of this spirit and of its injurious bearing on the United States strong proofs were afforded by the establishment at Amelia Island, and the purposes to which it was made instrumental by this band in 1817, and by the occurrences which took place in other parts of Florida in 1818, the details of which in both instances are too well known to require to be now recited. I am satisfied had a less decisive course been adopted that the worst consequences would have resulted from it. We have seen that these checks, decisive as they were, were not sufficient to crush that piratical spirit. Many culprits brought within our limits have been condemned to suffer death, the punishment due to that atrocious crime. The decisions of upright and enlightened tribunals fall equally on all whose crimes subject them, by a fair interpretation of the law, to its censure. It belongs to the executive not to suffer the executions under these decisions to transcend the great purpose for which punishment is necessary. The full benefit of example being secured, policy as well as humanity equally forbids that they should be carried further. I have acted on this principle, pardoning those who appear to have been led astray by ignorance of the criminality of the acts they had committed, and suffering the law to take effect on those only in whose favor no extenuating circumstances could be urged.

Great confidence is entertained that the late treaty with Spain, which has been ratified by both the parties, and the ratifications whereof have been exchanged, has placed the relations of the two countries on a basis of permanent friendship. The provision made by it for such of our citizens as have claims on Spain of the character described will, it is presumed, be very satisfactory to them, and the boundary which is established between the territories of the parties westward of the Mississippi, heretofore in dispute, has, it is thought, been settled on conditions just and advantageous to both. But to the acquisition of Florida too much importance can not be attached. It secures to the United States a territory important in itself, and whose importance is much increased by its bearing on many of the highest interests of the Union. It opens to several of the neighboring states a free passage to the ocean, through the province ceded, by several rivers, having their sources high up within their limits. It secures us against all future annoyance from powerful Indian tribes. It gives us several excellent harbors in the Gulf of Mexico for ships of war of the largest size. It covers by its position in the Gulf the Mississippi and other great waters within our extended limits, and thereby enables the United States to afford complete protection to the vast and very valuable productions of our whole western country, which find a market through those streams.

By a treaty with the British government, bearing date on the 20th of October 1818, the convention regulating the commerce between the United States and Great Britain, concluded on the 3rd of July 1815, which was about expiring, was revived and continued for the term of ten years from the time of its expiration. By that treaty, also, the differences which had arisen under the Treaty of Ghent respecting the right claimed by the United States for their citizens to take and cure fish on the coast of His Britannic Majesty's dominions in America, with other differences on important interests, were adjusted to the satisfaction of both parties. No agreement has yet been entered into respecting the commerce between the United States and the British dominions in the West Indies and on this continent. The restraints imposed on that commerce by Great Britain, and reciprocated by the United States on a principle of defense, continue still in force.

The negotiation with France for the regulation of the commercial relations between the two countries, which in the course of the last summer had been commenced at Paris, has since been transferred to this city, and will be pursued on the part of the United States in the spirit of conciliation, and with an earnest desire that it may terminate in an arrangement satisfactory to both parties.

Our relations with the Barbary powers are preserved in the same state and by the same means that were employed when I came into this office. As early as 1801 it was found necessary to send a squadron into the Mediterranean for the protection of our commerce and no period has intervened, a short term excepted, when it was thought advisable to withdraw it. The great interests which the United States have in the Pacific, in commerce and in the fisheries, have also made it necessary to maintain a naval force there. In disposing of this force in both instances the most effectual measures in our power have been taken, without interfering with its other duties, for the suppression of the slave trade and of piracy in the neighboring seas.

The situation of the United States in regard to their resources, the extent of their revenue, and the facility with which it is raised affords a most gratifying spectacle. The payment of nearly $67 million of the public debt, with the great progress made in measures of defense and in other improvements of various kinds since the late war, are conclusive proofs of this extraordinary prosperity, especially when it is recollected that these expenditures have been defrayed without a burthen on the people, the direct tax and excise having been repealed soon after the conclusion of the late war, and the revenue applied to these great objects having been raised in a manner not to be felt. Our great resources therefore remain untouched for any purpose which may affect the vital interests of the nation. For all such purposes they are inexhaustible. They are more especially to be found in the virtue, patriotism, and intelligence of our fellow citizens, and in the devotion with which they would yield up by any just measure of taxation all their property in support of the rights and honor of their country.

Under the present depression of prices, affecting all the productions of the country and every branch of industry, proceeding from causes explained on a former occasion, the revenue has considerably diminished, the effect of which has been to compel Congress either to abandon these great measures of defense or to resort to loans or internal taxes to supply the deficiency. On the presumption that this depression and the deficiency in the revenue arising from it would be temporary, loans were authorized for the demands of the last and present year. Anxious to relieve my fellow citizens in 1817 from every burthen which could be dispensed with and the state of the Treasury permitting it, I recommended the repeal of the internal taxes, knowing that such relief was then peculiarly necessary in consequence of the great exertions made in the late war. I made that recommendation under a pledge that should the public exigencies require a recurrence to them at any time while I remained in this trust, I would with equal promptitude perform the duty which would then be alike incumbent on me. By the experiment now making it will be seen by the next session of Congress whether the revenue shall have been so augmented as to be adequate to all these necessary purposes. Should the deficiency still continue, and especially should it be probable that it would be permanent, the course to be pursued appears to me to be obvious. I am satisfied that under certain circumstances loans may be resorted to with great advantage. I am equally well satisfied, as a general rule, that the demands of the current year, especially in time of peace, should be provided for by the revenue of that year.

I have never dreaded, nor have I ever shunned, in any situation in which I have been placed making appeals to the virtue and patriotism of my fellow citizens, well knowing that they could never be made in vain, especially in times of great emergency or for purposes of high national importance. Independently of the exigency of the case, many considerations of great weight urge a policy having in view a provision of revenue to meet to a certain extent the demands of the nation, without relying altogether on the precarious resource of foreign commerce. I am satisfied that internal duties and excises, with corresponding imposts on foreign articles of the same kind, would, without imposing any serious burdens on the people, enhance the price of produce, promote our manufactures, and augment the revenue, at the same time that they made it more secure and permanent.

The care of the Indian tribes within our limits has long been an essential part of our system, but, unfortunately, it has not been executed in a manner to accomplish all the objects intended by it. We have treated them as independent nations, without their having any substantial pretensions to that rank. The distinction has flattered their pride, retarded their improvement, and in many instances paved the way to their destruction. The progress of our settlements westward, supported as they are by a dense population, has constantly driven them back, with almost the total sacrifice of the lands which they have been compelled to abandon. They have claims on the magnanimity and, I may add, on the justice of this nation which we must all feel. We should become their real benefactors; we should perform the office of their Great Father, the endearing title which they emphatically give to the chief magistrate of our Union. Their sovereignty over vast territories should cease, in lieu of which the right of soil should be secured to each individual and his posterity in competent portions; and for the territory thus ceded by each tribe some reasonable equivalent should be granted, to be vested in permanent funds for the support of civil government over them and for the education of their children, for their instruction in the arts of husbandry, and to provide sustenance for them until they could provide it for themselves. My earnest hope is that Congress will digest some plan, founded on these principles, with such improvements as their wisdom may suggest, and carry it into effect as soon as it may be practicable.

Europe is again unsettled and the prospect of war increasing. Should the flame light up in any quarter, how far it may extend it is impossible to foresee. It is our peculiar felicity to be altogether unconnected with the causes which produce this menacing aspect elsewhere. With every power we are in perfect amity, and it is our interest to remain so if it be practicable on just conditions. I see no reasonable cause to apprehend variance with any power, unless it proceed from a violation of our maritime rights. In these contests, should they occur, and to whatever extent they may be carried, we shall be neutral; but as a neutral power we have rights which it is our duty to maintain. For like injuries it will be incumbent on us to seek redress in a spirit of amity, in full confidence that, injuring none, none would knowingly injure us. For more imminent dangers we should be prepared, and it should always be recollected that such preparation adapted to the circumstances and sanctioned by the judgment and wishes of our constituents cannot fail to have a good effect in averting dangers of every kind. We should recollect also that the season of peace is best adapted to these preparations.

If we turn our attention, fellow citizens, more immediately to the internal concerns of our country, and more especially to those on which its future welfare depends, we have every reason to anticipate the happiest results. It is now rather more than forty-four years since we declared our independence, and thirty-seven since it was acknowledged. The talents and virtues which were displayed in that great struggle were a sure presage of all that has since followed. A people who were able to surmount in their infant state such great perils would be more competent as they rose into manhood to repel any which they might meet in their progress. Their physical strength would be more adequate to foreign danger, and the practice of self-government, aided by the light of experience, could not fail to produce an effect equally salutary on all those questions connected with the internal organization. These favorable anticipations have been realized.

In our whole system, national and state, we have shunned all the defects which unceasingly preyed on the vitals and destroyed the ancient republics. In them there were distinct orders, a nobility and a people, or the people governed in one assembly. Thus, in the one instance there was a perpetual conflict between the orders in society for the ascendency, in which the victory of either terminated in the overthrow of the government and the ruin of the state; in the other, in which the people governed in a body, and whose dominions seldom exceeded the dimensions of a county in one of our states, a tumultuous and disorderly movement permitted only a transitory existence. In this great nation there is but one order, that of the people, whose power, by a peculiarly happy improvement of the representative principle, is transferred from them, without impairing in the slightest degree their sovereignty, to bodies of their own creation, and to persons elected by themselves, in the full extent necessary for all the purposes of free, enlightened and efficient government. The whole system is elective, the complete sovereignty being in the people, and every officer in every department deriving his authority from and being responsible to them for his conduct.

Our career has corresponded with this great outline. Perfection in our organization could not have been expected in the outset either in the national or state governments or in tracing the line between their respective powers. But no serious conflict has arisen, nor any contest but such as are managed by argument and by a fair appeal to the good sense of the people, and many of the defects which experience had clearly demonstrated in both governments have been remedied. By steadily pursuing this course in this spirit there is every reason to believe that our system will soon attain the highest degree of perfection of which human institutions are capable, and that the movement in all its branches will exhibit such a degree of order and harmony as to command the admiration and respect of the civilized world.

Our physical attainments have not been less eminent. Twenty-five years ago the river Mississippi was shut up and our western brethren had no outlet for their commerce. What has been the progress since that time? The river has not only become the property of the United States from its source to the ocean, with all its tributary streams (with the exception of the upper part of the Red River only), but Louisiana, with a fair and liberal boundary on the western side and the Floridas on the eastern, have been ceded to us. The United States now enjoy the complete and uninterrupted sovereignty over the whole territory from St. Croix to the Sabine. New states, settled from among ourselves in this and in other parts, have been admitted into our Union in equal participation in the national sovereignty with the original states. Our population has augmented in an astonishing degree and extended in every direction. We now, fellow citizens, comprise within our limits the dimensions and faculties of a great power under a government possessing all the energies of any government ever known to the Old World, with an utter incapacity to oppress the people.

Entering with these views the office which I have just solemnly sworn to execute with fidelity and to the utmost of my ability, I derive great satisfaction from a knowledge that I shall be assisted in the several departments by the very enlightened and upright citizens from whom I have received so much aid in the preceding term. With full confidence in the continuance of that candor and generous indulgence from my fellow citizens at large which I have heretofore experienced, and with a firm reliance on the protection of Almighty God, I shall forthwith commence the duties of the high trust to which you have called me.

John Quincy Adams

· INAUGURAL ADDRESS ·

Friday, March 4, 1825

John Quincy Adams, the son of the second President, perhaps the most well-prepared chief executive in the nation's history, is the only person to win the presidency despite losing both the popular vote and the electoral vote. Even the elections of 1876, when Rutherford B. Hayes lost the popular vote but prevailed in the Electoral College by a single vote, and 2000, when George W. Bush lost the popular vote but defeated Al Gore in the Electoral College after a long recount of votes in Florida, were not as bitter as the contest of 1824.

James Monroe had entered office in 1817 hoping to put an end to partisan divisions, and in a sense he succeeded. The Federalists expired during his watch. But sectional, ideological, and personal differences within the Democratic-Republican Party ensured that 1824 would be remembered as one of the country's most divisive elections.

Five well-known candidates, all of them Democratic-Republicans, competed with each other to succeed Monroe, including three members of his cabinet: Adams, the secretary of state; John C. Calhoun, the secretary of war; and William Crawford, secretary of the treasury. The field also included Senator Andrew Jackson, hero of the Battle of New Orleans, and Henry Clay, Speaker of the House of Representatives.

Sectional differences, which began to loom large during Monroe's second term, played an important role in defining the issues of 1824. The North was deeply resentful of Virginia's domination of the presidency, which worked to Adams's benefit and to the detriment of Crawford, a native of Virginia although a resident of Georgia. Calhoun, from South Carolina, and Clay, of Kentucky, had been outspoken advocates for American nationalism, emphasizing territorial expansion, internal improvements, protective tariffs, and a central bank of the United States. Jackson's views were nearly unknown.

The tariff question—that is, the taxes placed on foreign goods brought into the country—intersected with the growing economic divide between North and South. Northern manufacturers supported tariffs because they made foreign goods more expensive and so made their own products cheaper in the domestic market. Southern cotton planters, who had to purchase most of their manufactured goods, hated the tariff, regarding it as an example of federal policy being used to help one section at the expense of the other. With the support of northern and western states, Congress passed a tariff in 1824 that imposed duties on cheap manufactured goods from the nation's leading trading partner, Great Britain. The debate over the tariff added to sectional tensions that exploded during the debate over the admission of Missouri in 1820.

Calhoun dropped out of consideration in early 1824 as it became clear he didn't have enough support. When votes were cast in November, Jackson had the highest number of popular votes, 155,872, and the most electoral votes, 99. Adams had 105,321 popular votes and 84 electoral votes. The presence of Clay and Crawford in the race, with their combined total of 78 electoral votes, denied Jackson a majority in the Electoral College.

The election of 1824 was the first for which popular vote totals were available. In eighteen of the Union's twenty-four states, the popular vote decided which candidate received the state's electoral votes. Six states, however, still restricted the franchise to members of their respective state legislature, meaning that legislators, not the people, decided who would receive the state's electoral votes.

Because Jackson's electoral total fell short of a majority, the election was thrown to the House of Representatives, where the top-three vote getters—Jackson, Adams, and Crawford—were eligible, in accordance with the Twelfth Amendment. Significantly, the Speaker of that body, Clay, did not qualify, having finished fourth in the electoral vote. But if he could not be President, there was nothing in the Constitution that barred Clay from becoming kingmaker.

The contest in the House became a two-person race between Jackson and Adams, in part because of concerns over Crawford's health—he had suffered a stroke in 1823. Jackson was the favorite, but in the House one vote loomed above all others: that of Speaker Henry Clay. The Speaker's politics were more in tune with Adams than with Jackson. Clay's American System of protection and his support for internal improvements and a central bank of the United States were quasi-Federalist, as was Adams. What's more, Clay, a professional politician, was leery of Jackson, a newcomer. But Clay's decision was by no means an easy one. He and Adams had tangled in the past on foreign policy—Clay was a vocal supporter of Latin America's independence movements, but Adams was skeptical of them—so an alliance with the New Englander would require some fence-mending.

Mended they were. Clay visited Adams on January 9, 1825. According to Adams's account, the Speaker told him that in the contest "between General Jackson, Mr. Crawford, and myself, he had no hesitation in saying that his preference would be for me."[1] So, on February 9, 1825, the House chose Adams as President in a canvass of state delegations. Adams won thirteen states, with Clay helping to deliver several key western states. Jackson won seven states and Crawford four. When Adams later named Clay as his secretary of state, Jackson and his supporters denounced the alliance as a "corrupt bargain."

Given the controversy over Adams's election, it is hardly a wonder that the president-elect suffered through two sleepless nights before taking the oath of office on March 4. Adams, like his father, was a noted orator in his time, although his voice was high and his manner somewhat chilly. Wearing full-length pants rather than knee breeches, Adams candidly acknowledged the fraught process that led him to the House chamber on that day in March. "Fellow citizens," he said, "you are acquainted with the peculiar circumstances of the recent election, which have resulted in affording me the opportunity of addressing you at this time. . . . Less possessed of your confidence in advance than any of my predecessors, I am deeply conscious of the prospect that I shall stand more and oftener in need of your indulgence."

His embittered critics, who believed the presidency was stolen from Jackson, would offer him no quarter in the four years to come.

IN COMPLIANCE WITH AN USAGE COEVAL with the existence of our federal Constitution, and sanctioned by the example of my predecessors in the career upon which I am about to enter, I appear, my fellow citizens, in your presence and in that of heaven to bind myself by the solemnities of religious obligation to the faithful performance of the duties allotted to me in the station to which I have been called.

In unfolding to my countrymen the principles by which I shall be governed in the fulfillment of those duties my first resort will be to that Constitution which I shall swear to the best of my ability to preserve, protect, and defend. That revered instrument enumerates the powers and prescribes the duties of the executive magistrate, and in its first words declares the purposes to which these and the whole action of the government instituted by it should be invariably and sacredly devoted—to form a more perfect union, establish justice, insure domestic tranquility, provide for the common defense, promote the general welfare, and secure the blessings of liberty to the people of this Union in their successive generations. Since the adoption of this social compact one of these generations has passed away. It is the work of our forefathers. Administered by some of the most eminent men who contributed to its formation, through a most eventful period in the annals of the world, and through all the vicissitudes of peace and war incidental to the condition of associated man, it has not disappointed the hopes and aspirations of those illustrious benefactors of their age and nation. It has promoted the lasting welfare of that country so dear to us all; it has to an extent far beyond the ordinary lot of humanity secured the freedom and happiness of this people. We now receive it as a precious inheritance from those to whom we are indebted for its establishment, doubly bound by the examples which they have left us and by the blessings which we have enjoyed as the fruits of their labors to transmit the same unimpaired to the succeeding generation.

In the compass of thirty-six years since this great national covenant was instituted a body of laws enacted under its authority and in conformity with its provisions has unfolded its powers and carried into practical operation its effective energies. Subordinate departments have distributed the executive functions in their various relations to foreign affairs, to the revenue and expenditures, and to the military force of the Union by land and sea. A coordinate department of the judiciary has expounded the Constitution and the laws, settling in harmonious coincidence with the legislative will numerous weighty questions of construction which the imperfection of human language had rendered unavoidable. The year of jubilee since the first formation of our Union has just elapsed that of the declaration of our independence is at hand. The consummation of both was effected by this Constitution.

Since that period a population of four millions has multiplied to twelve. A territory bounded by the Mississippi has been extended from sea to sea. New states have been admitted to the Union in numbers nearly equal to those of the first confederation. Treaties of peace, amity, and commerce have been concluded with the principal dominions of the earth. The people of other nations, inhabitants of regions acquired not by conquest, but by compact, have been united with us in the participation of our rights and duties, of our burdens and blessings. The forest has fallen by the ax of our woodsmen; the soil has been made to teem by the tillage of our farmers; our commerce has whitened every ocean. The dominion of man over physical nature has been extended by the invention of our artists. Liberty and law have marched hand in hand. All the purposes of human association have been accomplished as effectively as under any other government on the globe, and at a cost little exceeding in a whole generation the expenditure of other nations in a single year.

Such is the unexaggerated picture of our condition under a Constitution founded upon the republican principle of equal rights. To admit that this picture has its shades is but to say that it is still the condition of men upon earth. From evil—physical, moral, and political—it is not our claim to be exempt. We have suffered sometimes by the visitation of heaven through disease; often by the wrongs and injustice of other nations, even to the extremities of war; and, lastly, by dissensions among ourselves— dissensions perhaps inseparable from the enjoyment of freedom, but which have more than once appeared to threaten the dissolution of the Union, and with it the overthrow of all the enjoyments of our present lot and all our earthly hopes of the future. The causes of these dissensions have been various, founded upon differences of speculation in the theory of republican government; upon conflicting views of policy in our relations with foreign nations; upon jealousies of partial and sectional interests, aggravated by prejudices and prepossessions which strangers to each other are ever apt to entertain.

It is a source of gratification and of encouragement to me to observe that the great result of this experiment upon the theory of human rights has at the close of that generation by which it was formed been crowned with success equal to the most sanguine expectations of its founders. Union, justice, tranquility, the common defense, the general welfare, and the blessings of liberty—all have been promoted by the government under which we have lived. Standing at this point of time, looking back to that generation which has gone by and forward to that which is advancing, we may at once indulge in grateful exultation and in cheering hope. From the experience of the past we derive instructive lessons for the future. Of the two great political parties which have divided the opinions and feelings of our country, the candid and the just will now admit that both have contributed splendid talents, spotless integrity, ardent patriotism, and disinterested sacrifices to the formation and administration of this government, and that both have required a liberal indulgence for a portion of human infirmity and error. The revolutionary wars of Europe, commencing precisely at the moment when the government of the United States first went into operation under this Constitution, excited a collision of sentiments and of sympathies which kindled all the passions and imbittered the conflict of parties till the nation was involved in war and the Union was shaken to its center. This time of trial embraced a period of five and twenty years, during which the policy of the Union in its relations with Europe constituted the principal basis of our political divisions and the most arduous part of the action of our federal government. With the catastrophe in which the wars of the French Revolution terminated, and our own subsequent peace with Great Britain, this baneful weed of party strife was uprooted. From that time no difference of principle, connected either with the theory of government or with our intercourse with foreign nations, has existed or been called forth in force sufficient to sustain a continued combination of parties or to give more than wholesome animation to public sentiment or legislative debate. Our political creed is, without a dissenting voice that can be heard, that the will of the people is the source and the happiness of the people the end of all legitimate government upon earth; that the best security for the beneficence and the best guaranty against the abuse of power consists in the freedom, the purity, and the frequency of popular elections; that the general government of the Union and the separate governments of the states are all sovereignties of limited powers, fellow servants of the same masters, uncontrolled within their respective spheres, uncontrollable by encroachments upon each other; that the firmest security of peace is the preparation during peace of the defenses of war; that a rigorous economy and accountability of public expenditures should guard against the aggravation and alleviate when possible the burden of taxation; that the military should be kept in strict subordination to the civil power; that the freedom of the press and of religious opinion should be inviolate; that the policy of our country is peace and the ark of our salvation union are articles of faith upon which we are all now agreed. If there have been those who doubted whether a confederated representative democracy were a government competent to the wise and orderly management of the common concerns of a mighty nation, those doubts have been dispelled; if there have been projects of partial confederacies to be erected upon the ruins of the Union, they have been scattered to the winds; if there have been dangerous attachments to one foreign nation and antipathies against another, they have been extinguished. Ten years of peace, at home and abroad, have assuaged the animosities of political contention and blended into harmony the most discordant elements of public opinion, There still remains one effort of magnanimity, one sacrifice of prejudice and passion, to be made by the individuals throughout the nation who have heretofore followed the standards of political party. It is that of discarding every remnant of rancor against each other, of embracing as countrymen and friends, and of yielding to talents and virtue alone that confidence which in times of contention for principle was bestowed only upon those who bore the badge of party communion.

The collisions of party spirit which originate in speculative opinions or in different views of administrative policy are in their nature transitory. Those which are founded on geographical divisions, adverse interests of soil, climate, and modes of domestic life are more permanent, and therefore, perhaps, more dangerous. It is this which gives inestimable value to the character of our government, at once federal and national. It holds out to us a perpetual admonition to preserve alike and with equal anxiety the rights of each individual state in its own government and the rights of the whole nation in that of the Union. Whatsoever is of domestic concernment, unconnected with the other members of the Union or with foreign lands, belongs exclusively to the administration of the state governments. Whatsoever directly involves the rights and interests of the federative fraternity or of foreign powers is of the resort of this general government. The duties of both are obvious in the general principle, though sometimes perplexed with difficulties in the detail. To respect the rights of the state governments is the inviolable duty of that of the Union; the government of every state will feel its own obligation to respect and preserve the rights of the whole. The prejudices everywhere too commonly entertained against distant strangers are worn away, and the jealousies of jarring interests are allayed by the composition and functions of the great national councils annually assembled from all quarters of the Union at this place. Here the distinguished men from every section of our country, while meeting to deliberate upon the great interests of those by whom they are deputed, learn to estimate the talents and do justice to the virtues of each other. The harmony of the nation is promoted and the whole Union is knit together by the sentiments of mutual respect, the habits of social intercourse, and the ties of personal friendship formed between the representatives of its several parts in the performance of their service at this metropolis.

Passing from this general review of the purposes and injunctions of the federal Constitution and their results as indicating the first traces of the path of duty in the discharge of my public trust, I turn to the administration of my immediate predecessor as the second. It has passed away in a period of profound peace, how much to the satisfaction of our country and to the honor of our country's name is known to you all. The great features of its policy, in general concurrence with the will of the legislature, have been to cherish peace while preparing for defensive war; to yield exact justice to other nations and maintain the rights of our own; to cherish the principles of freedom and of equal rights wherever they were proclaimed; to discharge with all possible promptitude the national debt; to reduce within the narrowest limits of efficiency the military force; to improve the organization and discipline of the army; to provide and sustain a school of military science; to extend equal protection to all the great interests of the nation; to promote the civilization of the Indian tribes, and to proceed in the great system of internal improvements within the limits of the constitutional power of the Union. Under the pledge of these promises, made by that eminent citizen at the time of his first induction to this office, in his career of eight years the internal taxes have been repealed; sixty millions of the public debt have been discharged; provision has been made for the comfort and relief of the aged and indigent among the surviving warriors of the Revolution; the regular armed force has been reduced and its constitution revised and perfected; the accountability for the expenditure of public moneys has been made more effective; the Floridas have been peaceably acquired, and our boundary has been extended to the Pacific Ocean; the independence of the southern nations of this hemisphere has been recognized, and recommended by example and by counsel to the potentates of Europe; progress has been made in the defense of the country by fortifications and the increase of the navy, toward the effectual suppression of the African traffic in slaves; in alluring the aboriginal hunters of our land to the cultivation of the soil and of the mind, in exploring the interior regions of the Union, and in preparing by scientific researches and surveys for the further application of our national resources to the internal improvement of our country.

In this brief outline of the promise and performance of my immediate predecessor the line of duty for his successor is clearly delineated. To pursue to their consummation those purposes of improvement in our common condition instituted or recommended by him will embrace the whole sphere of my obligations. To the topic of internal improvement, emphatically urged by him at his inauguration, I recur with peculiar satisfaction. It is that from which I am convinced that the unborn millions of our posterity who are in future ages to people this continent will derive their most fervent gratitude to the founders of the Union; that in which the beneficent action of its government will be most deeply felt and acknowledged. The magnificence and splendor of their public works are among the imperishable glories of the ancient republics. The roads and aqueducts of Rome have been the admiration of all after ages, and have survived thousands of years after all her conquests have been swallowed up in despotism or become the spoil of barbarians. Some diversity of opinion has prevailed with regard to the powers of Congress for legislation upon objects of this nature. The most respectful deference is due to doubts originating in pure patriotism and sustained by venerated authority. But nearly twenty years have passed since the construction of the first national road was commenced. The authority for its construction was then unquestioned. To how many thousands of our countrymen has it proved a benefit? To what single individual has it ever proved an injury? Repeated, liberal, and candid discussions in the legislature have conciliated the sentiments and approximated the opinions of enlightened minds upon the question of constitutional power. I cannot but hope that by the same process of friendly, patient, and persevering deliberation all constitutional objections will ultimately be removed. The extent and limitation of the powers of the general government in relation to this transcendently important interest will be settled and acknowledged to the common satisfaction of all, and every speculative scruple will be solved by a practical public blessing.

Fellow citizens, you are acquainted with the peculiar circumstances of the recent election, which have resulted in affording me the opportunity of addressing you at this time. You have heard the exposition of the principles which will direct me in the fulfillment of the high and solemn trust imposed upon me in this station. Less possessed of your confidence in advance than any of my predecessors, I am deeply conscious of the prospect that I shall stand more and oftener in need of your indulgence. Intentions upright and pure, a heart devoted to the welfare of our country, and the unceasing application of all the faculties allotted to me to her service are all the pledges that I can give for the faithful performance of the arduous duties I am to undertake. To the guidance of the legislative councils, to the assistance of the executive and subordinate departments, to the friendly cooperation of the respective state governments, to the candid and liberal support of the people so far as it may be deserved by honest industry and zeal, I shall look for whatever success may attend my public service; and knowing that "except the Lord keep the city the watchman waketh but in vain," with fervent supplications for His favor, to His overruling providence I commit with humble but fearless confidence my own fate and the future destinies of my country.

Andrew Jackson

· FIRST INAUGURAL ADDRESS ·

Wednesday, March 4, 1829

The presidential campaign of 1828 was without doubt the nastiest and filthiest in American history. John Quincy Adams, seeking reelection, was accused of pimping when he served as the U.S. minister to Russia, in that he procured an American girl to satisfy the lust of the czar. The charge was plainly ridiculous, but that did not stop his opponents from broadcasting the gossip. Worse, he was charged with engaging in a "corrupt bargain" with Henry Clay to obtain the presidency in 1825. This bargain involved the supposed fact that Speaker Clay provided Adams with the necessary votes in the House of Representatives to become the chief executive in return for which he promised to appoint Clay his secretary of state, an office that was considered at that time the stepping stone to the presidency. Shortly after his election as President, Adams did in fact appoint Clay his secretary of state, and that only confirmed what had long been rumored. It tended to prove that the two men did indeed conspire to "steal" the presidency from Andrew Jackson, Old Hickory, the man who had won a plurality of popular and electoral votes in the fall election of 1824.

If Adams was battered by shocking accusations in the 1828 election, Jackson, his rival, suffered worse treatment. He was accused of stealing another man's wife and living with her for two years, even though she was still married to someone else. She was condemned as a bigamist and he "her paramour lover." They were later legally married after her divorce from her first husband. Jackson's opponents also claimed that his mother was a prostitute who had been brought to this country to service British soldiers. He himself was lambasted as a ruffian and murderer who had assaulted dozens of individuals and even ordered the execution of innocent soldiers during the Creek War.

Far more important than these vicious stories was the fact that following the election of 1824–25 a number of leading Republicans who had backed William H. Crawford now turned to Jackson. Crawford had suffered a stroke and retired to his home in Georgia. His leading political operator, Senator Martin Van Buren of New York, now switched his allegiance to Old Hickory. He also began to draw other factions, such as the supporters of Vice President John C. Calhoun, into an alliance with the Jackson party. His actions represented a deliberate attempt to revive the two party system that had disappeared during the so-called Era of Good Feelings, following the collapse of the Federalist Party. Van Buren envisioned a party system consisting of Adams and Clay and their friends on one side, advocating Hamiltonian principles, and Jackson, Calhoun, and the friends of Crawford on the other side, committed to Jeffersonian principles. This latter group eventually became the Democratic Party. The Adams/Clay faction was known as the National Republican Party.

The 1828 election turned out to be a total victory for Jackson. He garnered 647,276 popular and 178 electoral votes compared to Adams's 508,064 popular and 83 electoral votes. Tragically, Jackson's wife, Rachel, died of a heart attack shortly after his victory became known. He believed it was the result of the attacks on her character by the opposition. He especially blamed Clay and Adams.

Before leaving his home, the Hermitage, in Tennessee, Jackson worked on what he called a "Rough Draft of the Inaugural Address." The document is remarkable for its language and ideas. In view of his limited education the address is rather outstanding. He declared that he had been called to this high office by the "voluntary suffrages of my country." He said he trusted in the smiles of an "overruling providence . . . for that animation . . . which shall enable us to steer, the Bark of liberty, through every difficulty." Then he outlined his reform program, stating that he wished to observe the strictest economy in governmental expenditures, liquidate the national debt, adjust the tariff rates to a more "judicious" level, and distribute the surplus in order to aid education and provide for internal improvements. It is truly a remarkable document for an indifferently educated frontiersman, but when he showed it to his friends in Washington, some of them declared that it was "absolutely disgraceful." Disgraceful it most assuredly was not. But their complaints about it convinced him to let them improve on it. So a second and a third address were cobbled together by an assortment of friends and advisers, such as James A. Hamilton, the son of Alexander Hamilton; Andrew Jackson Donelson, Old Hickory's nephew and ward; William B. Lewis, his friend and companion in the White House; and John H. Eaton, his secretary of war manqué.

The three drafts are different in many respects. What is most notable in the final version, the one he read at his inauguration—other than its lack of the original's vitality—is that Jackson mentions that he intends to remove the Indians from their home in the Southeast but do it by adhering to a "just and liberal policy." This version of the address also underscores the importance of reform. He says he wished to correct those "abuses that have brought the patronage of the federal government into conflict with the freedom of elections," an obvious reference to the "corrupt bargain" charge leveled at Adams and Clay. Furthermore, he would correct the "causes which have disturbed the rightful course of appointment" and placed power in "unfaithful or incompetent hands."

The day of the inauguration, March 4, 1829, proved to be bright and sunny. The capital awoke to the firing of cannons in a thirteen-gun salute. Immediately, the streets filled with people. By 10:00 A.M. the open area in front of the East Portico of the Capitol building was thronged with fifteen thousand to twenty thousand people. Francis Scott Key, author of "The Star-Spangled Banner," stared in wonder at the incredible spectacle of this surging, pulsating mass of humanity. "It is beautiful," he gasped, "it is sublime."

Jackson, with an escort, walked from his hotel to the Capitol. He headed first to the Senate chamber to witness the swearing-in of the reelected vice president, John C. Calhoun. At high noon, after the swearing-in, Jackson proceeded to the East Portico for the principal ceremony. When the crowd saw him, they burst out with repeated shouts of "huzza!" First he gave his inaugural address, and then Chief Justice John Marshall administered the oath of office.

When the ceremony concluded, Jackson reentered the Capitol, then walked down the hill where he found a horse waiting to take him to the White House. The mob followed, and they consisted of "country men, farmers, gentlemen, mounted and dismounted, boys, women and children, black and white. Carriages, wagons and carts all pursuing him to the President's house."

A "raving Democracy" poured through the White House. Pails of liquor and orange punch splashed to the floor when the mob spotted the waiters bringing them through the pantry doors. It was a "regular Saturnalia," laughed one senator. "The Majesty of the People had disappeared," said another, "and a rabble, a mob of boys, negros, women, children, scrambling, fighting, romping." To relieve the pressure in the house, tubs of punch and pails of liquor were carried to the lawn, outside and all the windows were opened to provide as many exits as possible. As anticipated, the rabble bolted after the liquor, using whatever exit was closest.

The "reign of KING MOB seemed triumphant."[1] Indeed, a new era had begun, one that would advance democracy throughout the country.

FELLOW CITIZENS:

About to undertake the arduous duties that I have been appointed to perform by the choice of a free people, I avail myself of this customary and solemn occasion to express the gratitude which their confidence inspires and to acknowledge the accountability which my situation enjoins. While the magnitude of their interests convinces me that no thanks can be adequate to the honor they have conferred, it admonishes me that the best return I can make is the zealous dedication of my humble abilities to their service and their good.

As the instrument of the federal Constitution it will devolve on me for a stated period to execute the laws of the United States, to superintend their foreign and their confederate relations, to manage their revenue, to command their forces, and, by communications to the legislature, to watch over and to promote their interests generally. And the principles of action by which I shall endeavor to accomplish this circle of duties it is now proper for me briefly to explain.

In administering the laws of Congress I shall keep steadily in view the limitations as well as the extent of the executive power, trusting thereby to discharge the functions of my office without transcending its authority. With foreign nations it will be my study to preserve peace and to cultivate friendship on fair and honorable terms, and in the adjustment of any differences that may exist or arise to exhibit the forbearance becoming a powerful nation rather than the sensibility belonging to a gallant people.

In such measures as I may be called on to pursue in regard to the rights of the separate states I hope to be animated by a proper respect for those

1. Joseph Story to Mrs. Joseph Story, March 7, 1829, in William W. Story, *Life and Letters of Joeseph Story* (Boston, 1851) I, 563.

sovereign members of our Union, taking care not to confound the powers they have reserved to themselves with those they have granted to the confederacy.

The management of the public revenue—that searching operation in all governments—is among the most delicate and important trusts in ours, and it will, of course, demand no inconsiderable share of my official solicitude. Under every aspect in which it can be considered it would appear that advantage must result from the observance of a strict and faithful economy. This I shall aim at the more anxiously both because it will facilitate the extinguishment of the national debt, the unnecessary duration of which is incompatible with real independence, and because it will counteract that tendency to public and private profligacy which a profuse expenditure of money by the government is but too apt to engender. Powerful auxiliaries to the attainment of this desirable end are to be found in the regulations provided by the wisdom of Congress for the specific appropriation of public money and the prompt accountability of public officers.

With regard to a proper selection of the subjects of impost with a view to revenue, it would seem to me that the spirit of equity, caution, and compromise in which the Constitution was formed requires that the great interests of agriculture, commerce, and manufactures should be equally favored, and that perhaps the only exception to this rule should consist in the peculiar encouragement of any products of either of them that may be found essential to our national independence.

Internal improvement and the diffusion of knowledge, so far as they can be promoted by the constitutional acts of the federal government, are of high importance.

Considering standing armies as dangerous to free governments in time of peace, I shall not seek to enlarge our present establishment, nor disregard that salutary lesson of political experience which teaches that the military should be held subordinate to the civil power. The gradual increase of our navy, whose flag has displayed in distant climes our skill in navigation and our fame in arms; the preservation of our forts, arsenals, and dockyards, and the introduction of progressive improvements in the discipline and science of both branches of our military service are so plainly prescribed by prudence that I should be excused for omitting their mention sooner than for enlarging on their importance. But the bulwark of our defense is the national militia, which in the present state of our intelligence and population must render us invincible. As long as our gov-

ernment is administered for the good of the people, and is regulated by
their will; as long as it secures to us the rights of person and of property,
liberty of conscience and of the press, it will be worth defending; and so
long as it is worth defending a patriotic militia will cover it with an im-
penetrable aegis. Partial injuries and occasional mortifications we may be
subjected to, but a million of armed freemen, possessed of the means of
war, can never be conquered by a foreign foe. To any just system, therefore, calculated to strengthen this natural safeguard of the country I shall
cheerfully lend all the aid in my power.

It will be my sincere and constant desire to observe toward the Indian
tribes within our limits a just and liberal policy, and to give that humane
and considerate attention to their rights and their wants which is consistent with the habits of our government and the feelings of our people.

The recent demonstration of public sentiment inscribes on the list of
executive duties, in characters too legible to be overlooked, the task of reform, which will require particularly the correction of those abuses that
have brought the patronage of the federal government into conflict with
the freedom of elections, and the counteraction of those causes which have
disturbed the rightful course of appointment and have placed or continued power in unfaithful or incompetent hands.

In the performance of a task thus generally delineated I shall endeavor
to select men whose diligence and talents will insure in their respective
stations able and faithful cooperation, depending for the advancement of
the public service more on the integrity and zeal of the public officers than
on their numbers.

A diffidence, perhaps too just, in my own qualifications will teach me
to look with reverence to the examples of public virtue left by my illustrious predecessors, and with veneration to the lights that flow from the
mind that founded and the mind that reformed our system. The same diffidence induces me to hope for instruction and aid from the coordinate
branches of the government, and for the indulgence and support of my fellow citizens generally. And a firm reliance on the goodness of that power
whose providence mercifully protected our national infancy, and has since
upheld our liberties in various vicissitudes, encourages me to offer up my
ardent supplications that He will continue to make our beloved country
the object of His divine care and gracious benediction.

Andrew Jackson

· SECOND INAUGURAL ADDRESS ·

Monday, March 4, 1833

*O*f the many reforms President Jackson hoped to see enacted, two of
them aroused bitter controversy. But they also defined who he was and
what he believed. In his first message to Congress, he asked that certain
changes be enacted with regard to the Second Bank of the United States
(BUS). Headquartered in Philadelphia, this institution had been established by the federal government in 1816 with a capital stock of thirty-five
million dollars, of which one-fifth was purchased by the government and
the remaining four-fifths sold to the public. The government's funds were
deposited in the Bank, which served as an agent in the collection and
transmittal of taxes. There were twenty-six branches of the BUS located in
the principal cities throughout the country. It was controlled by a board of
twenty-five directors, five of whom were appointed by the government and
twenty elected by the public stockholders. But the Bank was principally
managed by its very capable President, Nicholas Biddle.

Jackson's request for a reform of the Bank's operations resulted from
his belief that the institution's involvement in the political process, especially in assisting the election of persons who were friendly to the Bank,
threatened individual liberty. Indeed, he believed that the Bank had used
its money against him in the election of 1828. Congress paid no attention
to the President's request, since it reckoned that the institution was run efficiently and profitably.

The tense situation between Jackson and the Bank escalated when Biddle, goaded by Henry Clay, who now represented Kentucky in the Senate,
requested that Congress recharter the Bank four years before its present
charter was due to expire. Clay believed he could run against Jackson on
the National Republican ticket and defeat him in the presidential election
of 1832 if Old Hickory vetoed the recharter. Both Clay and Biddle were

certain that the American people would never tolerate the destruction of the Bank. Their action raised the conflict to all-out war. Jackson accepted the challenge, vetoed the recharter bill when Congress passed it, and went on to win reelection. He achieved a smashing victory over his opposition, proving once again how popular he was with the electorate. He received 688,242 popular and 219 electoral votes to Clay's 473,462 popular and 49 electoral votes.

The Bank War continued throughout Jackson's second term in office when the government's money was withdrawn from the BUS and the Senate censured the President for his action. But the charter for the institution was not renewed, and the Bank finally expired.

The second conflict arose over the attempt by Congress to enact a new tariff to replace the one passed in 1828 and known in the South as the Tariff of Abominations. In fact, Vice President Calhoun was so angered by the passage of the bill that he returned home to South Carolina and wrote an "Exposition and Protest" in which he advanced the doctrine of nullification, namely that when the federal government passes a law that acts in conflict with the interest of a particular state, that state may declare such action null and void within its borders. If three-fourths of all the states nullified the particular law, then it was repealed, just as if the Constitution itself had been amended. The South Carolina legislature adopted the "Exposition and Protest" and threatened nullification if the tariff was not adjusted fairly.

Congress did remove some of the "abominations" of the 1828 law by passing the Tariff of 1832. Unfortunately, it did not lower the tariff rates to any significant degree, so the governor of South Carolina called a special session of the legislature, which in turn ordered an elected convention to meet and take appropriate action. The convention did meet, and on November 24, 1832, it passed the Ordinance of Nullification declaring the tariffs of 1828 and 1832 null and void in South Carolina. It went on to threaten secession if force were used to coerce the state into compliance.

Jackson had just been reelected and felt the people of the country supported his view that no state could disobey the law and get away with it. He issued a proclamation on December 10, 1832, in which he said, "Disunion by armed force is *treason* . . . I consider, then, the power to annul a law of the United States, assumed by one State, *incompatible with the existence of the Union.*" Jackson then went on and asked Congress for a "Force Bill," authorizing him to deploy the military to put down armed rebellion if it should occur. The Force Bill passed along with the Compromise Tariff of 1833, which had been shepherded through the legislature by Senator Clay and which Calhoun pronounced acceptable to the nullifiers. This tariff provided a ten-year truce, during which time the rates would slowly fall, taking a sharper drop in the final years of the truce, when the duties would stand at a uniform 20 percent ad valorem rate and remain at that rate.

The crisis passed and Jackson prepared for his inauguration. Again he scribbled out his address, which ran to about a thousand words. It was a strong statement but unfortunately, like his first inaugural address, several editors laid rude hands upon it and softened its tone and tried to make it more conciliatory.

On Monday, March 4, 1833, Jackson and his new vice president, Martin Van Buren, were inaugurated. It was a cold and windy day and because of the inclement weather the swearing-in took place in the chamber of the House of Representatives. The President looked feeble and tired. When he rose from his seat to present his inaugural address he was greeted with cheers from the entire audience. The cheers continued for several minutes and when at last the noise level dissipated, Jackson began to read his address. He whipped through it very quickly since it was mercifully short. When he put down the last sheet, he was again saluted with prolonged cheers and applause. As always, he bowed before the "majesty of the people."

The old man looked on the verge of collapse. One observer ventured to bet that Jackson would "not outlive the present term of his office." But Old Hickory hung on and lived to see his successor, Martin Van Buren, sworn in as the eighth President of the United States.

FELLOW CITIZENS:

The will of the American people, expressed through their unsolicited suffrages, calls me before you to pass through the solemnities preparatory to taking upon myself the duties of President of the United States for another term. For their approbation of my public conduct through a period which has not been without its difficulties, and for this renewed expression of their confidence in my good intentions, I am at a loss for terms adequate to the expression of my gratitude. It shall be displayed to the extent of my humble abilities in continued efforts so to administer the government as to preserve their liberty and promote their happiness.

So many events have occurred within the last four years which have necessarily called forth—sometimes under circumstances the most delicate and painful—my views of the principles and policy which ought to be pursued by the general government that I need on this occasion but allude to a few leading considerations connected with some of them.

The foreign policy adopted by our government soon after the formation of our present Constitution, and very generally pursued by successive administrations, has been crowned with almost complete success, and has elevated our character among the nations of the earth. To do justice to all and to submit to wrong from none has been during my administration its governing maxim, and so happy have been its results that we are not only at peace with all the world, but have few causes of controversy, and those of minor importance, remaining unadjusted.

In the domestic policy of this government there are two objects which especially deserve the attention of the people and their representatives, and which have been and will continue to be the subjects of my increasing solicitude. They are the preservation of the rights of the several states and the integrity of the Union.

These great objects are necessarily connected, and can only be attained by an enlightened exercise of the powers of each within its appropriate sphere in conformity with the public will constitutionally expressed. To this end it becomes the duty of all to yield a ready and patriotic submission to the laws constitutionally enacted and thereby promote and strengthen a proper confidence in those institutions of the several states and of the United States which the people themselves have ordained for their own government.

My experience in public concerns and the observation of a life somewhat advanced confirm the opinions long since imbibed by me, that the destruction of our state governments or the annihilation of their control over the local concerns of the people would lead directly to revolution and anarchy, and finally to despotism and military domination. In proportion, therefore, as the general government encroaches upon the rights of the states, in the same proportion does it impair its own power and detract from its ability to fulfill the purposes of its creation. Solemnly impressed with these considerations, my countrymen will ever find me ready to exercise my constitutional powers in arresting measures which may directly or indirectly encroach upon the rights of the states or tend to consolidate all political power in the general government. But of equal and, indeed of incalculable, importance is the union of these states, and the sacred duty of all to contribute to its preservation by a liberal support of the general government in the exercise of its just powers. You have been wisely admonished to "accustom yourselves to think and speak of the Union as of the palladium of your political safety and prosperity, watching for its preservation with jealous anxiety, discountenancing whatever may suggest even a suspicion that it can in any event be abandoned, and indignantly frowning upon the first dawning of any attempt to alienate any portion of our country from the rest or to enfeeble the sacred ties which now link together the various parts." Without union our independence and liberty would never have been achieved; without union they never can be maintained. Divided into twenty-four, or even a smaller number, of separate communities, we shall see our internal trade burdened with numberless restraints and exactions; communication between distant points and sections obstructed or cut off; our sons made soldiers to deluge with blood the fields they now till in peace; the mass of our people borne down and impoverished by taxes to support armies and navies, and military leaders at the head of their victorious legions becoming our lawgivers and judges. The loss of liberty, of all good government, of peace, plenty, and happiness, must inevitably follow a dissolution of the Union. In supporting it, therefore, we support all that is dear to the freeman and the philanthropist.

The time at which I stand before you is full of interest. The eyes of all nations are fixed on our Republic. The event of the existing crisis will be decisive in the opinion of mankind of the practicability of our federal system of government. Great is the stake placed in our hands; great is the responsibility which must rest upon the people of the United States. Let us realize the importance of the attitude in which we stand before the world. Let us exercise forbearance and firmness. Let us extricate our country from the dangers which surround it and learn wisdom from the lessons they inculcate.

Deeply impressed with the truth of these observations, and under the obligation of that solemn oath which I am about to take, I shall continue to exert all my faculties to maintain the just powers of the Constitution and to transmit unimpaired to posterity the blessings of our federal Union. At the same time, it will be my aim to inculcate by my official acts the necessity of exercising by the general government those powers only that are clearly delegated; to encourage simplicity and economy in the expenditures of the government; to raise no more money from the people than may be requisite for these objects, and in a manner that will best promote the interests of all classes of the community and of all portions of the Union. Constantly bearing in mind that in entering into society "individuals must give up a share of liberty to preserve the rest," it will be my desire so to discharge my duties as to foster with our brethren in all parts of the country a spirit of liberal concession and compromise, and, by reconciling our fellow citizens to those partial sacrifices which they must unavoidably make for the preservation of a greater good, to recommend our invaluable government and Union to the confidence and affections of the American people.

Finally, it is my most fervent prayer to that Almighty Being before whom I now stand, and who has kept us in His hands from the infancy of our Republic to the present day, that He will so overrule all my intentions and actions and inspire the hearts of my fellow citizens that we may be preserved from dangers of all kinds and continue forever a united and happy people.

Martin Van Buren

Saturday, March 4, 1837

$\mathcal{M}$artin Van Buren was elected President of the United States in the fall of 1836 over Daniel Webster of Massachusetts, Hugh L. White of Tennessee, and William Henry Harrison of Ohio for one simple reason: President Andrew Jackson, the Hero of the Battle of New Orleans, had chosen him as his successor. Nevertheless, Van Buren had earned the right of succession if for no other reason than that he was unquestionably the chief architect in the creation of the Democratic Party that has continued as an important political force to this day.

He was born in Kinderhook, New York, and chose politics as the means to advance his career as a lawyer. A superb politician, he first built the Albany Regency, an organized group of skillful and powerful leaders of the New York Republican Party, to run the state in his absence after he was elected to the United States Senate. His success stemmed in large measure from his personal charm and ingratiating manner among his colleagues. A short man, he was courteous to all and possessed the "high art of blending dignity with ease and suavity." He was, commented one observer, "as polished and captivating a person in the social circle as America has ever known. . . ."[1] He was called the Little Magician because his political skills seemed at times to verge on the miraculous.

Once in Washington, he assumed control of the most conservative wing of the Democratic-Republican Party, labeled Radicals by many. That meant that he endorsed the republican doctrines of Thomas Jefferson, namely opposition to a strong central government and to a strict interpretation of the Constitution. He regarded the states as wholesome counterweights to the national government. In other words, he was a dedicated states-rights advocate.

---

1. Henry S. Foote, *Casket of Reminiscences* (Washington, D.C., 1874), 59.

Van Buren represented a new breed of American politicians following the War of 1812. He disagreed with the Founding Fathers, who looked upon political parties as little better than factions of greedy, designing, and dangerous adventurers. Rather, Van Buren regarded the party system as an essential ingredient for representative government. Modern efficient government, he argued, demanded well-functioning political parties openly arrayed against each other. According to Van Buren, a well-defined two-party system provided a balance of power between opposing forces, and this in turn safeguarded liberty and the institutions of republicanism. The era of a one-party system under James Monroe produced a concentration of power in Washington that necessarily generated corruption and led to the fraudulent election of John Quincy Adams in 1825.

In the election of 1824, Van Buren led the Radical Republicans to declare for William H. Crawford, the secretary of the treasury. When that effort ended in defeat, the Little Magician convinced the Radical Republicans and Vice President Calhoun and his friends to join the Jackson camp, thereby connecting Jackson's enormous personal popularity with Jeffersonian republican principles to form a new organization. It would revive a two-party system with Democrats (Jackson, Calhoun, and Van Buren) on one side and National Republicans (Adams and Clay), later Whigs, on the other.

Van Buren served as Jackson's secretary of state during Old Hickory's first term, and as vice president after the President broke with Calhoun. At their national nominating convention held in Baltimore on May 20, 1835, over six hundred delegates attended and unanimously nominated Van Buren, Jackson's choice for President. And in the general election in the fall of 1836, the Little Magician amassed 764,198 popular and 170 electoral votes over 73 electoral votes for Harrison, 26 for White, and 14 for Webster.

On March 4, 1837, Jackson and Van Buren rode a gleaming phaeton from the White House to the east front of the Capitol. They mounted the steps of the building to the cheers of an adoring crowd. They first attended the swearing-in of the new vice president, Richard Johnson of Kentucky, and then walked back to the East Portico where a wooden rostrum had been erected over the steps. Finally, Van Buren stepped forward and read his inaugural address, something that took over an hour to get through. He ended it by acknowledging his "illustrious predecessor's" brilliant success as President. Then the oath of office was administered by Chief Justice Roger B. Taney. Interestingly, both Clay and Webster attended the ceremony.

Jackson shook Van Buren's hand, turned, and descended the steps and headed for the phaeton waiting below. The crowd hushed. "This vast crowd," wrote Senator Thomas Hart Benton of Missouri, "remained riveted to their places, and profoundly silent. It was the stillness and silence of reverence and affection; and there was no room for mistake as to whom this mute and impressive homage was rendered. For once, the rising was eclipsed by the setting sun."

FELLOW CITIZENS:

The practice of all my predecessors imposes on me an obligation I cheerfully fulfill—to accompany the first and solemn act of my public trust with an avowal of the principles that will guide me in performing it and an expression of my feelings on assuming a charge so responsible and vast. In imitating their example I tread in the footsteps of illustrious men, whose superiors it is our happiness to believe are not found on the executive calendar of any country. Among them we recognize the earliest and firmest pillars of the Republic—those by whom our national independence was first declared, him who above all others contributed to establish it on the field of battle, and those whose expanded intellect and patriotism constructed, improved, and perfected the inestimable institutions under which we live. If such men in the position I now occupy felt themselves overwhelmed by a sense of gratitude for this the highest of all marks of their country's confidence, and by a consciousness of their inability adequately to discharge the duties of an office so difficult and exalted, how much more must these considerations affect one who can rely on no such claims for favor or forbearance! Unlike all who have preceded me, the Revolution that gave us existence as one people was achieved at the period of my birth; and whilst I contemplate with grateful reverence that memorable event, I feel that I belong to a later age and that I may not expect my countrymen to weigh my actions with the same kind and partial hand.

So sensibly, fellow citizens, do these circumstances press themselves upon me that I should not dare to enter upon my path of duty did I not look for the generous aid of those who will be associated with me in the various and coordinate branches of the government; did I not repose with unwavering reliance on the patriotism, the intelligence, and the kindness of a people who never yet deserted a public servant honestly laboring their cause; and, above all, did I not permit myself humbly to hope for the sustaining support of an ever-watchful and beneficent Providence.

To the confidence and consolation derived from these sources it would be ungrateful not to add those which spring from our present fortunate condition. Though not altogether exempt from embarrassments that disturb our tranquillity at home and threaten it abroad, yet in all the attributes of a great, happy, and flourishing people we stand without a parallel in the world. Abroad we enjoy the respect and, with scarcely an exception, the friendship of every nation; at home, while our government quietly but efficiently performs the sole legitimate end of political institutions—in doing the greatest good to the greatest number—we present an aggregate of human prosperity surely not elsewhere to be found.

How imperious, then, is the obligation imposed upon every citizen, in his own sphere of action, whether limited or extended, to exert himself in perpetuating a condition of things so singularly happy! All the lessons of history and experience must be lost upon us if we are content to trust alone to the peculiar advantages we happen to possess. Position and climate and the bounteous resources that nature has scattered with so liberal a hand— even the diffused intelligence and elevated character of our people—will avail us nothing if we fail sacredly to uphold those political institutions that were wisely and deliberately formed with reference to every circumstance that could preserve or might endanger the blessings we enjoy. The thoughtful framers of our Constitution legislated for our country as they found it. Looking upon it with the eyes of statesmen and patriots, they saw all the sources of rapid and wonderful prosperity; but they saw also that various habits, opinions, and institutions peculiar to the various portions of so vast a region were deeply fixed. Distinct sovereignties were in actual existence, whose cordial union was essential to the welfare and happiness of all. Between many of them there was, at least to some extent, a real diversity of interests, liable to be exaggerated through sinister designs; they differed in size, in population, in wealth, and in actual and prospective resources and power; they varied in the character of their industry and staple productions, and [in some] existed domestic institutions which, unwisely disturbed, might endanger the harmony of the whole. Most carefully were all these circumstances weighed, and the foundations of the new government laid upon principles of reciprocal concession and equitable compromise. The jealousies which the smaller states might entertain of the power of the rest were allayed by a rule of representation confessedly unequal at the time, and designed forever to remain so. A natural fear that the broad scope of general legislation might bear upon and unwisely control particular interests was counteracted by limits strictly drawn around the action of the federal authority, and to the people and the states was left unimpaired their sovereign power over the innumerable subjects embraced in the internal government of a just republic, excepting such only as necessarily appertain to the concerns of the whole confederacy or its intercourse as a united community with the other nations of the world.

This provident forecast has been verified by time. Half a century, teeming with extraordinary events, and elsewhere producing astonishing results, has passed along, but on our institutions it has left no injurious mark. From a small community we have risen to a people powerful in numbers and in strength; but with our increase has gone hand in hand the progress of just principles. The privileges, civil and religious, of the humblest individual are still sacredly protected at home, and while the valor and fortitude of our people have removed far from us the slightest apprehension of foreign power, they have not yet induced us in a single instance to forget what is right. Our commerce has been extended to the remotest nations; the value and even nature of our productions have been greatly changed; a wide difference has arisen in the relative wealth and resources of every portion of our country; yet the spirit of mutual regard and of faithful adherence to existing compacts has continued to prevail in our councils and never long been absent from our conduct. We have learned by experience a fruitful lesson—that an implicit and undeviating adherence to the principles on which we set out can carry us prosperously onward through all the conflicts of circumstances and vicissitudes inseparable from the lapse of years.

The success that has thus attended our great experiment is in itself a sufficient cause for gratitude, on account of the happiness it has actually conferred and the example it has unanswerably given. But to me, my fellow citizens, looking forward to the far-distant future with ardent prayers and confiding hopes, this retrospect presents a ground for still deeper delight. It impresses on my mind a firm belief that the perpetuity of our institutions depends upon ourselves; that if we maintain the principles on which they were established they are destined to confer their benefits on countless generations yet to come, and that America will present to every friend of mankind the cheering proof that a popular government, wisely formed, is wanting in no element of endurance or strength. Fifty years ago its rapid failure was boldly predicted. Latent and uncontrollable causes of dissolution were supposed to exist even by the wise and good, and not only did unfriendly or speculative theorists anticipate for us the fate of past republics, but the fears of many an honest patriot overbalanced his sanguine hopes. Look back on these forebodings, not hastily but reluctantly made, and see how in every instance they have completely failed.

An imperfect experience during the struggles of the Revolution was supposed to warrant the belief that the people would not bear the taxation requisite to discharge an immense public debt already incurred and to pay the necessary expenses of the government. The cost of two wars has been paid, not only without a murmur; but with unequaled alacrity. No one is now left to doubt that every burden will be cheerfully borne that may be necessary to sustain our civil institutions or guard our honor or welfare. Indeed, all experience has shown that the willingness of the people to contribute to these ends in cases of emergency has uniformly outrun the confidence of their representatives.

In the early stages of the new government, when all felt the imposing influence as they recognized the unequaled services of the first President, it was a common sentiment that the great weight of his character could alone bind the discordant materials of our government together and save us from the violence of contending factions. Since his death nearly forty years are gone. Party exasperation has been often carried to its highest point; the virtue and fortitude of the people have sometimes been greatly tried; yet our system, purified and enhanced in value by all it has encountered, still preserves its spirit of free and fearless discussion, blended with unimpaired fraternal feeling.

The capacity of the people for self-government, and their willingness, from a high sense of duty and without those exhibitions of coercive power so generally employed in other countries, to submit to all needful restraints and exactions of municipal law, have also been favorably exemplified in the history of the American states. Occasionally, it is true, the ardor of public sentiment, outrunning the regular progress of the judicial tribunals or seeking to reach cases not denounced as criminal by the existing law, has displayed itself in a manner calculated to give pain to the friends of free government and to encourage the hopes of those who wish for its overthrow. These occurrences, however, have been far less frequent in our country than in any other of equal population on the globe, and with the diffusion of intelligence it may well be hoped that they will constantly diminish in frequency and violence. The generous patriotism and sound common sense of the great mass of our fellow citizens will assuredly in time produce this result; for as every assumption of illegal power not only wounds the majesty of the law, but furnishes a pretext for abridging the liberties of the people, the latter have the most direct and permanent interest in preserving the landmarks of social order and maintaining on all occasions the inviolability of those constitutional and legal provisions which they themselves have made.

In a supposed unfitness of our institutions for those hostile emergencies which no country can always avoid their friends found a fruitful source of apprehension, their enemies of hope. While they foresaw less promptness of action than in governments differently formed, they overlooked the far more important consideration that with us war could never be the result of individual or irresponsible will, but must be a measure of redress for injuries sustained, voluntarily resorted to by those who were to bear the necessary sacrifice, who would consequently feel an individual interest in the contest, and whose energy would be commensurate with the difficulties to be encountered. Actual events have proved their error; the last war, far from impairing, gave new confidence to our government, and amid recent apprehensions of a similar conflict we saw that the energies of our country would not be wanting in ample season to vindicate its rights. We may not possess, as we should not desire to possess, the extended and ever-ready military organization of other nations; we may occasionally suffer in the outset for the want of it; but among ourselves all doubt upon this great point has ceased, while a salutary experience will prevent a contrary opinion from inviting aggression from abroad.

Certain danger was foretold from the extension of our territory, the multiplication of states, and the increase of population. Our system was supposed to be adapted only to boundaries comparatively narrow. These have been widened beyond conjecture; the members of our confederacy are already doubled, and the numbers of our people are incredibly augmented. The alleged causes of danger have long surpassed anticipation, but none of the consequences have followed. The power and influence of the Republic have arisen to a height obvious to all mankind; respect for its authority was not more apparent at its ancient than it is at its present limits; new and inexhaustible sources of general prosperity have been opened; the effects of distance have been averted by the inventive genius of our people, developed and fostered by the spirit of our institutions; and the enlarged variety and amount of interests, productions, and pursuits have strengthened the chain of mutual dependence and formed a circle of mutual benefits too apparent ever to be overlooked.

In justly balancing the powers of the federal and state authorities difficulties nearly insurmountable arose at the outset and subsequent collisions were deemed inevitable. Amid these it was scarcely believed possible that a scheme of government so complex in construction could remain uninjured. From time to time embarrassments have certainly occurred; but how just is the confidence of future safety imparted by the knowledge that each in succession has been happily removed! Overlooking partial and temporary evils as inseparable from the practical operation of all human institutions, and looking only to the general result, every patriot has reason to be satisfied. While the federal government has successfully performed its appropriate functions in relation to foreign affairs and concerns evidently national, that of every state has remarkably improved in protecting and developing local interests and individual welfare; and if the vibrations of authority have occasionally tended too much toward one or the other, it is unquestionably certain that the ultimate operation of the entire system has been to strengthen all the existing institutions and to elevate our whole country in prosperity and renown.

The last, perhaps the greatest, of the prominent sources of discord and disaster supposed to lurk in our political condition was the institution of domestic slavery. Our forefathers were deeply impressed with the delicacy of this subject, and they treated it with a forbearance so evidently wise that in spite of every sinister foreboding it never until the present period disturbed the tranquillity of our common country. Such a result is sufficient evidence of the justice and the patriotism of their course; it is evidence not to be mistaken that an adherence to it can prevent all embarrassment from this as well as from every other anticipated cause of difficulty or danger. Have not recent events made it obvious to the slightest reflection that the least deviation from this spirit of forbearance is injurious to every interest, that of humanity included? Amidst the violence of excited passions this generous and fraternal feeling has been sometimes disregarded; and standing as I now do before my countrymen, in this high place of honor and of trust, I can not refrain from anxiously invoking my fellow citizens never to be deaf to its dictates. Perceiving before my election the deep interest this subject was beginning to excite, I believed it a solemn duty fully to make known my sentiments in regard to it, and now, when every motive for misrepresentation has passed away, I trust that they will be candidly weighed and understood. At least they will be my standard of conduct in the path before me. I then declared that if the desire of those of my countrymen who were favorable to my election was gratified "I must go into the presidential chair the inflexible and uncompromising opponent of every attempt on the part of Congress to abolish slavery in the District of Columbia against the wishes of the slaveholding states, and also with a determination equally decided to resist the slightest interference with it in the states where it exists." I submitted also to my fellow citizens, with fullness and frankness, the reasons which led me to this determination. The result authorizes me to believe that they have been approved and are confided in by a majority of the people of the United States, including those whom they most immediately affect. It now only remains to add that no bill conflicting with these views can ever receive my constitutional sanction. These opinions have been adopted in the firm belief that they are in accordance with the spirit that actuated the venerated fathers of the Republic, and that succeeding experience has proved them to be humane, patriotic, expedient, honorable, and just. If the agitation of this subject was intended to reach the stability of our institutions, enough has occurred to show that it has signally failed, and that in this as in every other instance the apprehensions of the timid and the hopes of the wicked for the destruction of our government are again destined to be disappointed. Here and there, indeed, scenes of dangerous excitement have occurred, terrifying instances of local violence have been witnessed, and a reckless disregard of the consequences of their conduct has exposed individuals to popular indignation; but neither masses of the people nor sections of the country have been swerved from their devotion to the bond of union and the principles it has made sacred. It will be ever thus. Such attempts at dangerous agitation may periodically return, but with each the object will be better understood. That predominating affection for our political system which prevails throughout our territorial limits, that calm and enlightened judgment which ultimately governs our people as one vast body, will always be at hand to resist and control every effort, foreign or domestic, which aims or would lead to overthrow our institutions.

What can be more gratifying than such a retrospect as this? We look back on obstacles avoided and dangers overcome, on expectations more than realized and prosperity perfectly secured. To the hopes of the hostile, the fears of the timid, and the doubts of the anxious actual experience has given the conclusive reply. We have seen time gradually dispel every unfavorable foreboding and our Constitution surmount every adverse circumstance dreaded at the outset as beyond control. Present excitement will at all times magnify present dangers, but true philosophy must teach us that none more threatening than the past can remain to be overcome; and we ought (for we have just reason) to entertain an abiding confidence in the stability of our institutions and an entire conviction that if administered in the true form, character, and spirit in which they were established they are abundantly adequate to preserve to us and our children the rich blessings already derived from them, to make our beloved land for a thousand generations that chosen spot where happiness springs from a perfect equality of political rights.

For myself, therefore, I desire to declare that the principle that will govern me in the high duty to which my country calls me is a strict adherence to the letter and spirit of the Constitution as it was designed by those who framed it. Looking back to it as a sacred instrument carefully and not easily framed; remembering that it was throughout a work of concession and compromise; viewing it as limited to national objects; regarding it as leaving to the people and the states all power not explicitly parted with, I shall endeavor to preserve, protect, and defend it by anxiously referring to its provision for direction in every action. To matters of domestic concernment which it has intrusted to the federal government and to such as relate to our intercourse with foreign nations I shall zealously devote myself; beyond those limits I shall never pass.

To enter on this occasion into a further or more minute exposition of my views on the various questions of domestic policy would be as obtrusive as it is probably unexpected. Before the suffrages of my countrymen were conferred upon me I submitted to them, with great precision, my opinions on all the most prominent of these subjects. Those opinions I shall endeavor to carry out with my utmost ability.

Our course of foreign policy has been so uniform and intelligible as to constitute a rule of executive conduct which leaves little to my discretion, unless, indeed, I were willing to run counter to the lights of experience and the known opinions of my constituents. We sedulously cultivate the friendship of all nations as the conditions most compatible with our welfare and the principles of our government. We decline alliances as adverse to our peace. We desire commercial relations on equal terms, being ever willing to give a fair equivalent for advantages received. We endeavor to conduct our intercourse with openness and sincerity, promptly avowing our objects and seeking to establish that mutual frankness which is as beneficial in the dealings of nations as of men. We have no disposition and we disclaim all right to meddle in disputes, whether internal or foreign, that may molest other countries, regarding them in their actual state as social communities, and preserving a strict neutrality in all their controversies. Well knowing the tried valor of our people and our exhaustless resources, we neither anticipate nor fear any designed aggression; and in the consciousness of our own just conduct we feel a security that we shall never be called upon to exert our determination never to permit an invasion of our rights without punishment or redress.

In approaching, then, in the presence of my assembled countrymen, to make the solemn promise that yet remains, and to pledge myself that I will faithfully execute the office I am about to fill, I bring with me a settled purpose to maintain the institutions of my country, which I trust will atone for the errors I commit.

In receiving from the people the sacred trust twice confided to my illustrious predecessor, and which he has discharged so faithfully and so well, I know that I cannot expect to perform the arduous task with equal ability and success. But united as I have been in his counsels, a daily witness of his exclusive and unsurpassed devotion to his country's welfare, agreeing with him in sentiments which his countrymen have warmly supported, and permitted to partake largely of his confidence, I may hope that somewhat of the same cheering approbation will be found to attend upon my path. For him I but express with my own the wishes of all, that he may yet long live to enjoy the brilliant evening of his well-spent life; and for myself, conscious of but one desire, faithfully to serve my country, I throw myself without fear on its justice and its kindness. Beyond that I only look to the gracious protection of the Divine Being whose strengthening support I humbly solicit, and whom I fervently pray to look down upon us all. May it be among the dispensations of His providence to bless our beloved country with honors and with length of days. May her ways be ways of pleasantness and all her paths be peace!

William Henry Harrison

· INAUGURAL ADDRESS ·

Thursday, March 4, 1841

*P*oor William Henry Harrison! He is remembered today, if at all, as the man who gave the longest inaugural address in history, caught a cold in the process, and died a month later. He is the answer to a trivia question—which President served the shortest amount of time?—rather than a figure of historical significance.

His fate adds pathos to the very first words of his inaugural address, when he reminded his audience that he was called out of retirement to run for President as a candidate of the new Whig Party. He was sixty-eight years old when he became President, the oldest person to assume the office until Ronald Reagan in 1981. The years had brought him fame as a military leader and success as a political figure but perhaps not the wisdom that often accompanies age and accomplishment. Although his inauguration took place on an uncommonly cold March day in Washington, he wore neither an overcoat nor a hat. Worse, his speech of more than eight thousand words took him more than two hours to deliver. He became ill almost immediately but attempted to attend to his duties despite his worsening condition. He died on April 4, the first President to die in office.

Among the thousands of words he recited on that cold day in Washington, Harrison uttered a phrase never before heard in an inaugural address. On his first day as President, he pledged that "under no circumstances will I consent to serve a second term." This was probably intended as an act of statesmanship and as a concession, all too necessary as it turned out, to his age and health. But it was also an act of political foolishness, suggesting that Harrison would have had his hands full had he lived to make good on his pledge. The power of the presidency, Harry Truman observed, is the power to persuade. And the art of persuasion requires raw political power in addition to charm and perhaps even the occasional deployment of principle. Harrison made himself an instant lame duck in his first speech as President. It is hard to know what he might have been thinking, but it is fair to believe that the Whig lions in Congress, Henry Clay and Daniel Webster, were horrified, realizing that the President had just thrown away his main tool for implementing the party's program. But their horror might have given way to joy when the President announced that he would use his veto power sparingly, and asserted, in essence, that the true expression of the people's will was the Congress, not the President.

The election of 1840 saw the continued development of Jacksonian mass democracy, although it was the Whigs, rather than Jackson's Democrats, who benefited from the populism and spectacle of the nation's increasingly raucous style of politics. It helped that the incumbent, Martin Van Buren, was inept and distant. It helped even more that the country had plunged into a deep recession in 1837 and that Van Buren seemed out of touch with the nation's economic suffering .

Harrison was careful to avoid offering any specific solutions to the nation's economic problems. Instead, his campaign was about symbols and personality, one of the first recognizable modern campaigns that mobilized voters with mass literature, campaign rallies, slogans, and other spectacles. In contrast to the aristocratic Van Buren, Harrison, heir to Jacksonian populism, was cast by his supporters as an Everyman not unlike Old Hickory. It was on Harrison's behalf that the log cabin myth was created, and a myth it was. Harrison was born into a well-off Virginia family, but when a pro-Van Buren newspaper asserted that Harrison would be content to live on hard cider in a log cabin, the Whigs seized on this astounding bit of snobbery from, of all places, the party of Jackson. They embraced the fantasy that their planter–war hero candidate was a man of the people, a fantasy that would, in time, become a recurring theme in presidential politics.

What's more, Harrison's people managed to invent one of the best-remembered election slogans in American history: "Tippecanoe and Tyler, too." Harrison had won fame by defeating the great Indian chief Tecumseh at the Battle of Tippecanoe in 1811. John Tyler was Harrison's running mate, a renegade Democrat who sided with the Whigs, albeit not for long.

Harrison and Tyler easily defeated Van Buren and the Democrats, but the Whig victory was as short-lived as Harrison's tenure. When Tyler took office, the first vice president to fill a vacancy in the presidency, he reverted to his Democratic roots, a move that won him friends in neither party. He would become the first incumbent President denied his party's nomination for another term.

CALLED FROM A RETIREMENT which I had supposed was to continue for the residue of my life to fill the chief executive office of this great and free nation, I appear before you, fellow citizens, to take the oaths which the Constitution prescribes as a necessary qualification for the performance of its duties; and in obedience to a custom coeval with our government and what I believe to be your expectations I proceed to present to you a summary of the principles which will govern me in the discharge of the duties which I shall be called upon to perform.

It was the remark of a Roman consul in an early period of that celebrated republic that a most striking contrast was observable in the conduct of candidates for offices of power and trust before and after obtaining them, they seldom carrying out in the latter case the pledges and promises made in the former. However much the world may have improved in many respects in the lapse of upward of two thousand years since the remark was made by the virtuous and indignant Roman, I fear that a strict examination of the annals of some of the modern elective governments would develop similar instances of violated confidence.

Although the fiat of the people has gone forth proclaiming me the chief magistrate of this glorious Union, nothing upon their part remaining to be done, it may be thought that a motive may exist to keep up the delusion under which they may be supposed to have acted in relation to my principles and opinions; and perhaps there may be some in this assembly who have come here either prepared to condemn those I shall now deliver, or, approving them, to doubt the sincerity with which they are now uttered. But the lapse of a few months will confirm or dispel their fears. The outline of principles to govern and measures to be adopted by an administration not yet begun will soon be exchanged for immutable history, and I shall stand either exonerated by my countrymen or classed with the mass of those who promised that they might deceive and flattered with the intention to betray. However strong may be my present purpose to realize the expectations of a magnanimous and confiding people, I too well understand the dangerous temptations to which I shall be exposed from the magnitude of the power which it has been the pleasure of the people to commit to my hands not to place my chief confidence upon the aid of that Almighty Power which has hitherto protected me and enabled me to bring to favorable issues other important but still greatly inferior trusts heretofore confided to me by my country.

The broad foundation upon which our Constitution rests being the people—a breath of theirs having made, as a breath can unmake, change, or modify it—it can be assigned to none of the great divisions of government but to that of democracy. If such is its theory, those who are called upon to administer it must recognize as its leading principle the duty of shaping their measures so as to produce the greatest good to the greatest number. But with these broad admissions, if we would compare the sovereignty acknowledged to exist in the mass of our people with the power claimed by other sovereignties, even by those which have been considered most purely democratic, we shall find a most essential difference. All others lay claim to power limited only by their own will. The majority of our citizens, on the contrary, possess a sovereignty with an amount of power precisely equal to that which has been granted to them by the parties to the national compact, and nothing beyond. We admit of no government by divine right, believing that so far as power is concerned the Beneficent Creator has made no distinction amongst men; that all are upon an equality, and that the only legitimate right to govern is an express grant of power from the governed. The Constitution of the United States is the instrument containing this grant of power to the several departments composing the government. On an examination of that instrument it will be found to contain declarations of power granted and of power withheld. The latter is also susceptible of division into power which the majority had the right to grant, but which they do not think proper to intrust to their agents, and that which they could not have granted, not being possessed by themselves. In other words, there are certain rights possessed by each individual American citizen which in his compact with the others he has never surrendered. Some of them, indeed, he is unable to surrender, being, in the language of our system, unalienable. The boasted privilege of a Roman citizen was to him a shield only against a petty provincial ruler, whilst the proud democrat of Athens would console himself under a sentence of death for a supposed violation of the national faith—which no one understood and which at times was the subject of the mockery of all—or the banishment from his home, his family, and his country with or without an alleged cause, that it was the act not of a single tyrant or hated aristocracy, but of his assembled countrymen. Far different is the power of our sovereignty. It can interfere with no one's faith, prescribe forms of worship for no one's observance, inflict no punishment but after well-ascertained guilt, the result of investigation under rules prescribed by the Constitution itself. These precious privileges, and those scarcely less important of giving expression to his thoughts and opinions, either by writing or speaking, unrestrained but by the liability for injury to others, and that of a full participation in all the advantages which flow from the government, the acknowledged property of all, the American citizen derives from no charter granted by his fellow man. He claims them because he is himself a man, fashioned by the same Almighty hand as the rest of his species and entitled to a full share of the blessings with which He has endowed them. Notwithstanding the limited sovereignty possessed by the people of the United States and the restricted grant of power to the government which they have adopted, enough has been given to accomplish all the objects for which it was created. It has been found powerful in war, and hitherto justice has been administered, and intimate union effected, domestic tranquillity preserved, and personal liberty secured to the citizen. As was to be expected, however, from the defect of language and the necessarily sententious manner in which the Constitution is written, disputes have arisen as to the amount of power which it has actually granted or was intended to grant.

This is more particularly the case in relation to that part of the instrument which treats of the legislative branch, and not only as regards the exercise of powers claimed under a general clause giving that body the authority to pass all laws necessary to carry into effect the specified powers, but in relation to the latter also. It is, however, consolatory to reflect that most of the instances of alleged departure from the letter or spirit of the Constitution have ultimately received the sanction of a majority of the people. And the fact that many of our statesmen most distinguished for talent and patriotism have been at one time or other of their political career on both sides of each of the most warmly disputed questions forces upon us the inference that the errors, if errors there were, are attributable to the intrinsic difficulty in many instances of ascertaining the intentions of the framers of the Constitution rather than the influence of any sinister or unpatriotic motive. But the great danger to our institutions does not appear to me to be in a usurpation by the government of power not granted by the people, but by the accumulation in one of the departments of that which was assigned to others. Limited as are the powers which have been

granted, still enough have been granted to constitute a despotism if concentrated in one of the departments. This danger is greatly heightened, as
it has been always observable that men are less jealous of encroachments
of one department upon another than upon their own reserved rights.
When the Constitution of the United States first came from the hands of
the Convention which formed it, many of the sternest republicans of the
day were alarmed at the extent of the power which had been granted to the
federal government, and more particularly of that portion which had been
assigned to the executive branch. There were in it features which appeared
not to be in harmony with their ideas of a simple representative democracy or republic, and knowing the tendency of power to increase itself,
particularly when exercised by a single individual, predictions were made
that at no very remote period the government would terminate in virtual
monarchy. It would not become me to say that the fears of these patriots
have been already realized; but as I sincerely believe that the tendency of
measures and of men's opinions for some years past has been in that direction, it is, I conceive, strictly proper that I should take this occasion to repeat the assurances I have heretofore given of my determination to arrest
the progress of that tendency if it really exists and restore the government
to its pristine health and vigor, as far as this can be effected by any legitimate exercise of the power placed in my hands.

I proceed to state in as summary a manner as I can my opinion of the
sources of the evils which have been so extensively complained of and the
correctives which may be applied. Some of the former are unquestionably
to be found in the defects of the Constitution; others, in my judgment, are
attributable to a misconstruction of some of its provisions. Of the former
is the eligibility of the same individual to a second term of the presidency.
The sagacious mind of Mr. Jefferson early saw and lamented this error,
and attempts have been made, hitherto without success, to apply the amend-
atory power of the states to its correction. As, however, one mode of correction is in the power of every President, and consequently in mine, it
would be useless, and perhaps invidious, to enumerate the evils of which,
in the opinion of many of our fellow citizens, this error of the sages who
framed the Constitution may have been the source and the bitter fruits
which we are still to gather from it if it continues to disfigure our system.
It may be observed, however, as a general remark, that republics can commit no greater error than to adopt or continue any feature in their systems
of government which may be calculated to create or increase the lover of
power in the bosoms of those to whom necessity obliges them to commit

the management of their affairs; and surely nothing is more likely to produce such a state of mind than the long continuance of an office of high trust. Nothing can be more corrupting, nothing more destructive of all those noble feelings which belong to the character of a devoted republican patriot. When this corrupting passion once takes possession of the human mind, like the love of gold it becomes insatiable. It is the never-dying worm in his bosom, grows with his growth and strengthens with the declining years of its victim. If this is true, it is the part of wisdom for a republic to limit the service of that officer at least to whom she has intrusted the management of her foreign relations, the execution of her laws, and the command of her armies and navies to a period so short as to prevent his forgetting that he is the accountable agent, not the principal; the servant, not the master. Until an amendment of the Constitution can be effected public opinion may secure the desired object. I give my aid to it by renewing the pledge heretofore given that under no circumstances will I consent to serve a second term.

But if there is danger to public liberty from the acknowledged defects of the Constitution in the want of limit to the continuance of the executive power in the same hands, there is, I apprehend, not much less from a misconstruction of that instrument as it regards the powers actually given. I cannot conceive that by a fair construction any or either of its provisions would be found to constitute the President a part of the legislative power. It cannot be claimed from the power to recommend, since, although enjoined as a duty upon him, it is a privilege which he holds in common with every other citizen; and although there may be something more of confidence in the propriety of the measures recommended in the one case than in the other, in the obligations of ultimate decision there can be no difference. In the language of the Constitution, "all the legislative powers" which it grants "are vested in the Congress of the United States." It would be a solecism in language to say that any portion of these is not included in the whole.

It may be said, indeed, that the Constitution has given to the executive the power to annul the acts of the legislative body by refusing to them his assent. So a similar power has necessarily resulted from that instrument to the judiciary, and yet the judiciary forms no part of the legislature. There is, it is true, this difference between these grants of power: The executive can put his negative upon the acts of the legislature for other cause than that of want of conformity to the Constitution, whilst the judiciary can only declare void those which violate that instrument. But the decision of the judiciary is final in such a case, whereas in every instance where the veto of the executive is applied it may be overcome by a vote of two-thirds of both houses of Congress. The negative upon the acts of the legislative by the executive authority, and that in the hands of one individual, would seem to be an incongruity in our system. Like some others of a similar character, however, it appears to be highly expedient, and if used only with the forbearance and in the spirit which was intended by its authors it may be productive of great good and be found one of the best safeguards to the Union. At the period of the formation of the Constitution the principle does not appear to have enjoyed much favor in the state governments. It existed but in two, and in one of these there was a plural executive. If we would search for the motives which operated upon the purely patriotic and enlightened assembly which framed the Constitution for the adoption of a provision so apparently repugnant to the leading democratic principle that the majority should govern, we must reject the idea that they anticipated from it any benefit to the ordinary course of legislation. They knew too well the high degree of intelligence which existed among the people and the enlightened character of the state legislatures not to have the fullest confidence that the two bodies elected by them would be worthy representatives of such constituents, and, of course, that they would require no aid in conceiving and maturing the measures which the circumstances of the country might require. And it is preposterous to suppose that a thought could for a moment have been entertained that the President, placed at the capital, in the center of the country, could better understand the wants and wishes of the people than their own immediate representatives, who spend a part of every year among them, living with them, often laboring with them, and bound to them by the triple tie of interest, duty, and affection. To assist or control Congress, then, in its ordinary legislation could not, I conceive, have been the motive for conferring the veto power on the President. This argument acquires additional force from the fact of its never having been thus used by the first six Presidents—and two of them were members of the Convention, one presiding over its deliberations and the other bearing a larger share in consummating the labors of that august body than any other person. But if bills were never returned to Congress by either of the Presidents above referred to upon the ground of their being inexpedient or not as well adapted as they might be to the wants of the people, the veto was applied upon that of want of conformity to the Constitution or because errors had been committed from a too hasty enactment.

There is another ground for the adoption of the veto principle, which had probably more influence in recommending it to the Convention than any other. I refer to the security which it gives to the just and equitable action of the legislature upon all parts of the Union. It could not but have occurred to the Convention that in a country so extensive, embracing so great a variety of soil and climate, and consequently of products, and which from the same causes must ever exhibit a great difference in the amount of the population of its various sections, calling for a great diversity in the employments of the people, that the legislation of the majority might not always justly regard the rights and interests of the minority, and that acts of this character might be passed under an express grant by the words of the Constitution, and therefore not within the competency of the judiciary to declare void; that however enlightened and patriotic they might suppose from past experience the members of Congress might be, and however largely partaking, in the general, of the liberal feelings of the people, it was impossible to expect that bodies so constituted should not sometimes be controlled by local interests and sectional feelings. It was proper, therefore, to provide some umpire from whose situation and mode of appointment more independence and freedom from such influences might be expected. Such a one was afforded by the executive department constituted by the Constitution. A person elected to that high office, having his constituents in every section, state, and subdivision of the Union, must consider himself bound by the most solemn sanctions to guard, protect, and defend the rights of all and of every portion, great or small, from the injustice and oppression of the rest. I consider the veto power, therefore given by the Constitution to the executive of the United States solely as a conservative power, to be used only first, to protect the Constitution from violation; secondly, the people from the effects of hasty legislation where their will has been probably disregarded or not well understood, and, thirdly, to prevent the effects of combinations violative of the rights of minorities. In reference to the second of these objects I may observe that I consider it the right and privilege of the people to decide disputed points of the Constitution arising from the general grant of power to Congress to carry into effect the powers expressly given; and I believe with Mr. Madison that "repeated recognitions under varied circumstances in acts of the legislative, executive, and judicial branches of the government, accompanied by indications in different modes of the concurrence of the general will of the nation," as affording to the President sufficient authority for his considering such disputed points as settled.

Upward of half a century has elapsed since the adoption of the present form of government. It would be an object more highly desirable than the gratification of the curiosity of speculative statesmen if its precise situation could be ascertained, a fair exhibit made of the operations of each of its departments, of the powers which they respectively claim and exercise, of the collisions which have occurred between them or between the whole government and those of the states or either of them. We could then compare our actual condition after fifty years' trial of our system with what it was in the commencement of its operations and ascertain whether the predictions of the patriots who opposed its adoption or the confident hopes of its advocates have been best realized. The great dread of the former seems to have been that the reserved powers of the states would be absorbed by those of the federal government and a consolidated power established, leaving to the states the shadow only of that independent action for which they had so zealously contended and on the preservation of which they relied as the last hope of liberty. Without denying that the result to which they looked with so much apprehension is in the way of being realized, it is obvious that they did not clearly see the mode of its accomplishment. The general government has seized upon none of the reserved rights of the states. As far as any open warfare may have gone, the state authorities have amply maintained their rights. To a casual observer our system presents no appearance of discord between the different members which compose it. Even the addition of many new ones has produced no jarring. They move in their respective orbits in perfect harmony with the central head and with each other. But there is still an undercurrent at work by which, if not seasonably checked, the worst apprehensions of our anti-federal patriots will be realized, and not only will the state authorities be overshadowed by the great increase of power in the executive department of the general government, but the character of that government, if not its designation, be essentially and radically changed. This state of things has been in part effected by causes inherent in the Constitution and in part by the never-failing tendency of political power to increase itself. By making the President the sole distributer of all the patronage of the government the framers of the Constitution do not appear to have anticipated at how short a period it would become a formidable instrument to control the free operations of the state governments. Of trifling importance at first, it had early in Mr. Jefferson's administration become so powerful as to create great alarm in the mind of that patriot from the potent influence it might exert in controlling the freedom of the elective franchise. If such could have then been the effects of its influence, how much greater must be the danger at this time, quadrupled in amount as it certainly is and more completely under the control of the executive will than their construction of their powers allowed or the forbearing characters of all the early Presidents permitted them to make. But it is not by the extent of its patronage alone that the executive department has become dangerous, but by the use which it appears may be made of the appointing power to bring under its control the whole revenues of the country. The Constitution has declared it to be the duty of the President to see that the laws are executed, and it makes him the commander in chief of the armies and navy of the United States. If the opinion of the most approved writers upon that species of mixed government which in modern Europe is termed monarchy in contradistinction to despotism is correct, there was wanting no other addition to the powers of our chief magistrate to stamp a monarchical character on our government but the control of the public finances; and to me it appears strange indeed that anyone should doubt that the entire control which the President possesses over the officers who have the custody of the public money, by the power of removal with or without cause, does, for all mischievous purposes at least, virtually subject the treasure also to his disposal. The first Roman emperor, in his attempt to seize the sacred treasure, silenced the opposition of the officer to whose charge it had been committed by a significant allusion to his sword. By a selection of political instruments for the care of the public money a reference to their commissions by a President would be quite as effectual an argument as that of Caesar to the Roman knight. I am not insensible of the great difficulty that exists in drawing a proper plan for the safekeeping and disbursement of the public revenues, and I know the importance which has been attached by men of great abilities and patriotism to the divorce, as it is called, of the Treasury from the banking institutions. It is not the divorce which is complained of, but the unhallowed union of the Treasury with the executive department, which has created such extensive alarm. To this danger to our republican institutions and that created by the influence given to the executive through the instrumentality of the federal officers I propose to apply all the remedies which may be at my command. It was certainly a great error in the framers of the Constitution not to have made the officer at the head of the Treasury Department entirely independent of the executive. He should at least have been removable only upon the demand of the popular branch of the legislature. I have determined never to remove a secretary of the Treasury without communicating all the circumstances attending such removal to both houses of Congress.

The influence of the executive in controlling the freedom of the elective franchise through the medium of the public officers can be effectually checked by renewing the prohibition published by Mr. Jefferson forbidding their interference in elections further than giving their own votes, and their own independence secured by an assurance of perfect immunity in exercising this sacred privilege of freemen under the dictates of their own unbiased judgments. Never with my consent shall an officer of the people, compensated for his services out of their pockets, become the pliant instrument of executive will.

There is no part of the means placed in the hands of the executive which might be used with greater effect for unhallowed purposes than the control of the public press. The maxim which our ancestors derived from the mother country that "the freedom of the press is the great bulwark of civil and religious liberty" is one of the most precious legacies which they have left us. We have learned, too, from our own as well as the experience of other countries, that golden shackles, by whomsoever or by whatever pretense imposed, are as fatal to it as the iron bonds of despotism. The presses in the necessary employment of the government should never be used "to clear the guilty or to varnish crime." A decent and manly examination of the acts of the government should be not only tolerated, but encouraged.

Upon another occasion I have given my opinion at some length upon the impropriety of executive interference in the legislation of Congress— that the article in the Constitution making it the duty of the President to communicate information and authorizing him to recommend measures was not intended to make him the source in legislation, and, in particular, that he should never be looked to for schemes of finance. It would be very strange, indeed, that the Constitution should have strictly forbidden one branch of the legislature from interfering in the origination of such bills and that it should be considered proper that an altogether different department of the government should be permitted to do so. Some of our best political maxims and opinions have been drawn from our parent isle. There are others, however, which cannot be introduced in our system without singular incongruity and the production of much mischief, and this I conceive to be one. No matter in which of the houses of Parliament a bill may originate nor by whom introduced—a minister or a member of the opposition—by the fiction of law, or rather of constitutional principle, the sovereign is supposed to have prepared it agreeably to his will and then submitted it to Parliament for their advice and consent. Now the very reverse is the case here, not only with regard to the principle, but the forms prescribed by the Constitution. The principle certainly assigns to the only body constituted by the Constitution (the legislative body) the power to make laws, and the forms even direct that the enactment should be ascribed to them. The Senate, in relation to revenue bills, have the right to propose amendments, and so has the executive by the power given him to return them to the House of Representatives with his objections. It is in his power also to propose amendments in the existing revenue laws, suggested by his observations upon their defective or injurious operation. But the delicate duty of devising schemes of revenue should be left where the Constitution has placed it—with the immediate representatives of the people. For similar reasons the mode of keeping the public treasure should be prescribed by them, and the further removed it may be from the control of the executive the more wholesome the arrangement and the more in accordance with republican principle.

Connected with this subject is the character of the currency. The idea of making it exclusively metallic, however well intended, appears to me to be fraught with more fatal consequences than any other scheme having no relation to the personal rights of the citizens that has ever been devised. If any single scheme could produce the effect of arresting at once that mutation of condition by which thousands of our most indigent fellow citizens by their industry and enterprise are raised to the possession of wealth, that is the one. If there is one measure better calculated than another to produce that state of things so much deprecated by all true republicans, by which the rich are daily adding to their hoards and the poor sinking deeper into penury, it is an exclusive metallic currency. Or if there is a process by which the character of the country for generosity and nobleness of feeling may be destroyed by the great increase and neck toleration of usury, it is an exclusive metallic currency.

Amongst the other duties of a delicate character which the President is called upon to perform is the supervision of the government of the territories of the United States. Those of them which are destined to become members of our great political family are compensated by their rapid progress from infancy to manhood for the partial and temporary deprivation of their political rights. It is in this district only where American citizens are to be found who under a settled policy are deprived of many important political privileges without any inspiring hope as to the future.

Their only consolation under circumstances of such deprivation is that of the devoted exterior guards of a camp—that their sufferings secure tranquillity and safety within. Are there any of their countrymen, who would subject them to greater sacrifices, to any other humiliations than those essentially necessary to the security of the object for which they were thus separated from their fellow citizens? Are their rights alone not to be guaranteed by the application of those great principles upon which all our constitutions are founded? We are told by the greatest of British orators and statesmen that at the commencement of the War of the Revolution the most stupid men in England spoke of "their American subjects." Are there, indeed, citizens of any of our states who have dreamed of their subjects in the District of Columbia? Such dreams can never be realized by any agency of mine. The people of the District of Columbia are not the subjects of the people of the states, but free American citizens. Being in the latter condition when the Constitution was formed, no words used in that instrument could have been intended to deprive them of that character. If there is anything in the great principle of unalienable rights so emphatically insisted upon in our Declaration of Independence, they could neither make nor the United States accept a surrender of their liberties and become the subjects—in other words, the slaves—of their former fellow citizens. If this be true—and it will scarcely be denied by anyone who has a correct idea of his own rights as an American citizen—the grant to Congress of exclusive jurisdiction in the District of Columbia can be interpreted, so far as respects the aggregate people of the United States, as meaning nothing more than to allow to Congress the controlling power necessary to afford a free and safe exercise of the functions assigned to the general government by the Constitution. In all other respects the legislation of Congress should be adapted to their peculiar position and wants and be conformable with their deliberate opinions of their own interests.

I have spoken of the necessity of keeping the respective departments of the government, as well as all the other authorities of our country, within their appropriate orbits. This is a matter of difficulty in some cases, as the powers which they respectively claim are often not defined by any distinct lines. Mischievous, however, in their tendencies as collisions of this kind may be, those which arise between the respective communities which for certain purposes compose one nation are much more so, for no such nation can long exist without the careful culture of those feelings of confidence and affection which are the effective bonds to union between free and confederate states. Strong as is the tie of interest, it has been often found ineffectual. Men blinded by their passions have been known to adopt measures for their country in direct opposition to all the suggestions of policy. The alternative, then, is to destroy or keep down a bad passion by creating and fostering a good one, and this seems to be the cornerstone upon which our American political architects have reared the fabric of our government. The cement which was to bind it and perpetuate its existence was the affectionate attachment between all its members. To insure the continuance of this feeling, produced at first by a community of dangers, of sufferings, and of interests, the advantages of each were made accessible to all. No participation in any good possessed by any member of our extensive confederacy, except in domestic government, was withheld from the citizen of any other member. By a process attended with no difficulty, no delay, no expense but that of removal, the citizen of one might become the citizen of any other, and successively of the whole. The lines, too, separating powers to be exercised by the citizens of one state from those of another seem to be so distinctly drawn as to leave no room for misunderstanding. The citizens of each state unite in their persons all the privileges which that character confers and all that they may claim as citizens of the United States, but in no case can the same persons at the same time act as the citizen of two separate states, and he is therefore positively precluded from any interference with the reserved powers of any state but that of which he is for the time being a citizen. He may, indeed, offer to the citizens of other states his advice as to their management, and the form in which it is tendered is left to his own discretion and sense of propriety. It may be observed, however, that organized associations of citizens requiring compliance with their wishes too much resemble the recommendations of Athens to her allies, supported by an armed and powerful fleet. It was, indeed, to the ambition of the leading states of Greece to control the domestic concerns of the others that the destruction of that celebrated confederacy, and subsequently of all its members, is mainly to be attributed, and it is owing to the absence of that spirit that the Helvetic Confederacy has for so many years been preserved. Never has there been seen in the institutions of the separate members of any confederacy more elements of discord. In the principles and forms of government and religion, as well as in the circumstances of the several cantons, so marked a discrepancy was observable as to promise anything but harmony in their intercourse or permanency in their alliance, and yet for ages neither has been interrupted. Content with the positive benefits which their union produced, with the independence and safety from foreign aggression which it secured, these sagacious people respected the institutions of each other, however repugnant to their own principles and prejudices.

Our confederacy, fellow citizens, can only be preserved by the same forbearance. Our citizens must be content with the exercise of the powers with which the Constitution clothes them. The attempt of those of one state to control the domestic institutions of another can only result in feelings of distrust and jealousy, the certain harbingers of disunion, violence, and civil war, and the ultimate destruction of our free institutions. Our confederacy is perfectly illustrated by the terms and principles governing a common copartnership. There is a fund of power to be exercised under the direction of the joint councils of the allied members, but that which has been reserved by the individual members is intangible by the common government or the individual members composing it. To attempt it finds no support in the principles of our Constitution.

It should be our constant and earnest endeavor mutually to cultivate a spirit of concord and harmony among the various parts of our confederacy. Experience has abundantly taught us that the agitation by citizens of one part of the Union of a subject not confided to the general government, but exclusively under the guardianship of the local authorities, is productive of no other consequences than bitterness, alienation, discord, and injury to the very cause which is intended to be advanced. Of all the great interests which appertain to our country, that of union—cordial, confiding, fraternal union—is by far the most important, since it is the only true and sure guaranty of all others.

In consequence of the embarrassed state of business and the currency, some of the states may meet with difficulty in their financial concerns. However deeply we may regret anything imprudent or excessive in the engagements into which states have entered for purposes of their own, it does not become us to disparage the states' governments, nor to discourage them from making proper efforts for their own relief. On the contrary, it is our duty to encourage them to the extent of our constitutional authority to apply their best means and cheerfully to make all necessary sacrifices and submit to all necessary burdens to fulfill their engagements and maintain their credit, for the character and credit of the several states form a part of the character and credit of the whole country. The resources of the country are abundant, the enterprise and activity of our people proverbial, and we may well hope that wise legislation and prudent administration by the respective governments, each acting within its own sphere, will restore former prosperity.

Unpleasant and even dangerous as collisions may sometimes be between the constituted authorities of the citizens of our country in relation to the lines which separate their respective jurisdictions, the results can be of no vital injury to our institutions if that ardent patriotism, that devoted attachment to liberty, that spirit of moderation and forbearance for which our countrymen were once distinguished, continue to be cherished. If this continues to be the ruling passion of our souls, the weaker feeling of the mistaken enthusiast will be corrected, the utopian dreams of the scheming politician dissipated, and the complicated intrigues of the demagogue rendered harmless. The spirit of liberty is the sovereign balm for every injury which our institutions may receive. On the contrary, no care that can be used in the construction of our government, no division of powers, no distribution of checks in its several departments, will prove effectual to keep us a free people if this spirit is suffered to decay; and decay it will without constant nurture. To the neglect of this duty the best historians agree in attributing the ruin of all the republics with whose existence and fall their writings have made us acquainted. The same causes will ever produce the same effects, and as long as the love of power is a dominant passion of the human bosom, and as long as the understandings of men can be warped and their affections changed by operations upon their passions and prejudices, so long will the liberties of a people depend on their own constant attention to its preservation. The danger to all well-established free governments arises from the unwillingness of the people to believe in its existence or from the influence of designing men diverting their attention from the quarter whence it approaches to a source from which it can never come. This is the old trick of those who would usurp the government of their country. In the name of democracy they speak, warning the people against the influence of wealth and the danger of aristocracy. History, ancient and modern, is full of such examples. Caesar became the master of the Roman people and the senate under the pretense of supporting the democratic claims of the former against the aristocracy of the latter; Cromwell, in the character of protector of the liberties of the people, became the dictator of England, and Bolívar possessed himself of unlimited power with the title of his country's liberator. There is, on the contrary, no instance on record of an extensive and well-established republic being changed into an aristocracy. The tendencies of all such governments in their decline is to monarchy, and the antagonist principle to liberty there is the spirit of faction—a spirit which assumes the character and in times of great excitement imposes itself upon the people as the genuine spirit of freedom, and, like the false Christs whose coming was fore-told by the Savior, seeks to, and were it possible would, impose upon the true and most faithful disciples of liberty. It is in periods like this that it behooves the people to be most watchful of those to whom they have in-trusted power. And although there is at times much difficulty in distin-guishing the false from the true spirit, a calm and dispassionate investigation will detect the counterfeit, as well by the character of its op-erations as the results that are produced. The true spirit of liberty, al-though devoted, persevering, bold, and uncompromising in principle, that secured is mild and tolerant and scrupulous as to the means it employs, whilst the spirit of party, assuming to be that of liberty, is harsh, vindic-tive, and intolerant, and totally reckless as to the character of the allies which it brings to the aid of its cause. When the genuine spirit of liberty animates the body of a people to a thorough examination of their affairs, it leads to the excision of every excrescence which may have fastened itself upon any of the departments of the government, and restores the system to its pristine health and beauty. But the reign of an intolerant spirit of party amongst a free people seldom fails to result in a dangerous accession to the executive power introduced and established amidst unusual profes-sions of devotion to democracy.

The foregoing remarks relate almost exclusively to matters connected with our domestic concerns. It may be proper, however, that I should give some indications to my fellow citizens of my proposed course of conduct in the management of our foreign relations. I assure them, therefore, that it is my intention to use every means in my power to preserve the friendly intercourse which now so happily subsists with every foreign nation, and that although, of course, not well informed as to the state of pending ne-gotiations with any of them, I see in the personal characters of the sover-eigns, as well as in the mutual interests of our own and of the governments with which our relations are most intimate, a pleasing guaranty that the harmony so important to the interests of their subjects as well as of our citizens will not be interrupted by the advancement of any claim or pre-tension upon their part to which our honor would not permit us to yield. Long the defender of my country's rights in the field, I trust that my fellow citizens will not see in my earnest desire to preserve peace with foreign powers any indication that their rights will ever be sacrificed or the honor of the nation tarnished by any admission on the part of their chief magis-trate unworthy of their former glory. In our intercourse with our abor-iginal neighbors the same liberality and justice which marked the course prescribed to me by two of my illustrious predecessors when acting under their direction in the discharge of the duties of superintendent and commissioner shall be strictly observed. I can conceive of no more sublime spectacle, none more likely to propitiate an impartial and common Creator, than a rigid adherence to the principles of justice on the part of a powerful nation in its transactions with a weaker and uncivilized people whom circumstances have placed at its disposal.

Before concluding, fellow citizens, I must say something to you on the subject of the parties at this time existing in our country. To me it appears perfectly clear that the interest of that country requires that the violence of the spirit by which those parties are at this time governed must be greatly mitigated, if not entirely extinguished, or consequences will ensue which are appalling to be thought of.

If parties in a republic are necessary to secure a degree of vigilance sufficient to keep the public functionaries within the bounds of law and duty, at that point their usefulness ends. Beyond that they become destructive of public virtue, the parent of a spirit antagonist to that of liberty, and eventually its inevitable conqueror. We have examples of republics where the love of country and of liberty at one time were the dominant passions of the whole mass of citizens, and yet, with the continuance of the name and forms of free government, not a vestige of these qualities remaining in the bosoms of any one of its citizens. It was the beautiful remark of a distinguished English writer that "in the Roman senate Octavius had a party and Anthony a party, but the commonwealth had none." Yet the senate continued to meet in the temple of liberty to talk of the sacredness and beauty of the commonwealth and gaze at the statues of the elder Brutus and of the Curtii and Decii, and the people assembled in the forum, not, as in the days of Camillus and the Scipios, to cast their free votes for annual magistrates or pass upon the acts of the senate, but to receive from the hands of the leaders of the respective parties their share of the spoils and to shout for one or the other, as those collected in Gaul or Egypt and the lesser Asia would furnish the larger dividend. The spirit of liberty had fled, and, avoiding the abodes of civilized man, had sought protection in the wilds of Scythia or Scandinavia; and so under the operation of the same causes and influences it will fly from our Capitol and our forums. A calamity so awful, not only to our country, but to the world, must be deprecated by every patriot and every tendency to a state of things likely to produce it immediately checked. Such a tendency has existed—does exist. Always the friend of my countrymen, never their flatterer, it becomes my duty to say to them from this high place to which their partiality has exalted me that there exists in the land a spirit hostile to their best interests—hostile to liberty itself. It is a spirit contracted in its views, selfish in its objects. It looks to the aggrandizement of a few even to the destruction of the interests of the whole. The entire remedy is with the people. Something, however, may be effected by the means which they have placed in my hands. It is union that we want, not of a party for the sake of that party, but a union of the whole country for the sake of the whole country, for the defense of its interests and its honor against foreign aggression, for the defense of those principles for which our ancestors so gloriously contended. As far as it depends upon me it shall be accomplished. All the influence that I possess shall be exerted to prevent the formation at least of an executive party in the halls of the legislative body. I wish for the support of no member of that body to any measure of mine that does not satisfy his judgment and his sense of duty to those from whom he holds his appointment, nor any confidence in advance from the people but that asked for by Mr. Jefferson, "to give firmness and effect to the legal administration of their affairs."

I deem the present occasion sufficiently important and solemn to justify me in expressing to my fellow citizens a profound reverence for the Christian religion and a thorough conviction that sound morals, religious liberty, and a just sense of religious responsibility are essentially connected with all true and lasting happiness; and to that good Being who has blessed us by the gifts of civil and religious freedom, who watched over and prospered the labors of our fathers and has hitherto preserved to us institutions far exceeding in excellence those of any other people, let us unite in fervently commending every interest of our beloved country in all future time.

Fellow citizens, being fully invested with that high office to which the partiality of my countrymen has called me, I now take an affectionate leave of you. You will bear with you to your homes the remembrance of the pledge I have this day given to discharge all the high duties of my exalted station according to the best of my ability, and I shall enter upon their performance with entire confidence in the support of a just and generous people.

James K. Polk

Tuesday, March 4, 1845

James Knox Polk was an ardent expansionist in an age when expansion became part of American ideology. But in early 1844, he was a rather forgotten figure. He had been a Jacksonian member of Congress who served as Speaker of the House from 1835 to 1839. He left Washington when he won election as governor of Tennessee in 1839, but if he had hopes that he might use his new job as a springboard to greater success, they were quickly dashed. He lost his reelection bid in 1841 and a comeback bid in 1843. Nevertheless, he remained a presence in Democratic Party politics despite his losses. A protégé of Andrew Jackson, so much so that he was often called Young Hickory, Polk did not lack for influential friends and prospective allies.

In the election year of 1844, Polk's fellow Democrats appeared to be on the verge of renominating another Jacksonian, Martin Van Buren, who had lost reelection to William Henry Harrison in 1840. But Van Buren announced that he was opposed to the annexation of Texas, and he was not alone. Henry Clay, the Whig Party's candidate for President in 1844, also was opposed. Suddenly, the notion of American expansion seemed stalled.

When Democrats met to decide on their presidential candidate in 1844, Van Buren was the favorite, but his anti-annexation views kept him short of the two-thirds majority required to win the party's nomination. After multiple ballots, weary delegates began looking for an alternative and found one in Polk. It was the first time a deadlocked convention turned to a compromise candidate.

Beside Polk himself, nobody was more pleased about the party's choice than the nominee's mentor, Andrew Jackson. The former President abandoned Van Buren over Texas and designated Polk as his choice two weeks before the Democratic convention. The country now had a clear choice between an advocate of expansionism in Polk and a skeptic in Henry Clay.

Polk's enthusiasm for expansion was hardly limited to the South and Southwest. At a time when the United States and Britain were squabbling over the boundary of the Oregon Territory in the Pacific Northwest, Polk campaigned on a promise to seize the entire territory. His slogan, based on a latitude line that marked the U.S. claim, was "Fifty-four Forty or Fight!" The Americans believed Oregon's northern border ought to be the latitude line of 54° 40', and he made it clear he was willing to go to war over it.

The feisty Polk and the more deliberative Clay were not the only candidates in the race. In a sign of growing divisions over slavery, a third party entered the picture. The Liberty Party, founded in 1840, had one issue: opposition to the spread of slavery. Antislavery votes usually went to the Whigs, which made the Liberty Party a threat to Clay. Making matters worse for Clay, during the campaign he seemed to soften his anti-annexation stance, which pleased the South but hurt him in the North.

Polk, too, faced the possibility of a split vote when President Tyler tried to form an independent party. Although Tyler was vice president under the Whig administration of William Henry Harrison, he had been a Democrat and had hopes of rallying pro-Texas Democrats to his third-party campaign. Shrewdly, Polk worked over Tyler, calling in help from Jackson himself, until the President agreed to withdraw from the race.

On Election Day, the first when all states voted on the same day in November, Polk narrowly defeated Clay thanks to support for the Liberty Party candidate, James Birney, in New York. Without Birney in the race, Clay very likely would have won New York's thirty-six electoral votes and so defeated Polk in the Electoral College. As it was, Polk won less than 50 percent of the popular vote, losing his home state of Tennessee by a few hundred votes.

Regardless of the numbers, Polk regarded his election as evidence that Americans favored expansion. So did the outgoing President, who persuaded Congress to approve a resolution allowing for the admission of Texas in February 1845, before Polk took office. The move deftly avoided the pitfalls of annexing Texas by treaty, which had failed in the past because it required a two-thirds vote from the Senate.

In his inaugural address, Polk said that he regarded "the question of annexation as belonging exclusively to the United States and Texas."

The reference is an explicit rejection of Mexico's claims on Texas, despite Mexican threats to go to war if annexation went forward. War did indeed follow.

Polk was a highly regarded speaker and debater, best known for his straightforward style. His inaugural, once it moves beyond the expected nods to and praise of the Constitution, is nothing if not to the point. In addition to firing a shot across Mexico's bow, Polk laid out other goals for his administration: to continue a tariff, but to make it more equitable; to establish "the blessings of self-government" in the Oregon Territory, whose boundaries were a source of tension between Britain and the United States; and to encourage individual states to pay off their debts.

By the time he left office, voluntarily, in 1848, the expansion he so eagerly supported had been achieved through war (with Mexico) and treaties (with Britain and Mexico). The West was by no means settled, but it was open for business.

A personality modern Americans would describe as workaholic, Polk rarely took time off and so was exhausted after one term. His retirement, however, was short-lived. He died three months after leaving office, the shortest postpresidency in U.S. history.

Bibliographic Note

Paul H. Bergeron, *The Presidency of James K. Polk*
(Lawrence, KS: University Press of Kansas, 1987).

John H. Schroder, *Mr. Polk's War: American Opposition and Dissent, 1846–1848*
(Madison, WI: University of Wisconsin Press, 1973).

FELLOW CITIZENS:

Without solicitation on my part, I have been chosen by the free and voluntary suffrages of my countrymen to the most honorable and most responsible office on earth. I am deeply impressed with gratitude for the confidence reposed in me. Honored with this distinguished consideration at an earlier period of life than any of my predecessors, I can not disguise the diffidence with which I am about to enter on the discharge of my official duties.

If the more aged and experienced men who have filled the office of President of the United States even in the infancy of the Republic distrusted their ability to discharge the duties of that exalted station, what ought not to be the apprehensions of one so much younger and less endowed now that our domain extends from ocean to ocean, that our people have so greatly increased in numbers, and at a time when so great diversity of opinion prevails in regard to the principles and policy which should characterize the administration of our government? Well may the boldest fear and the wisest tremble when incurring responsibilities on which may depend our country's peace and prosperity, and in some degree the hopes and happiness of the whole human family.

In assuming responsibilities so vast I fervently invoke the aid of that Almighty Ruler of the Universe in whose hands are the destinies of nations and of men to guard this heaven-favored land against the mischiefs which without His guidance might arise from an unwise public policy. With a firm reliance upon the wisdom of Omnipotence to sustain and direct me in the path of duty which I am appointed to pursue, I stand in the presence of this assembled multitude of my countrymen to take upon myself the solemn obligation "to the best of my ability to preserve, protect, and defend the Constitution of the United States."

A concise enumeration of the principles which will guide me in the administrative policy of the government is not only in accordance with the examples set me by all my predecessors, but is eminently befitting the occasion.

The Constitution itself, plainly written as it is, the safeguard of our federative compact, the offspring of concession and compromise, binding together in the bonds of peace and union this great and increasing family of free and independent states, will be the chart by which I shall be directed.

It will be my first care to administer the government in the true spirit of that instrument, and to assume no powers not expressly granted or clearly implied in its terms. The government of the United States is one of delegated and limited powers, and it is by a strict adherence to the clearly granted powers and by abstaining from the exercise of doubtful or unauthorized implied powers that we have the only sure guaranty against the recurrence of those unfortunate collisions between the federal and state authorities which have occasionally so much disturbed the harmony of our system and even threatened the perpetuity of our glorious Union.

"To the States, respectively, or to the people" have been reserved "the powers not delegated to the United States by the Constitution nor prohibited by it to the States." Each state is a complete sovereignty within the sphere of its reserved powers. The government of the Union, acting within the sphere of its delegated authority, is also a complete sovereignty. While the general government should abstain from the exercise of authority not clearly delegated to it, the states should be equally careful that in the maintenance of their rights they do not overstep the limits of powers reserved to them. One of the most distinguished of my predecessors attached deserved importance to "the support of the state governments in all their rights, as the most competent administration for our domestic concerns and the surest bulwark against anti-republican tendencies," and to the "preservation of the general government in its whole constitutional vigor, as the sheet anchor of our peace at home and safety abroad."

To the government of the United States has been intrusted the exclusive management of our foreign affairs. Beyond that it wields a few general enumerated powers. It does not force reform on the states. It leaves individuals, over whom it casts its protecting influence, entirely free to improve their own condition by the legitimate exercise of all their mental and physical powers. It is a common protector of each and all the states; of every man who lives upon our soil, whether of native or foreign birth; of every religious sect, in their worship of the Almighty according to the dictates of their own conscience; of every shade of opinion, and the most free inquiry; of every art, trade, and occupation consistent with the laws of the states. And we rejoice in the general happiness, prosperity, and advancement of our country, which have been the offspring of freedom, and not of power.

This most admirable and wisest system of well-regulated self-government among men ever devised by human minds has been tested by its successful operation for more than half a century, and if preserved from the usurpations of the federal government on the one hand and the exercise by the states of powers not reserved to them on the other, will, I fervently hope and believe, endure for ages to come and dispense the blessings of civil and religious liberty to distant generations. To effect objects so dear to every patriot I shall devote myself with anxious solicitude. It will be my desire to guard against that most fruitful source of danger to the harmonious action of our system which consists in substituting the mere discretion and caprice of the executive or of majorities in the legislative department of the government for powers which have been withheld from the federal government by the Constitution. By the theory of our government majorities rule, but this right is not an arbitrary or unlimited one. It is a right to be exercised in subordination to the Constitution and in conformity to it. One great object of the Constitution was to restrain majorities from oppressing minorities or encroaching upon their just rights. Minorities have a right to appeal to the Constitution as a shield against such oppression.

That the blessings of liberty which our Constitution secures may be enjoyed alike by minorities and majorities, the executive has been wisely invested with a qualified veto upon the acts of the legislature. It is a negative power, and is conservative in its character. It arrests for the time hasty, inconsiderate, or unconstitutional legislation, invites reconsideration, and transfers questions at issue between the legislative and executive departments to the tribunal of the people. Like all other powers, it is subject to be abused. When judiciously and properly exercised, the Constitution itself may be saved from infraction and the rights of all preserved and protected.

The inestimable value of our federal Union is felt and acknowledged by all. By this system of united and confederated states our people are permitted collectively and individually to seek their own happiness in their own way, and the consequences have been most auspicious. Since the Union was formed the number of the states has increased from thirteen to twenty-eight; two of these have taken their position as members of the confederacy within the last week. Our population has increased from three to twenty millions. New communities and states are seeking protection under its aegis, and multitudes from the Old World are flocking to our shores to participate in its blessings. Beneath its benign sway peace and prosperity prevail. Freed from the burdens and miseries of war, our trade and intercourse have extended throughout the world. Mind, no longer tasked in devising means to accomplish or resist schemes of ambition, usurpation, or conquest, is devoting itself to man's true interests in developing his faculties and powers and the capacity of nature to minister to his enjoyments. Genius is free to announce its inventions and discoveries, and the hand is free to accomplish whatever the head conceives not incompatible with the rights of a fellow being. All distinctions of birth or of rank have been abolished. All citizens, whether native or adopted, are placed upon terms of precise equality. All are entitled to equal rights and equal protection. No union exists between church and state, and perfect freedom of opinion is guaranteed to all sects and creeds.

These are some of the blessings secured to our happy land by our federal Union. To perpetuate them it is our sacred duty to preserve it. Who shall assign limits to the achievements of free minds and free hands under the protection of this glorious Union? No treason to mankind since the organization of society would be equal in atrocity to that of him who would lift his hand to destroy it. He would overthrow the noblest structure of human wisdom, which protects himself and his fellow man. He would stop the progress of free government and involve his country either in anarchy or despotism. He would extinguish the fire of liberty, which warms and animates the hearts of happy millions and invites all the nations of the earth to imitate our example. If he say that error and wrong are committed in the administration of the government, let him remember that nothing human can be perfect, and that under no other system of government revealed by heaven or devised by man has reason been allowed so free and broad a scope to combat error. Has the sword of despots proved to be a safer or surer instrument of reform in government than enlightened reason? Does he expect to find among the ruins of this Union a happier abode for our swarming millions than they now have under it? Every lover of his country must shudder at the thought of the possibility of its dissolution, and will be ready to adopt the patriotic sentiment, "Our federal Union—it must be preserved." To preserve it the compromises which alone enabled our fathers to form a common Constitution for the government and protection of so many states and distinct communities, of such diversified habits, interests, and domestic institutions, must be sacredly and religiously observed. Any attempt to disturb or destroy these compromises, being terms of the compact of union, can lead to none other than the most ruinous and disastrous consequences.

It is a source of deep regret that in some sections of our country misguided persons have occasionally indulged in schemes and agitations whose object is the destruction of domestic institutions existing in other sections—institutions which existed at the adoption of the Constitution and were recognized and protected by it. All must see that if it were possible for them to be successful in attaining their object the dissolution of the Union and the consequent destruction of our happy form of government must speedily follow.

I am happy to believe that at every period of our existence as a nation there has existed, and continues to exist, among the great mass of our people a devotion to the Union of the states which will shield and protect it against the moral treason of any who would seriously contemplate its destruction. To secure a continuance of that devotion the compromises of the Constitution must not only be preserved, but sectional jealousies and heartburnings must be discountenanced, and all should remember that they are members of the same political family, having a common destiny.

To increase the attachment of our people to the Union, our laws should be just. Any policy which shall tend to favor monopolies or the peculiar interests of sections or classes must operate to the prejudice of the interest of their fellow citizens, and should be avoided. If the compromises of the Constitution be preserved, if sectional jealousies and heartburnings be discountenanced, if our laws be just and the government be practically administered strictly within the limits of power prescribed to it, we may discard all apprehensions for the safety of the Union.

With these views of the nature, character, and objects of the government and the value of the Union, I shall steadily oppose the creation of those institutions and systems which in their nature tend to pervert it from its legitimate purposes and make it the instrument of sections, classes, and individuals. We need no national banks or other extraneous institutions planted around the government to control or strengthen it in opposition to the will of its authors. Experience has taught us how unnecessary they are as auxiliaries of the public authorities—how impotent for good and how powerful for mischief.

Ours was intended to be a plain and frugal government, and I shall regard it to be my duty to recommend to Congress and, as far as the executive is concerned, to enforce by all the means within my power the strictest economy in the expenditure of the public money which may be compatible with the public interests.

A national debt has become almost an institution of European monarchies. It is viewed in some of them as an essential prop to existing governments. Melancholy is the condition of that people whose government can be sustained only by a system which periodically transfers large amounts from the labor of the many to the coffers of the few. Such a system is incompatible with the ends for which our republican government was instituted. Under a wise policy the debts contracted in our Revolution and during the War of 1812 have been happily extinguished. By a judicious application of the revenues not required for other necessary purposes, it is not doubted that the debt which has grown out of the circumstances of the last few years may be speedily paid off.

I congratulate my fellow citizens on the entire restoration of the credit of the general government of the Union and that of many of the states. Happy would it be for the indebted states if they were freed from their liabilities, many of which were incautiously contracted. Although the government of the Union is neither in a legal nor a moral sense bound for the debts of the states, and it would be a violation of our compact of union to assume them, yet we cannot but feel a deep interest in seeing all the states meet their public liabilities and pay off their just debts at the earliest practicable period. That they will do so as soon as it can be done without imposing too heavy burdens on their citizens there is no reason to doubt. The sound moral and honorable feeling of the people of the indebted states can not be questioned, and we are happy to perceive a settled disposition on their part, as their ability returns after a season of unexampled pecuniary embarrassment, to pay off all just demands and to acquiesce in any reasonable measures to accomplish that object.

One of the difficulties which we have had to encounter in the practical administration of the government consists in the adjustment of our revenue laws and the levy of the taxes necessary for the support of government. In the general proposition that no more money shall be collected than the necessities of an economical administration shall require all parties seem to acquiesce. Nor does there seem to be any material difference of opinion as to the absence of right in the government to tax one section of country, or one class of citizens, or one occupation, for the mere profit of another. "Justice and sound policy forbid the federal government to foster one branch of industry to the detriment of another, or to cherish the interests of one portion to the injury of another portion of our common country." I have heretofore declared to my fellow citizens that "in my judgment it is the duty of the government to extend, as far as it may be practicable to do so, by its revenue laws and all other means within its power, fair and just protection to all of the great interests of the whole Union, embracing agriculture, manufactures, the mechanic arts, commerce, and navigation." I have also declared my opinion to be "in favor of a tariff for revenue," and that "in adjusting the details of such a tariff I have sanctioned such moderate discriminating duties as would produce the amount of revenue needed and at the same time afford reasonable incidental protection to our home industry," and that I was "opposed to a tariff for protection merely, and not for revenue."

The power "to lay and collect taxes, duties, imposts, and excises" was an indispensable one to be conferred on the federal government, which without it would possess no means of providing for its own support. In executing this power by levying a tariff of duties for the support of government, the raising of revenue should be the object and protection the incident. To reverse this principle and make protection the object and revenue the incident would be to inflict manifest injustice upon all other than the protected interests. In levying duties for revenue it is doubtless proper to make such discriminations within the revenue principle as will afford incidental protection to our home interests. Within the revenue limit there is a discretion to discriminate; beyond that limit the rightful exercise of the power is not conceded. The incidental protection afforded to our home interests by discriminations within the revenue range it is believed will be ample. In making discriminations all our home interests should as far as practicable be equally protected. The largest portion of our people are agriculturists. Others are employed in manufactures, commerce, navigation, and the mechanic arts. They are all engaged in their respective pursuits and their joint labors constitute the national or home industry. To tax one branch of this home industry for the benefit of another would be unjust. No one of these interests can rightfully claim an advantage over the others, or to be enriched by impoverishing the others. All are equally entitled to the fostering care and protection of the government. In exercising a sound discretion in levying discriminating duties within the limit prescribed, care should be taken that it be done in a manner not to benefit the wealthy few at the expense of the toiling millions by taxing lowest the luxuries of life, or articles of superior quality and high price, which can only be consumed by the wealthy, and highest the necessaries of life, or articles of coarse quality and low price, which the poor and great mass of our people must consume. The burdens of government should as far as practicable be distributed justly and equally among all classes of our population. These general views, long entertained on this subject, I have deemed it proper to reiterate. It is a subject upon which conflicting interests of sections and occupations are supposed to exist, and a spirit of mutual concession and compromise in adjusting its details should be cherished by every part of our widespread country as the only means of preserving harmony and a cheerful acquiescence of all in the operation of our revenue laws. Our patriotic citizens in every part of the Union will readily submit to the payment of such taxes as shall be needed for the support of their government, whether in peace or in war, if they are so levied as to distribute the burdens as equally as possible among them.

The Republic of Texas has made known her desire to come into our Union, to form a part of our confederacy and enjoy with us the blessings of liberty secured and guaranteed by our Constitution. Texas was once a part of our country—was unwisely ceded away to a foreign power—is now independent, and possesses an undoubted right to dispose of a part or the whole of her territory and to merge her sovereignty as a separate and independent state in ours. I congratulate my country that by an act of the late

Congress of the United States the assent of this government has been given to the reunion, and it only remains for the two countries to agree upon the terms to consummate an object so important to both.

I regard the question of annexation as belonging exclusively to the United States and Texas. They are independent powers competent to contract, and foreign nations have no right to interfere with them or to take exceptions to their reunion. Foreign powers do not seem to appreciate the true character of our government. Our Union is a confederation of independent states, whose policy is peace with each other and all the world. To enlarge its limits is to extend the dominions of peace over additional territories and increasing millions. The world has nothing to fear from military ambition in our government. While the chief magistrate and the popular branch of Congress are elected for short terms by the suffrages of those millions who must in their own persons bear all the burdens and miseries of war, our government cannot be otherwise than pacific. Foreign powers should therefore look on the annexation of Texas to the United States not as the conquest of a nation seeking to extend her dominions by arms and violence, but as the peaceful acquisition of a territory once her own, by adding another member to our confederation, with the consent of that member, thereby diminishing the chances of war and opening to them new and ever-increasing markets for their products.

To Texas the reunion is important, because the strong protecting arm of our government would be extended over her, and the vast resources of her fertile soil and genial climate would be speedily developed, while the safety of New Orleans and of our whole southwestern frontier against hostile aggression, as well as the interests of the whole Union, would be promoted by it.

In the earlier stages of our national existence the opinion prevailed with some that our system of confederated states could not operate successfully over an extended territory, and serious objections have at different times been made to the enlargement of our boundaries. These objections were earnestly urged when we acquired Louisiana. Experience has shown that they were not well founded. The title of numerous Indian tribes to vast tracts of country has been extinguished; new states have been admitted into the Union; new territories have been created and our jurisdiction and laws extended over them. As our population has expanded, the Union has been cemented and strengthened. As our boundaries have been enlarged and our agricultural population has been spread over a large surface, our federative system has acquired additional strength and security.

It may well be doubted whether it would not be in greater danger of overthrow if our present population were confined to the comparatively narrow limits of the original thirteen states than it is now that they are sparsely settled over a more expanded territory. It is confidently believed that our system may be safely extended to the utmost bounds of our territorial limits, and that as it shall be extended the bonds of our Union, so far from being weakened, will become stronger.

None can fail to see the danger to our safety and future peace if Texas remains an independent state or becomes an ally or dependency of some foreign nation more powerful than herself. Is there one among our citizens who would not prefer perpetual peace with Texas to occasional wars, which so often occur between bordering independent nations? Is there one who would not prefer free intercourse with her to high duties on all our products and manufactures which enter her ports or cross her frontiers? Is there one who would not prefer an unrestricted communication with her citizens to the frontier obstructions which must occur if she remains out of the Union? Whatever is good or evil in the local institutions of Texas will remain her own whether annexed to the United States or not. None of the present states will be responsible for them any more than they are for the local institutions of each other. They have confederated together for certain specified objects. Upon the same principle that they would refuse to form a perpetual union with Texas because of her local institutions our forefathers would have been prevented from forming our present Union. Perceiving no valid objection to the measure and many reasons for its adoption vitally affecting the peace, the safety, and the prosperity of both countries, I shall on the broad principle which formed the basis and produced the adoption of our Constitution, and not in any narrow spirit of sectional policy, endeavor by all constitutional, honorable, and appropriate means to consummate the expressed will of the people and government of the United States by the reannexation of Texas to our Union at the earliest practicable period.

Nor will it become in a less degree my duty to assert and maintain by all constitutional means the right of the United States to that portion of our territory which lies beyond the Rocky Mountains. Our title to the country of the Oregon is "clear and unquestionable," and already are our people preparing to perfect that title by occupying it with their wives and children. But eighty years ago our population was confined on the west by the ridge of the Alleghenies. Within that period—within the lifetime, I might say, of some of my hearers—our people, increasing to many millions, have filled the eastern valley of the Mississippi, adventurously ascended the Missouri to its headsprings, and are already engaged in establishing the blessings of self-government in valleys of which the rivers flow to the Pacific. The world beholds the peaceful triumphs of the industry of our emigrants. To us belongs the duty of protecting them adequately wherever they may be upon our soil. The jurisdiction of our laws and the benefits of our republican institutions should be extended over them in the distant regions which they have selected for their homes. The increasing facilities of intercourse will easily bring the states, of which the formation in that part of our territory cannot be long delayed, within the sphere of our federative Union. In the meantime every obligation imposed by treaty or conventional stipulations should be sacredly respected.

In the management of our foreign relations it will be my aim to observe a careful respect for the rights of other nations, while our own will be the subject of constant watchfulness. Equal and exact justice should characterize all our intercourse with foreign countries. All alliances having a tendency to jeopard the welfare and honor of our country or sacrifice any one of the national interests will be studiously avoided, and yet no opportunity will be lost to cultivate a favorable understanding with foreign governments by which our navigation and commerce may be extended and the ample products of our fertile soil, as well as the manufactures of our skillful artisans, find a ready market and remunerating prices in foreign countries.

In taking "care that the laws be faithfully executed," a strict performance of duty will be exacted from all public officers. From those officers, especially, who are charged with the collection and disbursement of the public revenue will prompt and rigid accountability be required. Any culpable failure or delay on their part to account for the moneys intrusted to them at the times and in the manner required by law will in every instance terminate the official connection of such defaulting officer with the government.

Although in our country the chief magistrate must almost of necessity be chosen by a party and stand pledged to its principles and measures, yet in his official action he should not be the President of a part only, but of the whole people of the United States. While he executes the laws with an impartial hand, shrinks from no proper responsibility, and faithfully carries out in the executive department of the government the principles and policy of those who have chosen him, he should not be unmindful that our fellow citizens who have differed with him in opinion are entitled to the full and free exercise of their opinions and judgments, and that the rights of all are entitled to respect and regard.

Confidently relying upon the aid and assistance of the coordinate departments of the government in conducting our public affairs, I enter upon the discharge of the high duties which have been assigned me by the people, again humbly supplicating that Divine Being who has watched over and protected our beloved country from its infancy to the present hour to continue His gracious benedictions upon us, that we may continue to be a prosperous and happy people.

Zachary Taylor

Monday March 5, 1849

Zachary Taylor won the presidency in 1848 based on his role in the Mexican War. Ironically enough, he won as the candidate of the Whig Party, which had opposed the war. Among the conflict's more outspoken critics was a Whig congressman from Illinois named Abraham Lincoln, who roundly criticized President Polk's justification for invading Mexico.

Like the Federalists who opposed the War of 1812 only to be put on the defensive when public opinion shifted dramatically at war's end, the Whigs found themselves in an awkward position after the decisive U.S. victory brought with it a huge territorial expansion. Under the Treaty of Guadalupe Hidalgo, which brought the war to an end in early 1848, the United States gained control of a huge swath of today's Southwest, including parts of Arizona and New Mexico, and all of California.

Taylor emerged as one of the war's heroes. Troops under his command invaded Mexico in 1846 and won several victories, but his fame grew out of his victory at the Battle of Buena Vista in Februrary 1847. Taylor was vastly outnumbered, with just forty-five hundred troops to twenty thousand under the command of General Antonio López de Santa Anna. Taylor refused Santa Anna's demand that he surrender and took personal command of the battlefield. Those actions and his victory made him a national hero.

As the election of 1848 approached, Whigs and Democrats alike courted Taylor, who had never held any elected office. As would be the case with Dwight Eisenhower a century later, Taylor was a political novice whose views and party affiliations were unknown but whose fame was irresistible. He declared himself a Whig, and the party nominated him over three contenders: Henry Clay, trying yet again for the White House; another war hero, Winfield Scott; and veteran senator Daniel Webster.

In nominating a man about whom they knew almost nothing, the Whigs avoided the nation's increasingly deep divisions over slavery and its expansion. The Democrats, however, could not paper over their differences. In 1846, a Democratic congressman from Pennsylvania, David Wilmot, proposed that slavery be banned from the new territories won from Mexico. Known as the Wilmot Proviso, the proposal shattered any semblance of election-year unity in the Democratic Party. When the party's national convention rejected a pro-proviso position, a group of dissidents split to form the Free-Soil Party, nominating former President Martin Van Buren for another term as President. The Free-Soil Party opposed the extension of slavery into new territories, arguing in favor of "free soil, free speech, free labor, and free men."

As was the case with the Louisiana Purchase, the territorial expansion in the 1840s placed slavery in the forefront of American politics. Parties might try to evade the issue, or they could rely on compromise as they did in 1820 and would again in 1850. But there was no denying the enormous fissure in American society over slavery.

Thanks to Taylor's military fame and his blank record, the 1848 election did not necessarily reflect the nation's sectional divide. The Whig ticket, which teamed a Southern slaveholder (Taylor) with a Northern supporter of the Wilmont Proviso (Millard Fillmore), captured a half dozen Southern and border states while nearly sweeping the North. The Free-Soilers took no electoral votes but with 10 percent of the popular vote prevented Taylor from winning an absolute majority of the raw vote.

Taylor delivered his inaugural address on March 5, because March 4 fell on a Sunday. Compared with Polk's straightforward, goal-oriented speech in 1845, Taylor's was a listless collection of platitudes that never came close to acknowledging the nation's sharp divisions. The challenge posed by the addition of the Southwest and of the Oregon Territory—another transaction that Polk completed—was unmentioned. The future of slavery was the issue that dared not speak its name.

Once in office, Taylor forthrightly faced the issues he had evaded as a candidate. He supported the quick admission of California, which wished to enter the Union as a free state. The discovery of gold at Sutter's Mill in 1849 and the sudden influx of fortune seekers gave some urgency to the question of statehood. But Taylor's fellow Southerners were furious. Henry Clay proposed a compromise that would allow the admission of California as a free state, the abolition of the slave trade—but not slavery itself—

in the District of Columbia, reliance on popular sovereignty to determine whether a new state would be free or slave, and passage of a fugitive-slave law requiring all American citizens to assist in tracking down runaway slaves and denying slaves a trial by jury. The proposed compromise prompted one of Capitol Hill's great debates, with the lions of antebellum America—Daniel Webster, John C. Calhoun, Stephen Douglas, William Seward, Henry Clay, and Jefferson Davis—all playing prominent roles. Calhoun was in such poor health that his speech had to be read for him. In it, he argued that the Union could be saved only if the North ceased its antislavery agitation and allowed the South to deal with slavery as it saw fit. Failing that, he said, both sides ought to simply separate.

Calhoun died shortly after his ominous speech. Taylor watched the debate continue to unfold on Capitol Hill. Although a Southerner and a slave owner, he was determined to crush any movement toward the disunion that Calhoun believed was inevitable.

On July 4, 1850, a hot summer's day in Washington, D.C., an overworked Zachary Taylor sat through an hour-long speech at the new Washington Monument. Many of his fellow Washingtonians were ill that summer with maladies that seemed related to heat and poor sanitation. Later that day, the President was taken ill with an intestinal ailment. He died on July 9, and was succeeded by Vice President Fillmore.

Zachary Taylor is remembered today as one of several Presidents who seemed helpless as the issue of slavery slashed away at the bonds of Union. But his position was clear at the time: he would do whatever he could to preserve the Union. But he did not live long enough to put that view to the ultimate test.

Bibliographic Note
Elbert B. Smith, *The Presidencies of Zachary Taylor and Millard Fillmore*
(Lawrence, KS: University Press of Kansas, 1988).

ELECTED BY THE AMERICAN people to the highest office known to our laws, I appear here to take the oath prescribed by the Constitution, and, in compliance with a time-honored custom, to address those who are now assembled.

The confidence and respect shown by my countrymen in calling me to be the chief magistrate of a Republic holding a high rank among the nations of the earth have inspired me with feelings of the most profound gratitude; but when I reflect that the acceptance of the office which their partiality has bestowed imposes the discharge of the most arduous duties and involves the weightiest obligations, I am conscious that the position which I have been called to fill, though sufficient to satisfy the loftiest ambition, is surrounded by fearful responsibilities. Happily, however, in the performance of my new duties I shall not be without able cooperation. The legislative and judicial branches of the government present prominent examples of distinguished civil attainments and matured experience, and it shall be my endeavor to call to my assistance in the executive departments individuals whose talents, integrity, and purity of character will furnish ample guaranties for the faithful and honorable performance of the trusts to be committed to their charge. With such aids and an honest purpose to do whatever is right, I hope to execute diligently, impartially, and for the best interests of the country the manifold duties devolved upon me.

In the discharge of these duties my guide will be the Constitution, which I this day swear to "preserve, protect, and defend." For the interpretation of that instrument I shall look to the decisions of the judicial tribunals established by its authority and to the practice of the government under the earlier Presidents, who had so large a share in its formation. To the example of those illustrious patriots I shall always defer with reverence, and especially to his example who was by so many titles "the Father of His Country."

To command the army and navy of the United States; with the advice and consent of the Senate, to make treaties and to appoint ambassadors and other officers; to give to Congress information of the state of the Union and recommend such measures as he shall judge to be necessary; and to take care that the laws shall be faithfully executed—these are the most important functions intrusted to the President by the Constitution, and it may be expected that I shall briefly indicate the principles which will control me in their execution.

Chosen by the body of the people under the assurance that my administration would be devoted to the welfare of the whole country, and not to the support of any particular section or merely local interest, I this day renew the declarations I have heretofore made and proclaim my fixed determination to maintain to the extent of my ability the government in its original purity and to adopt as the basis of my public policy those great republican doctrines which constitute the strength of our national existence.

In reference to the army and navy, lately employed with so much distinction on active service, care shall be taken to insure the highest condition of efficiency, and in furtherance of that object the military and naval schools, sustained by the liberality of Congress, shall receive the special attention of the executive.

As American freemen we cannot but sympathize in all efforts to extend the blessings of civil and political liberty, but at the same time we are warned by the admonitions of history and the voice of our own beloved Washington to abstain from entangling alliances with foreign nations. In all disputes between conflicting governments it is our interest not less than our duty to remain strictly neutral, while our geographical position, the genius of our institutions and our people, the advancing spirit of civilization, and, above all, the dictates of religion direct us to the cultivation of peaceful and friendly relations with all other powers. It is to be hoped that no international question can now arise which a government confident in its own strength and resolved to protect its own just rights may not settle by wise negotiation; and it eminently becomes a government like our own, founded on the morality and intelligence of its citizens and upheld by their affections, to exhaust every resort of honorable diplomacy before appealing to arms. In the conduct of our foreign relations I shall conform to these views, as I believe them essential to the best interests and the true honor of the country.

The appointing power vested in the President imposes delicate and onerous duties. So far as it is possible to be informed, I shall make honesty, capacity, and fidelity indispensable prerequisites to the bestowal of office, and the absence of either of these qualities shall be deemed sufficient cause for removal.

It shall be my study to recommend such constitutional measures to Congress as may be necessary and proper to secure encouragement and protection to the great interests of agriculture, commerce, and manufactures, to improve our rivers and harbors, to provide for the speedy extinguishment of the public debt, to enforce a strict accountability on the part of all officers of the government and the utmost economy in all public expenditures; but it is for the wisdom of Congress itself, in which all legislative powers are vested by the Constitution, to regulate these and other matters of domestic policy. I shall look with confidence to the enlightened patriotism of that body to adopt such measures of conciliation as may harmonize conflicting interests and tend to perpetuate that Union which should be the paramount object of our hopes and affections. In any action calculated to promote an object so near the heart of everyone who truly loves his country I will zealously unite with the coordinate branches of the government.

In conclusion I congratulate you, my fellow citizens, upon the high state of prosperity to which the goodness of Divine Providence has conducted our common country. Let us invoke a continuance of the same protecting care which has led us from small beginnings to the eminence we this day occupy, and let us seek to deserve that continuance by prudence and moderation in our councils, by well-directed attempts to assuage the bitterness which too often marks unavoidable differences of opinion, by the promulgation and practice of just and liberal principles, and by an enlarged patriotism, which shall acknowledge no limits but those of our own widespread Republic.

Franklin Pierce

Friday, March 4, 1853

A light morning snow turned heavy as Franklin Pierce took the oath of office from Chief Justice Roger Taney on Friday, March 4, 1853. The temperature was just above freezing, but a steady wind made it feel colder.

No President has ever taken the oath of office with a heavier heart, not even those who succeeded Presidents who died in office. Just six weeks earlier, on January 16, 1853, Pierce, his wife, Jane Appleton, and their only surviving child, eleven-year-old Bennie, were traveling by train from Boston when their car derailed near Andover, Massachusetts. The president-elect and his wife sustained only minor injuries, but their son died before their eyes, crushed to death and nearly decapitated.

It was, sadly, only the latest family tragedy for the Pierces. The couple already had suffered through the deaths of two other children. One died three days after his birth, and the other died of typhus at age four.

Jane Appleton Pierce was not with her husband on Inauguration Day, too grief stricken to participate in public ceremonies. Also absent was the new vice president, William King of Alabama, who was gravely ill with tuberculosis. He went to Cuba in vain hopes of recovery and actually took the oath of office on Cuban soil on March 24. He died a short time later, after returning to Alabama.

Pierce's ascent to the nation's highest office was both unlikely and not surprising. He was very much a young man in a hurry in his native New Hampshire and was elected to the state's legislature in 1829 at the age of twenty-five. (His father was elected governor that same year.) He was elected to the House of Representatives in 1833 and to the U.S. Senate in 1836, when he was just thirty-two. He seemed well on his way to a brilliant career.

In 1842, however, he resigned from the Senate, in part because his wife disliked politics and Washington, and returned to New Hampshire. He established a thriving law practice and remained active in local politics. When war broke out with Mexico in 1848, he joined the army as a private and finished the war as a brigadier general.

He was by no means a stranger to the upper echelons of the Democratic Party. Still, he was hardly considered a presidential prospect when the Democrats gathered for their nominating convention in 1852. The contenders were Stephen A. Douglas of Illinois, James Buchanan of Pennsylvania, and Lewis Cass of Michigan, among others. After forty-eight ballots, the party was hopelessly deadlocked. In search of a compromise candidate who had alienated nobody, the Democrats turned to Pierce. One can only imagine his wife's reaction upon learning that she might have to return to a life she loathed and believed she had left behind.

The election of 1852 featured two veterans of the Mexican War, Pierce and the more celebrated Winfield Scott, who ran as a Whig. Of the two generals, Pierce was the more experienced politician, and a good one by all accounts. He won 254 of the Electoral College's 296 votes to become, at age forty-eight, the nation's youngest President thus far.

In taking the oath of office, Pierce chose to affirm, rather than swear, that he would faithfully execute the powers of his office and uphold the Constitution. He did not use a Bible for the ceremony, choosing a law book instead. He began his speech with an implicit and moving acknowledgement of his grief. "It is a relief . . . ," he said, "that no heart but my own can know the personal regret and bitter sorrow over which I have been borne to a position so suitable for others rather than desirable for myself."

Pierce's speech today seems tragically out of touch with the tensions that would explode when the Kansas-Nebraska Act repealed the Missouri Compromise in 1854. Pierce asserted that his administration "will not be controlled by any timid forebodings of evil from expansion." Indeed, he promised more expansion, a vow he would deliver in 1853 in the form of the Gadsden Purchase, which added territory from Mexico to the southern parts of today's New Mexico and Arizona.

Expansion, of course, always was freighted with implications about slavery. Pierce supported the Compromise of 1850, but if he believed it postponed a day of reckoning, he was gravely mistaken. Under the Kansas-Nebraska Act, settlers in those territories, which were above the Missouri Compromise line, were allowed to decide for themselves whether they would have slavery or not. Kansas bled as slavery's opponents and advocates battled each other for control of the territory. A new political faction, the Republican Party, rose to challenge the spread of slavery in the West and to argue against Pierce's enforcement of the Fugitive Slave Law.

Although he was a Northerner, Pierce certainly was not looking to confront the South over the extension of slavery. He said he would offer "a ready and stern resistance" to "every theory of society or government . . . calculated to dissolve the bonds of law and affection which unite us," but he also attempted to mollify Southern slave owners. "I believe that involuntary servitude, as it exists in different states of this confederacy, is recognized by the Constitution," he said. "I believe that it stands like any other admitted right, and that the states where it exists are entitled to efficient remedies to enforce the constitutional provisions." He contended that "the rights of the South" deserved the respect of all Americans, "not with a reluctance encouraged by abstract opinions as to their propriety in a different state of society, but cheerfully . . ."

Pierce's balancing act was doomed to failure. The problems he faced surely were overwhelming, but for Pierce, personal grief made matters all the more tragic. His wife was despondent and unwilling to play the role of First Lady. The President himself sought comfort in the bottle.

Forebodings, timid or not, certainly did come to control the presidency of Franklin Pierce. He would be deserted by his party after four years, a lonely, solitary, and beaten man.

My countrymen:

It a relief to feel that no heart but my own can know the personal regret and bitter sorrow over which I have been borne to a position so suitable for others rather than desirable for myself.

The circumstances under which I have been called for a limited period to preside over the destinies of the Republic fill me with a profound sense of responsibility, but with nothing like shrinking apprehension. I repair to the post assigned me not as to one sought, but in obedience to the unsolicited expression of your will, answerable only for a fearless, faithful, and diligent exercise of my best powers. I ought to be, and am, truly grateful for the rare manifestation of the nation's confidence; but this, so far from lightening my obligations, only adds to their weight. You have summoned me in my weakness; you must sustain me by your strength. When looking for the fulfillment of reasonable requirements, you will not be unmindful of the great changes which have occurred, even within the last quarter of a century, and the consequent augmentation and complexity of duties imposed in the administration both of your home and foreign affairs.

Whether the elements of inherent force in the Republic have kept pace with its unparalleled progression in territory, population, and wealth has been the subject of earnest thought and discussion on both sides of the ocean. Less than sixty-four years ago the Father of His Country made the then "recent accession of the important state of North Carolina to the Constitution of the United States" one of the subjects of his special congratulation. At that moment, however, when the agitation consequent upon the Revolutionary struggle had hardly subsided, when we were just emerging from the weakness and embarrassments of the confederation, there was an evident consciousness of vigor equal to the great mission so wisely and bravely fulfilled by our fathers. It was not a presumptuous assurance, but a calm faith, springing from a clear view of the sources of power in a government constituted like ours. It is no paradox to say that although comparatively weak the newborn nation was intrinsically strong. Inconsiderable in population and apparent resources, it was upheld by a broad and intelligent comprehension of rights and an all-pervading purpose to maintain them, stronger than armaments. It came from the furnace of the Revolution, tempered to the necessities of the times. The thoughts of the men of that day were as practical as their sentiments were patriotic. They wasted no portion of their energies upon idle and delusive speculations, but with a firm and fearless step advanced beyond the governmental landmarks which had hitherto circumscribed the limits of human freedom and planted their standard, where it has stood against dangers which have threatened from abroad, and internal agitation, which has at times fearfully menaced at home. They proved themselves equal to the solution of the great problem, to understand which their minds had been illuminated by the dawning lights of the Revolution. The object sought was not a thing dreamed of; it was a thing realized. They had exhibited not only the power to achieve, but, what all history affirms to be so much more unusual, the capacity to maintain. The oppressed throughout the world from that day to the present have turned their eyes hitherward, not to find those lights extinguished or to fear lest they should wane, but to be constantly cheered by their steady and increasing radiance.

In this our country has, in my judgment, thus far fulfilled its highest duty to suffering humanity. It has spoken and will continue to speak, not only by its words, but by its acts, the language of sympathy, encouragement, and hope to those who earnestly listen to tones which pronounce for the largest rational liberty. But after all, the most animating encouragement and potent appeal for freedom will be its own history—its trials and its triumphs. Preeminently, the power of our advocacy reposes in our example; but no example, be it remembered, can be powerful for lasting good, whatever apparent advantages may be gained, which is not based upon eternal principles of right and justice. Our fathers decided for themselves, both upon the hour to declare and the hour to strike. They were their own judges of the circumstances under which it became them to pledge to each other "their lives, their fortunes, and their sacred honor" for the acquisition of the priceless inheritance transmitted to us. The energy with which that great conflict was opened and, under the guidance of a manifest and beneficent Providence, the uncomplaining endurance with which it was prosecuted to its consummation were only surpassed by the wisdom and patriotic spirit of concession which characterized all the counsels of the early fathers.

One of the most impressive evidences of that wisdom is to be found in the fact that the actual working of our system has dispelled a degree of solicitude which at the outset disturbed bold hearts and far-reaching intellects. The apprehension of dangers from extended territory, multiplied states, accumulated wealth, and augmented population has proved to be unfounded. The stars upon your banner have become nearly threefold their original number; your densely populated possessions skirt the shores of the two great oceans; and yet this vast increase of people and territory has not only shown itself compatible with the harmonious action of the states and federal government in their respective constitutional spheres, but has afforded an additional guaranty of the strength and integrity of both.

With an experience thus suggestive and cheering, the policy of my administration will not be controlled by any timid forebodings of evil from expansion. Indeed, it is not to be disguised that our attitude as a nation and our position on the globe render the acquisition of certain possessions not within our jurisdiction eminently important for our protection, if not in the future essential for the preservation of the rights of commerce and the peace of the world. Should they be obtained, it will be through no grasping spirit, but with a view to obvious national interest and security, and in a manner entirely consistent with the strictest observance of national faith. We have nothing in our history or position to invite aggression; we have everything to beckon us to the cultivation of relations of peace and amity with all nations. Purposes, therefore, at once just and pacific will be significantly marked in the conduct of our foreign affairs. I intend that my administration shall leave no blot upon our fair record, and trust I may safely give the assurance that no act within the legitimate scope of my constitutional control will be tolerated on the part of any portion of our citizens which cannot challenge a ready justification before the tribunal of the civilized world. An administration would be unworthy of confidence at home or respect abroad should it cease to be influenced by the conviction that no apparent advantage can be purchased at a price so dear as that of national wrong or dishonor. It is not your privilege as a nation to speak of a distant past. The striking incidents of your history, replete with instruction and furnishing abundant grounds for hopeful confidence, are comprised in a period comparatively brief. But if your past is limited, your future is boundless. Its obligations throng the unexplored pathway of advancement, and will be limitless as duration. Hence a sound and comprehensive policy should embrace not less the distant future than the urgent present.

The great objects of our pursuit as a people are best to be attained by peace, and are entirely consistent with the tranquillity and interests of the rest of mankind. With the neighboring nations upon our continent we should cultivate kindly and fraternal relations. We can desire nothing in regard to them so much as to see them consolidate their strength and pursue the paths of prosperity and happiness. If in the course of their growth we should open new channels of trade and create additional facilities for friendly intercourse, the benefits realized will be equal and mutual. Of the complicated European systems of national polity we have heretofore been independent. From their wars, their tumults, and anxieties we have been, happily, almost entirely exempt. Whilst these are confined to the nations which gave them existence, and within their legitimate jurisdiction, they cannot affect us except as they appeal to our sympathies in the cause of human freedom and universal advancement. But the vast interests of commerce are common to all mankind, and the advantages of trade and international intercourse must always present a noble field for the moral influence of a great people.

With these views firmly and honestly carried out, we have a right to expect, and shall under all circumstances require, prompt reciprocity. The rights which belong to us as a nation are not alone to be regarded, but those which pertain to every citizen in his individual capacity, at home and abroad, must be sacredly maintained. So long as he can discern every star in its place upon that ensign, without wealth to purchase for him preferment or title to secure for him place, it will be his privilege, and must be his acknowledged right, to stand unabashed even in the presence of princes, with a proud consciousness that he is himself one of a nation of sovereigns and that he cannot in legitimate pursuit wander so far from home that the agent whom he shall leave behind in the place which I now occupy will not see that no rude hand of power or tyrannical passion is laid upon him with impunity. He must realize that upon every sea and on every soil where our enterprise may rightfully seek the protection of our flag American citizenship is an inviolable panoply for the security of American rights. And in this connection it can hardly be necessary to reaffirm a principle which should now be regarded as fundamental. The rights, security, and repose of this confederacy reject the idea of interference or colonization on this side of the ocean by any foreign power beyond present jurisdiction as utterly inadmissible.

The opportunities of observation furnished by my brief experience as a soldier confirmed in my own mind the opinion, entertained and acted upon by others from the formation of the government, that the maintenance of large standing armies in our country would be not only dangerous, but unnecessary. They also illustrated the importance—I might well say the absolute necessity—of the military science and practical skill furnished in such an eminent degree by the institution which has made your army what it is, under the discipline and instruction of officers not more distinguished for their solid attainments, gallantry, and devotion to the public service than for unobtrusive bearing and high moral tone. The army as organized must be the nucleus around which in every time of need the strength of your military power, the sure bulwark of your defense—a national militia—may be readily formed into a well-disciplined and efficient organization. And the skill and self-devotion of the navy assure you that you may take the performance of the past as a pledge for the future, and may confidently expect that the flag which has waved its untarnished folds over every sea will still float in undiminished honor. But these, like many other subjects, will be appropriately brought at a future time to the attention of the coordinate branches of the government, to which I shall always look with profound respect and with trustful confidence that they will accord to me the aid and support which I shall so much need and which their experience and wisdom will readily suggest.

In the administration of domestic affairs you expect a devoted integrity in the public service and an observance of rigid economy in all departments, so marked as never justly to be questioned. If this reasonable expectation be not realized, I frankly confess that one of your leading hopes is doomed to disappointment, and that my efforts in a very important particular must result in a humiliating failure. Offices can be properly regarded only in the light of aids for the accomplishment of these objects, and as occupancy can confer no prerogative nor importunate desire for preferment any claim, the public interest imperatively demands that they be considered with sole reference to the duties to be performed. Good citizens may well claim the protection of good laws and the benign influence of good government, but a claim for office is what the people of a republic should never recognize. No reasonable man of any party will expect the administration to be so regardless of its responsibility and of the obvious elements of success as to retain persons known to be under the influence of political hostility and partisan prejudice in positions which will require not only severe labor, but cordial cooperation. Having no implied engagements to ratify, no rewards to bestow, no resentments to remember, and no personal wishes to consult in selections for official station, I shall fulfill this difficult and delicate trust, admitting no motive as worthy either of my character or position which does not contemplate an efficient discharge of duty and the best interests of my country. I acknowledge my obligations to the masses of my countrymen, and to them alone. Higher objects than personal aggrandizement gave direction and energy to their exertions in the late canvass, and they shall not be disappointed. They require at my hands diligence, integrity, and capacity wherever there are duties to be performed. Without these qualities in their public servants, more stringent laws for the prevention or punishment of fraud, negligence, and peculation will be vain. With them they will be unnecessary.

But these are not the only points to which you look for vigilant watchfulness. The dangers of a concentration of all power in the general government of a confederacy so vast as ours are too obvious to be disregarded. You have a right, therefore, to expect your agents in every department to regard strictly the limits imposed upon them by the Constitution of the United States. The great scheme of our constitutional liberty rests upon a proper distribution of power between the state and federal authorities, and experience has shown that the harmony and happiness of our people must depend upon a just discrimination between the separate rights and responsibilities of the states and your common rights and obligations under the general government; and here, in my opinion, are the considerations which should form the true basis of future concord in regard to the questions which have most seriously disturbed public tranquillity. If the federal government will confine itself to the exercise of powers clearly granted by the Constitution, it can hardly happen that its action upon any question should endanger the institutions of the states or interfere with their right to manage matters strictly domestic according to the will of their own people.

In expressing briefly my views upon an important subject [which] has recently agitated the nation to almost a fearful degree, I am moved by no other impulse than a most earnest desire for the perpetuation of that Union which has made us what we are, showering upon us blessings and conferring a power and influence which our fathers could hardly have anticipated, even with their most sanguine hopes directed to a far-off future. The sentiments I now announce were not unknown before the expression of the voice which called me here. My own position upon this subject was clear and unequivocal, upon the record of my words and my acts, and it is only recurred to at this time because silence might perhaps be misconstrued. With the Union my best and dearest earthly hopes are entwined. Without it what are we individually or collectively? What becomes of the noblest field ever opened for the advancement of our race in religion, in government, in the arts, and in all that dignifies and adorns mankind? From that radiant constellation which both illumines our own way and points out to struggling nations their course, let but a single star be lost, and, if these be not utter darkness, the luster of the whole is dimmed. Do my countrymen need any assurance that such a catastrophe is not to overtake them while I possess the power to stay it? It is with me an earnest and vital belief that as the Union has been the source, under Providence, of our prosperity to this time, so it is the surest pledge of a continuance of the blessings we have enjoyed, and which we are sacredly bound to transmit undiminished to our children. The field of calm and free discussion in our country is open, and will always be so, but never has been and never can be traversed for good in a spirit of sectionalism and uncharitableness. The founders of the Republic dealt with things as they were presented to them, in a spirit of self-sacrificing patriotism, and, as time has proved, with a comprehensive wisdom which it will always be safe for us to consult. Every measure tending to strengthen the fraternal feelings of all the members of our Union has had my heartfelt approbation. To every theory of society or government, whether the offspring of feverish ambition or of morbid enthusiasm, calculated to dissolve the bonds of law and affection which unite us, I shall interpose a ready and stern resistance. I believe that involuntary servitude, as it exists in different states of this confederacy, is recognized by the Constitution. I believe that it stands like any other admitted right, and that the states where it exists are entitled to efficient remedies to enforce the constitutional provisions. I hold that the laws of 1850, commonly called the "compromise measures," are strictly constitutional and to be unhesitatingly carried into effect. I believe that the constituted authorities of this Republic are bound to regard the rights of the South in this respect as they would view any other legal and constitutional right, and that the laws to enforce them should be respected and obeyed, not with a reluctance encouraged by abstract opinions as to their propriety in a different state of society, but cheerfully and according to the decisions of the tribunal to which their exposition belongs. Such have been, and are, my convictions, and upon them I shall act. I fervently hope that the question is at rest, and that no sectional or ambitious or fanatical excitement may again threaten the durability of our institutions or obscure the light of our prosperity.

But let not the foundation of our hope rest upon man's wisdom. It will not be sufficient that sectional prejudices find no place in the public deliberations. It will not be sufficient that the rash counsels of human passion are rejected. It must be felt that there is no national security but in the nation's humble, acknowledged dependence upon God and His overruling providence.

We have been carried in safety through a perilous crisis. Wise counsels, like those which gave us the Constitution, prevailed to uphold it. Let the period be remembered as an admonition, and not as an encouragement, in any section of the Union, to make experiments where experiments are fraught with such fearful hazard. Let it be impressed upon all hearts that, beautiful as our fabric is, no earthly power or wisdom could ever reunite its broken fragments. Standing, as I do, almost within view of the green slopes of Monticello, and, as it were, within reach of the tomb of Washington, with all the cherished memories of the past gathering around me like so many eloquent voices of exhortation from heaven, I can express no better hope for my country than that the kind Providence which smiled upon our fathers may enable their children to preserve the blessings they have inherited.

James Buchanan

· INAUGURAL ADDRESS ·

Wednesday, March 4, 1857

*I*n one of his earliest speeches as a candidate for President in 1960, John F. Kennedy described his vision of the presidency. It was, not surprisingly, an expansive vision, taking as its models Abraham Lincoln, Andrew Jackson, and the two Roosevelts. These strong leaders, Kennedy conceded, often were criticized because they used their powers to the fullest. But Kennedy argued that it was better "to have a Roosevelt or a Wilson than to have another James Buchanan, cringing in the White House, afraid to move."

Kennedy's contempt for the fifteenth President was and remains shared by most historians. But before his presidency collapsed in disappointment, Buchanan believed he had a plan to avoid a suicidal separation between North and South. Though he was a Northerner himself and personally opposed slavery, he believed that Northern abolitionists were the cause of sectional discord. As he prepared to enter the office that would prove to be his ruin, he said he would seek to stamp out antislavery agitation in the North. The threat to the Union, he believed, was in the North, not the South.

That insight is more helpful than Kennedy's image of a President who cringed rather than decided, a weak man unsuited for hard decisions. Buchanan actually believed he had the formula, a secret plan, so to speak, to head off disaster. He would rely on the Supreme Court to take the lead on slavery, allowing the other two branches of government—far more susceptible to public pressure—to follow rather than lead. Perhaps not the most courageous formula, but a plan nevertheless.

James Buchanan did not win the presidency because he offered bold solutions to the nation's bitter divisions. His fellow Democrats turned to him, after rejecting Franklin Pierce, in part because he had played no role in the divisive debate over Kansas. He had been in London, as Pierce's ambassador to Great Britain. So, for the same reason the party turned to Pierce in 1852—he was a blank slate—it turned to Buchanan as Pierce's prospective replacement in 1856.

The Whigs went the way of the Federalists during the Pierce years; the party fractured beyond repair after the Kansas-Nebraska Act. Southern Whigs supported the measure, leading many Northern antislavery Whigs to join the new Republican Party, among them Abraham Lincoln. Other Whigs turned against the wave of immigrants, mostly German and Irish Catholics, who were transforming Northern cities. They backed the new American Party, better known as the Know-Nothings, which was the voice of the anti-immigrant nativist movement.

The Republicans nominated John C. Fremont in 1856 as their first presidential candidate. Fremont was a famed military officer and explorer who was nicknamed the Pathfinder in recognition of his expeditions to the Oregon Territory and California. In 1850, the new state of California sent him to the U.S. Senate.

Fremont was a devout abolitionist. The American Party's candidate, former President Millard Fillmore, was a firm opponent of immigration. Buchanan, then, was the moderate in the race, and in the three-way campaign he captured 45 percent of the popular vote and 174 electoral votes to win with relative ease.

While Buchanan prepared to assume his new responsibilities, the Supreme Court was considering a fateful case, which Buchanan carefully monitored. A slave named Dred Scott had sued for his freedom because his owner had taken him from a slave state, Missouri, to the free state of Illinois and to the Wisconsin Territory, which was free as well. Scott's abolitionist attorneys argued that he was now a free man.

In the days before Buchanan took the oath of office, he learned—improperly—that the Court, headed by Chief Justice Roger Taney, was about to decide against Scott on the grounds that as a slave he was not a U.S. citizen and so had no standing to sue. In addition, the Court would rule that Congress had no authority to prohibit slavery from new territories, declaring the Missouri Compromise, which already had been repealed by the Kansas-Nebraska Act, to be unconstitutional. In Buchanan's view, the Court's ruling decided the issue.

Inauguration Day 1857 was a warmer and brighter day than it was for Pierce. The nation's troubles seemed, for a moment, distant as official

Washington paraded and marched in solemn ceremony. Buchanan himself was not in the best of moods. He had developed dysentery during the train journey from Pennsylvania to Washington. Not long before he set out to the Capitol to take the oath, he was given a bit of brandy and some medicine to settle his stomach. Fortunately, the combination worked.

Buchanan's inaugural address contained surprisingly few of the opening platitudes that cluttered the speeches of his most immediate predecessors. But that was because he had news to deliver. Taking note of the bloody situation in Kansas, Buchanan noted that a "difference of opinion" had arisen about whether Kansas could be admitted as a free or slave state. "This is, happily, a matter of but little practical importance," he announced. "Besides, it is a judicial question, which legitimately belongs to the Supreme Court of the United States, before whom it is now pending, and will, it is understood, be speedily and finally settled." A strict interpreter of the Constitution, Buchanan noted that he would "cheerfully submit" to whatever the Court decided.

After dispensing with the related issues of slavery and new territories, Buchanan launched into a polite tirade against the "wild schemes of expenditure" that had produced a group of "speculators and jobbers, whose ingenuity is exerted in contriving and promoting expedients to obtain public money." Buchanan clearly was no advocate for a latter-day American System of internal improvements. He promised to ride herd on the nation's finances, a rather moot point considering the expenditure in lives and treasure that lay just around the corner.

Buchanan's speech was not that of a man who feared the worst. It was of a man who believed he had the answers. When that illusion was shattered, he had nothing more to offer.

Bibliographic Note

Kenneth M. Stampp, *America in 1857: A Nation on the Brink*
(New York: Oxford University Press, 1990).

Philip Shriver Klein, *President James Buchanan: A Biography*
(University Park, PA: Pennsylvania State University Press, 1962).

FELLOW CITIZENS:

I appear before you this day to take the solemn oath "that I will faithfully execute the office of President of the United States and will to the best of my ability preserve, protect, and defend the Constitution of the United States."

In entering upon this great office I must humbly invoke the God of our fathers for wisdom and firmness to execute its high and responsible duties in such a manner as to restore harmony and ancient friendship among the people of the several states and to preserve our free institutions throughout many generations. Convinced that I owe my election to the inherent love for the Constitution and the Union which still animates the hearts of the American people, let me earnestly ask their powerful support in sustaining all just measures calculated to perpetuate these, the richest political blessings which heaven has ever bestowed upon any nation. Having determined not to become a candidate for reelection, I shall have no motive to influence my conduct in administering the government except the desire ably and faithfully to serve my country and to live in grateful memory of my countrymen.

We have recently passed through a presidential contest in which the passions of our fellow citizens were excited to the highest degree by questions of deep and vital importance; but when the people proclaimed their will the tempest at once subsided and all was calm.

The voice of the majority, speaking in the manner prescribed by the Constitution, was heard, and instant submission followed. Our own country could alone have exhibited so grand and striking a spectacle of the capacity of man for self-government.

What a happy conception, then, was it for Congress to apply this simple rule, that the will of the majority shall govern, to the settlement of the question of domestic slavery in the territories. Congress is neither "to legislate slavery into any territory or state nor to exclude it therefrom, but to leave the people thereof perfectly free to form and regulate their domestic institutions in their own way, subject only to the Constitution of the United States."

As a natural consequence, Congress has also prescribed that when the territory of Kansas shall be admitted as a state it "shall be received into the Union with or without slavery, as their constitution may prescribe at the time of their admission." A difference of opinion has arisen in regard to the point of time when the people of a territory shall decide this question for themselves.

This is, happily, a matter of but little practical importance. Besides, it is a judicial question, which legitimately belongs to the Supreme Court of the United States, before whom it is now pending, and will, it is understood, be speedily and finally settled. To their decision, in common with all good citizens, I shall cheerfully submit, whatever this may be, though it has ever been my individual opinion that under the Nebraska-Kansas Act the appropriate period will be when the number of actual residents in the territory shall justify the formation of a constitution with a view to its admission as a state into the Union. But be this as it may, it is the imperative and indispensable duty of the government of the United States to secure to every resident inhabitant the free and independent expression of his opinion by his vote. This sacred right of each individual must be preserved. That being accomplished, nothing can be fairer than to leave the people of a territory free from all foreign interference to decide their own destiny for themselves, subject only to the Constitution of the United States.

The whole territorial question being thus settled upon the principle of popular sovereignty—a principle as ancient as free government itself—everything of a practical nature has been decided. No other question remains for adjustment, because all agree that under the Constitution slavery in the states is beyond the reach of any human power except that of the respective states themselves wherein it exists. May we not, then, hope that the long agitation on this subject is approaching its end, and that the geographical parties to which it has given birth, so much dreaded by the Father of His Country, will speedily become extinct? Most happy will it be for the country when the public mind shall be diverted from this question to others of more pressing and practical importance. Throughout the whole progress of this agitation, which has scarcely known any intermission for more than twenty years, whilst it has been productive of no positive good to any human being it has been the prolific source of great evils to the master, to the slave, and to the whole country. It has alienated and estranged the people of the sister states from each other, and has even seriously endangered the very existence of the Union. Nor has the danger yet entirely ceased. Under our system there is a remedy for all mere political evils in the sound sense and sober judgment of the people. Time is a great corrective. Political subjects which but a few years ago excited and exasperated the public mind have passed away and are now nearly forgotten. But this question of domestic slavery is of far graver importance than any mere political question, because should the agitation continue it may eventually endanger the personal safety of a large portion of our countrymen where the institution exists. In that event no form of government, however admirable in itself and however productive of material benefits, can compensate for the loss of peace and domestic security around the family altar. Let every Union-loving man, therefore, exert his best influence to suppress this agitation, which since the recent legislation of Congress is without any legitimate object.

It is an evil omen of the times that men have undertaken to calculate the mere material value of the Union. Reasoned estimates have been presented of the pecuniary profits and local advantages which would result to different states and sections from its dissolution and of the comparative injuries which such an event would inflict on other states and sections. Even descending to this low and narrow view of the mighty question, all such calculations are at fault. The bare reference to a single consideration will be conclusive on this point. We at present enjoy a free trade throughout our extensive and expanding country such as the world has never witnessed. This trade is conducted on railroads and canals, on noble rivers and arms of the sea, which bind together the North and the South, the East and the West, of our confederacy. Annihilate this trade, arrest its free progress by the geographical lines of jealous and hostile states, and you destroy the prosperity and onward march of the whole and every part and involve all in one common ruin. But such considerations, important as they are in themselves, sink into insignificance when we reflect on the terrific evils which would result from disunion to every portion of the confederacy—to the North, not more than to the South, to the East, not more than to the West. These I shall not attempt to portray, because I feel an humble confidence that the kind Providence which inspired our fathers with wisdom to frame the most perfect form of government and union ever devised by man will not suffer it to perish until it shall have been peacefully instrumental by its example in the extension of civil and religious liberty throughout the world.

Next in importance to the maintenance of the Constitution and the Union is the duty of preserving the government free from the taint or even the suspicion of corruption. Public virtue is the vital spirit of republics, and history proves that when this has decayed and the love of money has usurped its place, although the forms of free government may remain for a season, the substance has departed forever.

Our present financial condition is without a parallel in history. No nation has ever before been embarrassed from too large a surplus in its Treasury. This almost necessarily gives birth to extravagant legislation. It produces wild schemes of expenditure and begets a race of speculators and jobbers, whose ingenuity is exerted in contriving and promoting expedients to obtain public money. The purity of official agents, whether rightfully or wrongfully, is suspected, and the character of the government suffers in the estimation of the people. This is in itself a very great evil.

The natural mode of relief from this embarrassment is to appropriate the surplus in the Treasury to great national objects for which a clear warrant can be found in the Constitution. Among these I might mention the extinguishment of the public debt, a reasonable increase of the navy, which is at present inadequate to the protection of our vast tonnage afloat, now greater than that of any other nation, as well as to the defense of our extended seacoast.

It is beyond all question the true principle that no more revenue ought to be collected from the people than the amount necessary to defray the expenses of a wise, economical, and efficient administration of the government. To reach this point it was necessary to resort to a modification of the tariff, and this has, I trust, been accomplished in such a manner as to do as little injury as may have been practicable to our domestic manufactures, especially those necessary for the defense of the country. Any discrimination against a particular branch for the purpose of benefiting favored corporations, individuals, or interests would have been unjust to the rest of the community and inconsistent with that spirit of fairness and equality which ought to govern in the adjustment of a revenue tariff.

But the squandering of the public money sinks into comparative insignificance as a temptation to corruption when compared with the squandering of the public lands.

No nation in the tide of time has ever been blessed with so rich and noble an inheritance as we enjoy in the public lands. In administering this important trust, whilst it may be wise to grant portions of them for the improvement of the remainder, yet we should never forget that it is our cardinal policy to reserve these lands, as much as may be, for actual settlers, and this at moderate prices. We shall thus not only best promote the prosperity of the new states and territories, by furnishing them a hardy and independent race of honest and industrious citizens, but shall secure homes for our children and our children's children, as well as for those exiles from foreign shores who may seek in this country to improve their condition and to enjoy the blessings of civil and religious liberty. Such emigrants have done much to promote the growth and prosperity of the country. They have proved faithful both in peace and in war. After becoming citizens they are entitled, under the Constitution and laws, to be placed on a perfect equality with native-born citizens, and in this character they should ever be kindly recognized.

The federal Constitution is a grant from the states to Congress of certain specific powers, and the question whether this grant should be liberally or strictly construed has more or less divided political parties from the beginning. Without entering into the argument, I desire to state at the commencement of my administration that long experience and observation have convinced me that a strict construction of the powers of the government is the only true, as well as the only safe, theory of the Constitution. Whenever in our past history doubtful powers have been exercised by Congress, these have never failed to produce injurious and unhappy consequences. Many such instances might be adduced if this were the proper occasion. Neither is it necessary for the public service to strain the language of the Constitution, because all the great and useful powers required for a successful administration of the government, both in peace and in war, have been granted, either in express terms or by the plainest implication.

Whilst deeply convinced of these truths, I yet consider it clear that under the war-making power Congress may appropriate money toward the construction of a military road when this is absolutely necessary for the defense of any state or territory of the Union against foreign invasion. Under the Constitution Congress has power "to declare war," "to raise and support armies," "to provide and maintain a navy," and to call forth the militia to "repel invasions." Thus endowed, in an ample manner, with the war-making power, the corresponding duty is required that "the United States shall protect each of them [the states] against invasion." Now, how is it possible to afford this protection to California and our Pacific possessions except by means of a military road through the territories of the United States, over which men and munitions of war may be speedily transported from the Atlantic states to meet and to repel the invader? In the event of a war with a naval power much stronger than our own we should then have no other available access to the Pacific Coast, because such a power would instantly close the route across the isthmus of Central

America. It is impossible to conceive that whilst the Constitution has expressly required Congress to defend all the states it should yet deny to them, by any fair construction, the only possible means by which one of these states can be defended. Besides, the government, ever since its origin, has been in the constant practice of constructing military roads. It might also be wise to consider whether the love for the Union which now animates our fellow citizens on the Pacific coast may not be impaired by our neglect or refusal to provide for them, in their remote and isolated condition, the only means by which the power of the states on this side of the Rocky Mountains can reach them in sufficient time to "protect" them "against invasion." I forbear for the present from expressing an opinion as to the wisest and most economical mode in which the government can lend its aid in accomplishing this great and necessary work. I believe that many of the difficulties in the way, which now appear formidable, will in a great degree vanish as soon as the nearest and best route shall have been satisfactorily ascertained.

It may be proper that on this occasion I should make some brief remarks in regard to our rights and duties as a member of the great family of nations. In our intercourse with them there are some plain principles, approved by our own experience, from which we should never depart. We ought to cultivate peace, commerce, and friendship with all nations, and this not merely as the best means of promoting our own material interests, but in a spirit of Christian benevolence toward our fellow men, wherever their lot may be cast. Our diplomacy should be direct and frank, neither seeking to obtain more nor accepting less than is our due. We ought to cherish a sacred regard for the independence of all nations, and never attempt to interfere in the domestic concerns of any unless this shall be imperatively required by the great law of self-preservation. To avoid entangling alliances has been a maxim of our policy ever since the days of Washington, and its wisdom no one will attempt to dispute. In short, we ought to do justice in a kindly spirit to all nations and require justice from them in return.

It is our glory that whilst other nations have extended their dominions by the sword we have never acquired any territory except by fair purchase or, as in the case of Texas, by the voluntary determination of a brave, kindred, and independent people to blend their destinies with our own. Even our acquisitions from Mexico form no exception. Unwilling to take advantage of the fortune of war against a sister republic, we purchased these possessions under the treaty of peace for a sum which was considered at the time a fair equivalent. Our past history forbids that we shall in the future acquire territory unless this be sanctioned by the laws of justice and honor. Acting on this principle, no nation will have a right to interfere or to complain if in the progress of events we shall still further extend our possessions. Hitherto in all our acquisitions the people, under the protection of the American flag, have enjoyed civil and religious liberty, as well as equal and just laws, and have been contented, prosperous, and happy. Their trade with the rest of the world has rapidly increased, and thus every commercial nation has shared largely in their successful progress.

I shall now proceed to take the oath prescribed by the Constitution, whilst humbly invoking the blessing of Divine Providence on this great people.

Abraham Lincoln

· FIRST INAUGURAL ADDRESS ·

March 4, 1861

*D*ecades of patched-together compromises and bitter debates had postponed a national reckoning over slavery. But on November 6, 1860, a plurality of the nation's voters chose Abraham Lincoln, candidate of the antislavery Republican Party, as the sixteenth President. The reckoning, feared by some and welcomed by others, had arrived.

Lincoln's sizable victory in the Electoral College—he won 180 votes, more than the combined total of his three opponents—masked the divisions in the popular vote. Some 1.8 million voters supported Lincoln, but 2.8 million more did not, preferring either Stephen Douglas of Illinois, John Breckinridge of Kentucky, or John Bell of Tennessee. Lincoln won, not simply because of antislavery sentiment in the North, although that was part of it, but also because the Democrats were hopelessly divided between Douglas and Breckinridge, with Bell further splitting the anti-Republican vote as the voice of nativist ex-Whigs.

Throughout the divisive campaign, Lincoln was careful to point out that he was not an abolitionist. His assertions that Southerners had nothing to fear from a Republican victory did not silence talk about secession. The candidate chose kindness rather than confrontation in dealing with the Southern threat. "The good people of the South have too much good sense and good temper to attempt the ruin of the government,"[1] he said. This appeal earned Lincoln little in the way of good will and nothing in the way of Southern electoral votes.

The four-month interlude between Lincoln's election in November and his inauguration in March was the most dramatic such interregnum in U.S. history. Outgoing President James Buchanan believed the Southern

---

1. Stephen B. Oates, *With Malice Toward None* (New York: Harper and Row, 1977), 188.

states had no legal right to leave the Union. But state after state did so, beginning in early January. Within a month, secessionists from six states met in Alabama and declared themselves members of a new confederation.

In Springfield, Illinois, Lincoln made no public statements about the crisis. Instead, he began work on the most important speech of his life. The president-elect sequestered himself in a small room above a storefront and across the street from the state capitol. He wrote, revised, and wrote some more, with a copy of the U.S. Constitution nearby, along with copies of Andrew Jackson's proclamation against nullification, Henry Clay's speech in favor of the Compromise of 1850, and Daniel Webster's magisterial reply to Senator Robert Hayne's argument in favor of nullification in 1829.[2]

When he finished a complete draft, he shared copies with two key aides, Senator William Seward and Orville Browning, a lawyer and friend of the president-elect. Meanwhile, Lincoln set out for Washington from Springfield on February 11, embarking on the most dangerous journey any president-elect has ever taken. Rumors of assassination plots abounded. Along the way, Lincoln gave a series of conciliatory speeches as he traveled through Ohio, Pennsylvania, New York, and New Jersey. Even as he did so, the new Southern confederation announced that it had elected its first President, Jefferson Davis, who had resigned as a U.S. Senator from Mississippi when his home state seceded in January.

So serious were the threats to Lincoln's life that his security detail urged him to leave his official train in Philadelphia, disguise himself, and board another train that would take him to dangerous Baltimore with its large population of Southern sympathizers. From Baltimore, he would board another train for Washington. Lincoln reluctantly agreed. The president-elect entered the nation's capital not in triumph but in secret, at daybreak on February 23.

The extraordinary circumstances of Lincoln's election and pending inauguration, the very need for secrecy and security during the final stage of his journey to Washington, suggested a moment unlike any other in the nation's history. But for those who wished jobs and favors from the new administration, it was, amazingly, business as usual. Patronage seekers filled the president-elect's calendar, even as he met with representatives from Virginia, which remained outside the new confederation, and other

---

2. Roy P. Basler, ed., *Abraham Lincoln: His Speeches and Writings* (New York: De Capo Press, 1988), 588–89.

wavering states along the borderland between North and South. He spent his final days as president-elect trying to save the Union and keep the party's favor seekers happy.

On the eve of the inauguration, Lincoln and Seward revised the inaugural address one final time. Seward, a onetime Whig from New York who bitterly opposed slavery and who was about to become Lincoln's secretary of state, persuaded the President to tone down his threat to retake the fortifications already seized by the rebel confederation. The critical audience for this speech would not be those states that had already seceded but those on the border that were wavering. The prospective rebellion might be stopped in its tracks if Virginia, Maryland, Tennessee, Delaware, Kentucky, Missouri, and, farther south, North Carolina remained outside the Confederacy. They might respond to reassurance, but threats surely would only make matters worse.

Washington was an armed camp on Inauguration Day 1861. Following precedent, Lincoln and the outgoing chief executive, Buchanan, rode to the Capitol together. Absent from the city's streets was the unbroken sea of smiling faces that usually accompanied inaugural festivities. Inauguration Day 1861 was not a festive occasion.

Lincoln had grown a beard during the interval between his election and his swearing-in. He wore a black suit and, before taking the oath from the withered Chief Justice Roger Taney, author of the Dred Scott decision, Lincoln took off his hat, put on his spectacles, and addressed not only the dignitaries in attendance but also those who would eagerly read his speech in the public prints.

No President before or since has had to deliver so fateful an inaugural address. Lincoln's greatest speeches are works of art, but his manner of delivering them was hardly in the tradition of era's great orators. Lincoln had a high, clear voice, effective enough but not especially notable compared with the thundering instruments wielded by some of his contemporaries.

His speech was designed for Southern ears that might yet be willing to hear, Southern minds that might yet be open. He offered his support for one of the several desperate compromises under discussion: a constitutional amendment, already passed by Congress, that would prohibit the federal government from interfering with slavery where it existed.

Mixed in with these reassurances was an overt appeal to Southerners, especially those in border and wavering states, who might have gazed into the abyss and saw nothing but fear, anxiety, and worse. The Union, Lincoln said, was intact because separation was impossible. Those who believed and acted otherwise were "insurrectionary or revolutionary" and could be dealt with accordingly.

His long brief on behalf of Union and conciliation ended with one of the most famous passages in American rhetoric. The moving imagery ("every . . . patriot grave" and "every living heart and hearthstone"), the poetic touches ("mystic chords of memory") and appeals to decency ("the better angels of our nature") earned this speech a place not only in the annals of American political rhetoric but in American letters as well.

FELLOW CITIZENS OF THE UNITED STATES:

In compliance with a custom as old as the government itself, I appear before you to address you briefly and to take in your presence the oath prescribed by the Constitution of the United States to be taken by the President "before he enters on the execution of this office."

I do not consider it necessary at present for me to discuss those matters of administration about which there is no special anxiety or excitement.

Apprehension seems to exist among the people of the Southern states that by the accession of a Republican administration their property and their peace and personal security are to be endangered. There has never been any reasonable cause for such apprehension. Indeed, the most ample evidence to the contrary has all the while existed and been open to their inspection. It is found in nearly all the published speeches of him who now addresses you. I do but quote from one of those speeches when I declare that I have no purpose, directly or indirectly, to interfere with the institution of slavery in the states where it exists. I believe I have no lawful right to do so, and I have no inclination to do so.

Those who nominated and elected me did so with full knowledge that I had made this and many similar declarations and had never recanted them; and more than this, they placed in the platform for my acceptance, and as a law to themselves and to me, the clear and emphatic resolution which I now read: "Resolved, that the maintenance inviolate of the rights of the states, and especially the right of each state to order and control its own domestic institutions according to its own judgment exclusively, is essential to that balance of power on which the perfection and endurance of our political fabric depend; and we denounce the lawless invasion by armed force of the soil of any state or territory, no matter what pretext, as among the gravest of crimes."

I now reiterate these sentiments, and in doing so I only press upon the public attention the most conclusive evidence of which the case is susceptible that the property, peace, and security of no section are to be in any wise endangered by the now incoming administration. I add, too, that all the protection which, consistently with the Constitution and the laws, can be given will be cheerfully given to all the states when lawfully demanded, for whatever cause—as cheerfully to one section as to another.

There is much controversy about the delivering up of fugitives from service or labor. The clause I now read is as plainly written in the Constitution as any other of its provisions:

"No person held to service or labor in one state, under the laws thereof, escaping into another, shall in consequence of any law or regulation therein be discharged from such service or labor, but shall be delivered up on claim of the party to whom such service or labor may be due."

It is scarcely questioned that this provision was intended by those who made it for the reclaiming of what we call fugitive slaves; and the intention of the lawgiver is the law. All members of Congress swear their support to the whole Constitution—to this provision as much as to any other. To the proposition, then, that slaves whose cases come within the terms of this clause "shall be delivered up" their oaths are unanimous. Now, if they would make the effort in good temper, could they not with nearly equal unanimity frame and pass a law by means of which to keep good that unanimous oath?

There is some difference of opinion whether this clause should be enforced by national or by state authority, but surely that difference is not a very material one. If the slave is to be surrendered, it can be of but little consequence to him or to others by which authority it is done. And should anyone in any case be content that his oath shall go unkept on a merely unsubstantial controversy as to how it shall be kept?

Again: in any law upon this subject ought not all the safeguards of liberty known in civilized and humane jurisprudence to be introduced, so that a free man be not in any case surrendered as a slave? And might it not be well at the same time to provide by law for the enforcement of that clause in the Constitution which guarantees that "the citizens of each state shall be entitled to all privileges and immunities of citizens in the several states"?

I take the official oath today with no mental reservations and with no purpose to construe the Constitution or laws by any hypercritical rules;

and while I do not choose now to specify particular acts of Congress as proper to be enforced, I do suggest that it will be much safer for all, both in official and private stations, to conform to and abide by all those acts which stand unrepealed than to violate any of them trusting to find impunity in having them held to be unconstitutional.

It is seventy-two years since the first inauguration of a President under our national Constitution. During that period fifteen different and greatly distinguished citizens have in succession administered the executive branch of the government. They have conducted it through many perils, and generally with great success. Yet, with all this scope of precedent, I now enter upon the same task for the brief constitutional term of four years under great and peculiar difficulty. A disruption of the federal Union, heretofore only menaced, is now formidably attempted.

I hold that in contemplation of universal law and of the Constitution the Union of these states is perpetual. Perpetuity is implied, if not expressed, in the fundamental law of all national governments. It is safe to assert that no government proper ever had a provision in its organic law for its own termination. Continue to execute all the express provisions of our national Constitution, and the Union will endure forever, it being impossible to destroy it except by some action not provided for in the instrument itself.

Again: If the United States be not a government proper, but an association of states in the nature of contract merely, can it, as a contract, be peaceably unmade by less than all the parties who made it? One party to a contract may violate it—break it, so to speak—but does it not require all to lawfully rescind it?

Descending from these general principles, we find the proposition that in legal contemplation the Union is perpetual confirmed by the history of the Union itself. The Union is much older than the Constitution. It was formed, in fact, by the Articles of Association in 1774. It was matured and continued by the Declaration of Independence in 1776. It was further matured, and the faith of all the then thirteen states expressly plighted and engaged that it should be perpetual, by the Articles of Confederation in 1778. And finally, in 1787, one of the declared objects for ordaining and establishing the Constitution was "to form a more perfect Union."

But if destruction of the Union by one or by a part only of the states be lawfully possible, the Union is less perfect than before the Constitution, having lost the vital element of perpetuity.

It follows from these views that no state upon its own mere motion can lawfully get out of the Union; that resolves and ordinances to that effect are legally void, and that acts of violence within any state or states against the authority of the United States are insurrectionary or revolutionary, according to circumstances.

I therefore consider that in view of the Constitution and the laws the Union is unbroken, and to the extent of my ability, I shall take care, as the Constitution itself expressly enjoins upon me, that the laws of the Union be faithfully executed in all the states. Doing this I deem to be only a simple duty on my part, and I shall perform it so far as practicable unless my rightful masters, the American people, shall withhold the requisite means or in some authoritative manner direct the contrary. I trust this will not be regarded as a menace, but only as the declared purpose of the Union that it will constitutionally defend and maintain itself.

In doing this there needs to be no bloodshed or violence, and there shall be none unless it be forced upon the national authority. The power confided to me will be used to hold, occupy, and possess the property and places belonging to the government and to collect the duties and imposts; but beyond what may be necessary for these objects, there will be no invasion, no using of force against or among the people anywhere. Where hostility to the United States in any interior locality shall be so great and universal as to prevent competent resident citizens from holding the federal offices, there will be no attempt to force obnoxious strangers among the people for that object. While the strict legal right may exist in the government to enforce the exercise of these offices, the attempt to do so would be so irritating and so nearly impracticable withal that I deem it better to forego for the time the uses of such offices.

The mails, unless repelled, will continue to be furnished in all parts of the Union. So far as possible the people everywhere shall have that sense of perfect security which is most favorable to calm thought and reflection. The course here indicated will be followed unless current events and experience shall show a modification or change to be proper, and in every case and exigency my best discretion will be exercised, according to circumstances actually existing and with a view and a hope of a peaceful solution of the national troubles and the restoration of fraternal sympathies and affections.

That there are persons in one section or another who seek to destroy the Union at all events and are glad of any pretext to do it I will neither affirm nor deny; but if there be such, I need address no word to them. To those, however, who really love the Union may I not speak?

Before entering upon so grave a matter as the destruction of our national fabric, with all its benefits, its memories, and its hopes, would it not be wise to ascertain precisely why we do it? Will you hazard so desperate a step while there is any possibility that any portion of the ills you fly from have no real existence? Will you, while the certain ills you fly to are greater than all the real ones you fly from, will you risk the commission of so fearful a mistake?

All profess to be content in the Union if all constitutional rights can be maintained. Is it true, then, that any right plainly written in the Constitution has been denied? I think not. Happily, the human mind is so constituted that no party can reach to the audacity of doing this. Think, if you can, of a single instance in which a plainly written provision of the Constitution has ever been denied. If by the mere force of numbers a majority should deprive a minority of any clearly written constitutional right, it might in a moral point of view justify revolution; certainly would if such right were a vital one. But such is not our case. All the vital rights of minorities and of individuals are so plainly assured to them by affirmations and negations, guaranties and prohibitions, in the Constitution that controversies never arise concerning them. But no organic law can ever be framed with a provision specifically applicable to every question which may occur in practical administration. No foresight can anticipate nor any document of reasonable length contain express provisions for all possible questions. Shall fugitives from labor be surrendered by national or by state authority? The Constitution does not expressly say. May Congress prohibit slavery in the territories? The Constitution does not expressly say. Must Congress protect slavery in the territories? The Constitution does not expressly say.

From questions of this class spring all our constitutional controversies, and we divide upon them into majorities and minorities. If the minority will not acquiesce, the majority must, or the government must cease. There is no other alternative, for continuing the government is acquiescence on one side or the other. If a minority in such case will secede rather than acquiesce, they make a precedent which in turn will divide and ruin them, for a minority of their own will secede from them whenever a majority refuses to be controlled by such minority. For instance, why may not any portion of a new confederacy a year or two hence arbitrarily secede again, precisely as portions of the present Union now claim to secede from it? All who cherish disunion sentiments are now being educated to the exact temper of doing this.

Is there such perfect identity of interests among the states to compose a new union as to produce harmony only and prevent renewed secession?

Plainly the central idea of secession is the essence of anarchy. A majority held in restraint by constitutional checks and limitations, and always changing easily with deliberate changes of popular opinions and sentiments, is the only true sovereign of a free people. Whoever rejects it does of necessity fly to anarchy or to despotism. Unanimity is impossible. The rule of a minority, as a permanent arrangement, is wholly inadmissible; so that, rejecting the majority principle, anarchy or despotism in some form is all that is left.

I do not forget the position assumed by some that constitutional questions are to be decided by the Supreme Court, nor do I deny that such decisions must be binding in any case upon the parties to a suit as to the object of that suit, while they are also entitled to very high respect and consideration in all parallel cases by all other departments of the government. And while it is obviously possible that such decision may be erroneous in any given case, still the evil effect following it, being limited to that particular case, with the chance that it may be overruled and never become a precedent for other cases, can better be borne than could the evils of a different practice. At the same time, the candid citizen must confess that if the policy of the government upon vital questions affecting the whole people is to be irrevocably fixed by decisions of the Supreme Court, the instant they are made in ordinary litigation between parties in personal actions the people will have ceased to be their own rulers, having to that extent practically resigned their government into the hands of that eminent tribunal. Nor is there in this view any assault upon the court or the judges. It is a duty from which they may not shrink to decide cases properly brought before them, and it is no fault of theirs if others seek to turn their decisions to political purposes.

One section of our country believes slavery is right and ought to be extended, while the other believes it is wrong and ought not to be extended. This is the only substantial dispute. The fugitive-slave clause of the Constitution and the law for the suppression of the foreign slave trade are each as well enforced, perhaps, as any law can ever be in a community where the moral sense of the people imperfectly supports the law itself. The great body of the people abide by the dry legal obligation in both cases, and a few break over in each. This, I think, cannot be perfectly cured, and it would be worse in both cases after the separation of the sections than be-

fore. The foreign slave trade, now imperfectly suppressed, would be ultimately revived without restriction in one section, while fugitive slaves,
now only partially surrendered, would not be surrendered at all by the
other.

Physically speaking, we cannot separate. We cannot remove our respective sections from each other nor build an impassable wall between
them. A husband and wife may be divorced and go out of the presence
and beyond the reach of each other, but the different parts of our country
cannot do this. They cannot but remain face to face, and intercourse, either amicable or hostile, must continue between them. Is it possible, then,
to make that intercourse more advantageous or more satisfactory after
separation than before? Can aliens make treaties easier than friends can
make laws? Can treaties be more faithfully enforced between aliens than
laws can among friends? Suppose you go to war, you cannot fight always;
and when, after much loss on both sides and no gain on either, you cease
fighting, the identical old questions, as to terms of intercourse, are again
upon you.

This country, with its institutions, belongs to the people who inhabit
it. Whenever they shall grow weary of the existing government, they can
exercise their constitutional right of amending it or their revolutionary
right to dismember or overthrow it. I cannot be ignorant of the fact that
many worthy and patriotic citizens are desirous of having the national
Constitution amended. While I make no recommendation of amendments, I fully recognize the rightful authority of the people over the whole
subject, to be exercised in either of the modes prescribed in the instrument itself; and I should, under existing circumstances, favor rather than
oppose a fair opportunity being afforded the people to act upon it. I will
venture to add that to me the convention mode seems preferable, in that it
allows amendments to originate with the people themselves, instead of
only permitting them to take or reject propositions originated by others,
not especially chosen for the purpose, and which might not be precisely
such as they would wish to either accept or refuse. I understand a proposed amendment to the Constitution—which amendment, however, I
have not seen—has passed Congress, to the effect that the federal government shall never interfere with the domestic institutions of the states, including that of persons held to service. To avoid misconstruction of what I
have said, I depart from my purpose not to speak of particular amendments so far as to say that, holding such a provision to now be implied

constitutional law, I have no objection to its being made express and irrevocable.

The chief magistrate derives all his authority from the people, and they have referred none upon him to fix terms for the separation of the states. The people themselves can do this if also they choose, but the executive as such has nothing to do with it. His duty is to administer the present government as it came to his hands and to transmit it unimpaired by him to his successor.

Why should there not be a patient confidence in the ultimate justice of the people? Is there any better or equal hope in the world? In our present differences, is either party without faith of being in the right? If the Almighty Ruler of Nations, with His eternal truth and justice, be on your side of the North, or on yours of the South, that truth and that justice will surely prevail by the judgment of this great tribunal of the American people.

By the frame of the government under which we live this same people have wisely given their public servants but little power for mischief, and have with equal wisdom provided for the return of that little to their own hands at very short intervals. While the people retain their virtue and vigilance no administration by any extreme of wickedness or folly can very seriously injure the government in the short space of four years.

My countrymen, one and all, think calmly and well upon this whole subject. Nothing valuable can be lost by taking time. If there be an object to hurry any of you in hot haste to a step which you would never take deliberately, that object will be frustrated by taking time; but no good object can be frustrated by it. Such of you as are now dissatisfied still have the old Constitution unimpaired, and, on the sensitive point, the laws of your own framing under it; while the new administration will have no immediate power, if it would, to change either. If it were admitted that you who are dissatisfied hold the right side in the dispute, there still is no single good reason for precipitate action. Intelligence, patriotism, Christianity, and a firm reliance on Him who has never yet forsaken this favored land are still competent to adjust in the best way all our present difficulty.

In your hands, my dissatisfied fellow countrymen, and not in mine, is the momentous issue of civil war. The government will not assail you. You can have no conflict without being yourselves the aggressors. You have no oath registered in heaven to destroy the government, while I shall have the most solemn one to "preserve, protect, and defend it."

I am loath to close. We are not enemies, but friends. We must not be enemies. Though passion may have strained it must not break our bonds of affection. The mystic chords of memory, stretching from every battle-field and patriot grave to every living heart and hearthstone all over this broad land, will yet swell the chorus of the Union, when again touched, as surely they will be, by the better angels of our nature.

Abraham Lincoln

Saturday, March 4, 1865

The war between the Northern and Southern states raged on for three years. By 1864 it had begun to turn in favor of the Union forces. General Ulysses S. Grant assumed command of the Army of the Potomac, which now numbered over one hundred thousand. Then, on May 5, 1864, he began his monthlong Wilderness Campaign in which he expected to decimate General Robert E. Lee's army of sixty thousand. The slaughter on both sides was horrendous. In the battles of Spotsylvania and Cold Harbor, Grant's losses rose to nearly sixty thousand as against half that number suffered by Lee. But there was no way the Confederates could replace their losses. The Confederacy was slowly bleeding to death.

Meanwhile, on May 7, General William T. Sherman set out with one hundred thousand troops from Chattanooga, Tennessee, and headed toward Atlanta, Georgia. On September 2 he occupied Atlanta, and after destroying whatever supplies or materials that might prove helpful to the enemy, he headed for the sea.

In Washington, Lincoln had already devised what is known as the "10 percent plan" for reuniting the country. This plan required 10 percent of the 1860 electorate to take an oath of loyalty to the Union; and those state governments that agreed to the emancipation of their slaves could be restored to the Union by executive decree. President Lincoln assumed he had the authority to direct Reconstruction and had in fact already recognized Arkansas and Louisiana as restored to the Union. But when representatives from those states appeared in Congress they were denied their seats.

Although most Radical Republicans in Congress chafed at what Lincoln was doing, they knew that he would be nominated for a second term as President at their convention in Baltimore on June 7, 1864. But some of them were offended by the selection as vice president of Andrew Johnson, the loyalist military governor of Tennessee, in an attempt to appeal to Democrats. In late August, the Democratic Party held its convention in Chicago and nominated the popular general George B. McClellan for President and George H. Pendleton of Ohio for vice president.

Radical Republicans realized that they had to come up with their own plan to restore the Union or forfeit all direction of Reconstruction to the President. So, on July 4, 1864, they enacted the Wade-Davis Bill, named after Senator Ben Wade of Ohio and Representative Henry Winter Davis of Maryland. The measure required that a majority of the electorate, not 10 percent, swear to past and present loyalty before they could form a government. It also demanded the abolishment of slavery and decreed that no one could vote who held a Confederate office or carried arms against the United States. Until Congress recognized state governments organized under its auspices, said Davis, "there is no government in the rebel states except the authority of Congress."

Lincoln pocket vetoed the bill, declaring that he was not committed to any single plan of Reconstruction. He also expressed his doubt that Congress had the authority to abolish slavery. The Radicals responded with the Wade-Davis Manifesto, which asserted Congress's absolute authority to deal with the rebellious states and instructed Lincoln to execute the laws of the country, not legislate them.

Lincoln's political position was strengthened with the capture of Atlanta by the Union forces under General Sherman in early September, and he won reelection on November 8, 1864, by an electoral count of 212 to 21 and a popular majority vote of four hundred thousand out of four million who cast ballots.

When Congress reconvened in December, it passed the Thirteenth Amendment to the Constitution, which would end slavery. It was declared ratified on December 18, 1865. At the same time, General Sherman, having reached the sea, then swung northward through South Carolina. The state capital, Columbia, was burned and Charleston occupied. The war was finally winding down.

Early in March 1865, General Robert E. Lee attempted to engage General Grant in negotiations to attend a military convention. Grant notified Lincoln of this overture, but the President instructed his general against any conference with Lee unless it be to receive the capitulation of the Confederate army under Lee's command.

Then, on March 4, Lincoln was inaugurated for a second term. It was a cloudy, uncomfortable day with sudden gusts of wind, but a large crowd turned out and cheered the President when he and the official party emerged from the Capitol and took their assigned places in the East Portico.

Lincoln rose and read a very short address and acknowledged that this terrible war now seemed to be headed for termination. "Fondly do we hope," he said, "fervently do we pray, that this mighty scourge of war may speedily pass away." Then he closed with some of the most memorable words in all American history, words so eloquent and so splendid and appropriate that they still have the power to move us deeply: "With malice toward none, with charity for all, with firmness in the right as God gives us to see the right, let us strive on . . . to bind up the nation's wounds, to care for him who shall have borne the battle and for his widow and his orphan, to do all which may achieve and cherish a just and lasting peace among ourselves and with all nations."

When he finished, Chief Justice Salmon P. Chase administered the oath of office. Lincoln kissed the Bible, bowed to the people witnessing the ceremony, turned, and left the platform. On April 9, Generals Grant and Lee met at Appomattox Court House and Lee surrendered. A few days later, on April 14, 1865, at Ford's Theater, President Lincoln was assassinated by John Wilkes Booth. He had saved the Union, only to die before he could "bind up the nation's wounds" and bring about a "just and lasting peace."

FELLOW COUNTRYMEN:

At this second appearing to take the oath of the Presidential office there is less occasion for an extended address than there was at the first. Then a statement somewhat in detail of a course to be pursued seemed fitting and proper. Now, at the expiration of four years, during which public declarations have been constantly called forth on every point and phase of the great contest which still absorbs the attention and engrosses the energies of the nation, little that is new could be presented. The progress of our arms, upon which all else chiefly depends, is as well known to the public as to myself, and it is, I trust, reasonably satisfactory and encouraging to all. With high hope for the future, no prediction in regard to it is ventured.

On the occasion corresponding to this four years ago all thoughts were anxiously directed to an impending civil war. All dreaded it, all sought to avert it. While the inaugural address was being delivered from this place, devoted altogether to saving the Union without war, insurgent agents were in the city seeking to destroy it without war—seeking to dissolve the Union and divide effects by negotiation. Both parties deprecated war, but one of them would make war rather than let the nation survive, and the other would accept war rather than let it perish, and the war came.

One-eighth of the whole population were colored slaves, not distributed generally over the Union, but localized in the southern part of it. These slaves constituted a peculiar and powerful interest. All knew that this interest was somehow the cause of the war. To strengthen, perpetuate, and extend this interest was the object for which the insurgents would rend the Union even by war, while the government claimed no right to do more than to restrict the territorial enlargement of it. Neither party expected for the war the magnitude or the duration which it has already attained. Neither anticipated that the cause of the conflict might cease with or even before the conflict itself should cease. Each looked for an easier triumph, and a result less fundamental and astounding. Both read the same Bible and pray to the same God, and each invokes His aid against the other. It may seem strange that any men should dare to ask a just God's assistance in wringing their bread from the sweat of other men's faces, but let us judge not, that we be not judged. The prayers of both could not be answered. That of neither has been answered fully. The Almighty has His own purposes. "Woe unto the world because of offenses; for it must needs be that offenses come, but woe to that man by whom the offense cometh." If we shall suppose that American slavery is one of those offenses which, in the providence of God, must needs come, but which, having continued through His appointed time, He now wills to remove, and that He gives to both North and South this terrible war as the woe due to those by whom the offense came, shall we discern therein any departure from those divine attributes which the believers in a living God always ascribe to Him? Fondly do we hope, fervently do we pray, that this mighty scourge of war may speedily pass away. Yet, if God wills that it continue until all the wealth piled by the bondsman's two hundred and fifty years of unrequited toil shall be sunk, and until every drop of blood drawn with the lash shall be paid by another drawn with the sword, as was said three thousand years ago, so still it must be said "the judgments of the Lord are true and righteous altogether."

With malice toward none, with charity for all, with firmness in the right as God gives us to see the right, let us strive on to finish the work we are in, to bind up the nation's wounds, to care for him who shall have borne the battle and for his widow and his orphan, to do all which may achieve and cherish a just and lasting peace among ourselves and with all nations.

Ulysses S. Grant

· FIRST INAUGURAL ADDRESS ·

Thursday, March 4, 1869

As he became the first elected President of post–Civil War America, Ulysses Grant said little about the enormous and violent conflict in which he played so large a part as a general. Instead, Grant talked mostly about the nation's creditworthiness and the need to pay down debt. These were significant issues for a nation that spent so much during the war, but to eyes accustomed to Abraham Lincoln's vivid imagery and broad strokes Grant's speech reads like a corporate ledger.

There was in Grant's first inaugural no soulful reflection on the war and its resolution. Instead, he simply asked that his fellow citizens act "without prejudice, hate, or sectional pride" in resolving the issues of the day.

If Grant's first inaugural is notable only because of its straightforward, businesslike tone, perhaps that should not be surprising. Grant, after all, won his reputation on the battlefield by forging ahead with brute force rather than dancing around his opponents with complicated maneuvers. His inaugural address was a reflection of a man who smashed rather than bloodied the Army of Northern Virginia, a general who wore a private's uniform when he accepted Robert E. Lee's surrender. There was nothing ornate about Ulysses S. Grant.

That said, his speech contained one bon mot that ought to be better known to students of political science and American rhetoric. In a brief reflection on the rule of law, Grant noted, "I know no method to secure the repeal of bad or obnoxious laws so effective as their stringent execution." Grant's faith in the democratic process was either touching or naïve. The people, he implied, surely will rise to challenge a bad law, but only if that law is actually enforced. This was a counterpoint to the dead letter of nullification, which asserted that bad laws could be declared null and void before they were put into practice. The wisdom of the people, he suggested, could take care of bad lawmaking.

Ulysses Grant's first inaugural address was also his first speech as an elected officeholder. Although Grant worked with congressional Republicans on Reconstruction issues, he was a political neophyte. And it would show, in time.

Grant had been reluctant to convert his military fame to political power. In a postwar letter to Ohio senator John Sherman, brother of General William T. Sherman, Grant insisted he had no interest in pursuing the presidency. "All that I can say to discourage the idea of my ever being a candidate for an office I do say,"[1] Grant wrote. Sherman's brother certainly believed his old comrade wanted nothing to do with politics. "I think that if Grant can avoid the nomination, he will,"[2] General Sherman wrote in late 1867.

But Grant never shut the door entirely, and that was fine for the Republican Party. As a military hero who appeared to be above partisan politics, Grant seemed like an ideal choice for a country divided by Reconstruction and the upcoming impeachment of Andrew Johnson.

According to historian Brooks D. Simpson's study of Grant, the impeachment of Andrew Johnson in early 1868, brought on by Johnson's attempt to fire Secretary of War Edwin Stanton in August 1867, was an object lesson for Grant in the perils of Reconstruction politics. After Johnson suspended Stanton from his duties, Grant accepted the job as interim secretary of war, but Stanton refused to relinquish the office. After months of being caught between the President and Congress, Grant resigned in February when the Senate reinstated Stanton as secretary. Johnson believed Grant had betrayed him.

Grant certainly did not sympathize with Johnson, but he loathed the harsh tone of the President's critics and supporters alike. Nevertheless, on May 21 the Republican convention in Chicago nominated Ulysses Grant for President by acclamation. Grant learned the news while at his desk in Washington, where he continued to serve as the army's top general. He accepted the nomination, and, in keeping with tradition, did little else. Candidates still did not personally campaign for the presidency. Their

---

1. Brooks D. Simpson, *Let Us Have Peace: Ulysses S. Grant and the Politics of War and Reconstruction* (Chapel Hill, NC: University of North Carolina Press, 1991), 206.

2. Ibid., 214.

surrogates did the speaking for them, while they did their best to appear above the battle. But his letter of acceptance contained a simple phrase that spoke to the nation's mood. "Let us have peace," he wrote. It became the Republican Party's mantra in 1868.

Opposing Grant, and representing a badly divided Democratic Party, was another reluctant candidate, Horatio Seymour, a former governor of New York. When the Democrats met in July in Seymour's home state, the ex-governor urged them to nominate the chief justice of the Supreme Court, Salmon P. Chase, who had presided over the Senate's impeachment trial of Andrew Johnson. But Chase was a firm advocate for black voting rights, which won him the enmity of Southern Democrats. With the convention approaching a stalemate, Democrats turned to Seymour, who accepted the nomination.

The candidates' surrogates waged a particularly nasty campaign, befitting a nation still torn over issues like black suffrage and Reconstruction. Democrats spread rumors about Grant's fondness for the bottle, while Republicans reminded the nation that under Seymour's watch, New York City had fought a mini–civil war of its own in 1863, when rioters burned the city to protest the draft.

Grant won easily, compiling 214 electoral votes to Seymour's 80. But his share of the popular vote was far less impressive and perhaps closer to the temper of the still-divided nation. Grant captured just under 53 percent of the vote, and Democrats posted significant gains on Capitol Hill.

Inauguration Day 1869 was not a bipartisan celebration of shared values. Andrew Johnson chose not to attend the ceremony after Grant declined to follow tradition by sharing a ride with the outgoing President from the White House to the ceremony on Capitol Hill.

In his speech, Grant promised that "all laws will be faithfully executed," which his listeners knew was no mere banality but a reference to Johnson's refusal to abide by the Tenure of Office Act. It was just after making this vow that Grant unveiled his formula for changing bad laws: enforce them, stringently, and they will be changed.

At the end of his speech, after declaring that "the original occupants of this land" deserved "proper treatment," Grant announced his support for the third and last of the Reconstruction amendments. The Fifteenth Amendment outlawed racial discrimination in the voting booth, and it passed during Grant's first year in office.

It would take nearly a century before it was vigorously enforced.

YOUR SUFFRAGES HAVING ELECTED ME to the office of President of the United States, I have, in conformity to the Constitution of our country, taken the oath of office prescribed therein. I have taken this oath without mental reservation and with the determination to do to the best of my ability all that is required of me. The responsibilities of the position I feel, but accept them without fear. The office has come to me unsought; I commence its duties untrammeled. I bring to it a conscious desire and determination to fill it to the best of my ability to the satisfaction of the people.

On all leading questions agitating the public mind I will always express my views to Congress and urge them according to my judgment, and when I think it advisable will exercise the constitutional privilege of interposing a veto to defeat measures which I oppose; but all laws will be faithfully executed, whether they meet my approval or not.

I shall on all subjects have a policy to recommend, but none to enforce against the will of the people. Laws are to govern all alike—those opposed as well as those who favor them. I know no method to secure the repeal of bad or obnoxious laws so effective as their stringent execution.

The country having just emerged from a great rebellion, many questions will come before it for settlement in the next four years which preceding administrations have never had to deal with. In meeting these it is desirable that they should be approached calmly, without prejudice, hate, or sectional pride, remembering that the greatest good to the greatest number is the object to be attained.

This requires security of person, property, and free religious and political opinion in every part of our common country, without regard to local prejudice. All laws to secure these ends will receive my best efforts for their enforcement.

A great debt has been contracted in securing to us and our posterity the Union. The payment of this, principal and interest, as well as the return to a specie basis as soon as it can be accomplished without material detriment to the debtor class or to the country at large, must be provided for. To protect the national honor, every dollar of government indebtedness should be paid in gold, unless otherwise expressly stipulated in the contract. Let it be understood that no repudiator of one farthing of our public debt will be trusted in public place, and it will go far toward strengthening a credit which ought to be the best in the world, and will ultimately enable us to replace the debt with bonds bearing less interest than we now pay. To this should be added a faithful collection of the revenue, a strict accountability to the Treasury for every dollar collected, and the greatest practicable retrenchment in expenditure in every department of government.

When we compare the paying capacity of the country now, with the ten states in poverty from the effects of war, but soon to emerge, I trust, into greater prosperity than ever before, with its paying capacity twenty-five years ago, and calculate what it probably will be twenty-five years hence, who can doubt the feasibility of paying every dollar then with more ease than we now pay for useless luxuries? Why, it looks as though Providence had bestowed upon us a strong box in the precious metals locked up in the sterile mountains of the far West, and which we are now forging the key to unlock, to meet the very contingency that is now upon us.

Ultimately it may be necessary to insure the facilities to reach these riches and it may be necessary also that the general government should give its aid to secure this access; but that should only be when a dollar of obligation to pay secures precisely the same sort of dollar to use now, and not before. Whilst the question of specie payments is in abeyance the prudent businessman is careful about contracting debts payable in the distant future. The nation should follow the same rule. A prostrate commerce is to be rebuilt and all industries encouraged.

The young men of the country—those who from their age must be its rulers twenty-five years hence—have a peculiar interest in maintaining the national honor. A moment's reflection as to what will be our commanding influence among the nations of the earth in their day, if they are only true to themselves, should inspire them with national pride. All divisions—geographical, political, and religious—can join in this common sentiment. How the public debt is to be paid or specie payments resumed is not so important as that a plan should be adopted and acquiesced in. A united determination to do is worth more than divided counsels upon the method of doing. Legislation upon this subject may not be necessary now, or even advisable, but it will be when the civil law is more fully restored in all parts of the country and trade resumes its wonted channels.

It will be my endeavor to execute all laws in good faith, to collect all revenues assessed, and to have them properly accounted for and economically disbursed. I will to the best of my ability appoint to office those only who will carry out this design.

In regard to foreign policy, I would deal with nations as equitable law requires individuals to deal with each other, and I would protect the law-abiding citizen, whether of native or foreign birth, wherever his rights are jeopardized or the flag of our country floats. I would respect the rights of all nations, demanding equal respect for our own. If others depart from this rule in their dealings with us, we may be compelled to follow their precedent.

The proper treatment of the original occupants of this land—the Indians—is one deserving of careful study. I will favor any course toward them which tends to their civilization and ultimate citizenship.

The question of suffrage is one which is likely to agitate the public so long as a portion of the citizens of the nation are excluded from its privileges in any state. It seems to me very desirable that this question should be settled now, and I entertain the hope and express the desire that it may be by the ratification of the fifteenth article of amendment to the Constitution.

In conclusion I ask patient forbearance one toward another throughout the land, and a determined effort on the part of every citizen to do his share toward cementing a happy union; and I ask the prayers of the nation to Almighty God in behalf of this consummation.

Ulysses S. Grant

Tuesday, March 4, 1873

*U*lysses S. Grant was no more poetic in his second inaugural than he was in his first. But as he addressed the nation and his colleagues in 1873, he once again demonstrated that there is art in unadorned language and direct address.

In denouncing violent attempts to deny freed slaves their civil rights, arguing that expansion and democratic rule were compatible, framing better treatment of Indians as a moral issue, and firing back at his critics, Grant showed flashes of the spare but effective prose style that would earn his memoirs a place in American literature. Historians rate Grant among the worst Presidents, thanks largely to corruption, which plagued his second term, and his failure to act on some of the sentiments voiced in his second inaugural. But the speech itself deserves a better fate for its clarity of prose and of its purpose.

The nation was four more years removed from the Civil War when Grant took the oath of office for a second time. The states of the former Confederacy were now returned to full membership in the Union, meaning they all had recognized state governments and were once again sending representatives and senators to Congress. Railroads and the telegraph shortened the distance between East and West, North and South. Cities were becoming crowded with wage earners, native born and immigrant alike, who toiled in mass production at factories in the North and Midwest.

And yet, even as the nation returned to a peacetime economy and resumed its relentless westward expansion, the conflicts and divisions which plunged the nation into civil war remained unresolved. The Reconstruction amendments had put an end to slavery, promised equal treatment for all people under the law, and guaranteed the franchise to adult males regardless of skin color. But those rights existed only in the abstract throughout the old Confederacy. Groups of armed white supremacists terrorized free blacks who dared to take advantage of the rights they were promised. The formal war between the states was over, but a low-intensity insurgency in the South mocked any claim that the nation was at peace. White guerillas could not bring back slavery, but, with the aid of white politicians, they sought to keep the South's black population politically and economically enslaved.

Ulysses Grant saw the white supremacist campaign in the South as an effort to undo the victories his armies had won during the war. His continued support for black civil rights and for a federal presence in the South to uphold the Reconstruction amendments was spelled out in his second inaugural. In a characteristically blunt passage early in the speech, he noted that the South's former slaves were "not possessed of the civil rights which citizenship should carry with it.

"This," he continued, "is wrong, and should be corrected." The phrase is striking not because of its elaborate construction but because of its straightforward simplicity. It is perhaps too passive—"must be corrected" would have been stronger—and lacked both stick and carrot. But in calling wrong by its name, Grant exhibited no small amount of moral courage. After all, a white supremacist had murdered Lincoln. Equally violent men were killing black people by the scores in the South. It surely was not hard to imagine a similar strike against the man held responsible for the Confederacy's military defeat.

That is not to say that Grant's views on race reached beyond his time and place. After pledging his support for civil rights, Grant drew the line at enforcing other forms of equality. "Social quality," he said, "is not a subject to be legislated upon, nor shall I ask that anything be done to advance the social status of the colored man, except to give him a fair chance to develop what there is good in him, give him access to the schools, and when he travels let him feel assured that his conduct will regulate the treatment and fare he will receive." Even that minimalist approach to equality was doomed to shameful failure.

Grant's support for a continued federal presence in the South split his party during the 1872 campaign. A group of dissidents, calling themselves Liberal Republicans, believed that Reconstruction was over and so federal troops ought to be withdrawn from the South. They nominated Horace Greeley, noted editor of the *New York Tribune* and himself once a Radical Republican. Democrats, often the target of Greeley's editorial abuse, put aside pride and endorsed him, too. Democrats shared Greeley's newfound belief that the South should be left to conduct its affairs as it—or, more to the point, as its unreconstructed white population—saw fit.

The main contenders for President, then, were political amateurs, although Grant's four years as President obviously gave him an edge over his journalist-rival. But they were not the only political newcomers in the race. A new entity called the Equal Rights Party nominated Victoria Woodhull, a crusader for women's suffrage, for President, and Frederick Douglass, a onetime slave who won fame as an abolitionist, public speaker, and author, for vice president. Their campaign may have been symbolic only—Woodhull was actually too young, at thirty-four, to become President—but their very presence in the race foreshadowed broader social movements to come. On Election Day, women's suffrage leaders were arrested for attempting to cast a vote.

Grant won an easy victory, with more then 55 percent of the popular vote and 286 electoral votes. His main opponent, Greeley, died in late November, so the editor's electoral votes were cast for other candidates when the Electoral College met in February.

The ease with which Grant won reelection masked legitimate and, in the end, justifiable concerns about political corruption, which Greeley and the Liberal Republicans blamed on the President. What's more, the nation's booming postwar economy was about to crash, and when it did, in 1873, thousands of businesses closed and the unemployment rate rose to 14 percent.

So scandals and hard times were near at hand but not yet visible as Grant began his second term. The sentiments in his speech—of the Indian's plight, he said, "The wrong inflicted upon him should be taken into account and the balance placed to his credit"—would soon become casualties of postwar politics. Likewise, Grant's very reputation. The final words of his inaugural rank among the bitterest in American history. Noting his long service to his country as a soldier and a President, he said, "I have been the subject of abuse and slander scarcely ever equaled in political history."

All these years later, the words still apply.

FELLOW CITIZENS:

Under Providence I have been called a second time to act as executive over this great nation. It has been my endeavor in the past to maintain all the laws, and, so far as lay in my power, to act for the best interests of the whole people. My best efforts will be given in the same direction in the future, aided, I trust, by my four years' experience in the office.

When my first term of the office of chief executive began, the country had not recovered from the effects of a great internal revolution, and three of the former states of the Union had not been restored to their federal relations.

It seemed to me wise that no new questions should be raised so long as that condition of affairs existed. Therefore the past four years, so far as I could control events, have been consumed in the effort to restore harmony, public credit, commerce, and all the arts of peace and progress. It is my firm conviction that the civilized world is tending toward republicanism, or government by the people through their chosen representatives, and that our own great Republic is destined to be the guiding star to all others.

Under our Republic we support an army less than that of any European power of any standing and a navy less than that of either of at least five of them. There could be no extension of territory on the continent which would call for an increase of this force, but rather might such extension enable us to diminish it.

The theory of government changes with general progress. Now that the telegraph is made available for communicating thought, together with rapid transit by steam, all parts of a continent are made contiguous for all purposes of government, and communication between the extreme limits of the country made easier than it was throughout the old thirteen states at the beginning of our national existence.

The effects of the late civil strife have been to free the slave and make him a citizen. Yet he is not possessed of the civil rights which citizenship should carry with it. This is wrong, and should be corrected. To this correction I stand committed, so far as executive influence can avail.

Social equality is not a subject to be legislated upon, nor shall I ask that anything be done to advance the social status of the colored man, except to give him a fair chance to develop what there is good in him, give him access to the schools, and when he travels let him feel assured that his conduct will regulate the treatment and fare he will receive.

The states lately at war with the general government are now happily rehabilitated, and no executive control is exercised in any one of them that would not be exercised in any other state under like circumstances.

In the first year of the past administration the proposition came up for the admission of Santo Domingo as a territory of the Union. It was not a question of my seeking, but was a proposition from the people of Santo Domingo, and which I entertained. I believe now, as I did then, that it was for the best interest of this country, for the people of Santo Domingo, and all concerned that the proposition should be received favorably. It was, however, rejected constitutionally, and therefore the subject was never brought up again by me.

In future, while I hold my present office, the subject of acquisition of territory must have the support of the people before I will recommend any proposition looking to such acquisition. I say here, however, that I do not share in the apprehension held by many as to the danger of governments becoming weakened and destroyed by reason of their extension of territory. Commerce, education, and rapid transit of thought and matter by telegraph and steam have changed all this. Rather do I believe that our Great Maker is preparing the world, in His own good time, to become one nation, speaking one language, and when armies and navies will be no longer required.

My efforts in the future will be directed to the restoration of good feeling between the different sections of our common country; to the restoration of our currency to a fixed value as compared with the world's standard of values—gold—and, if possible, to a par with it; to the construction of cheap routes of transit throughout the land, to the end that the products of all may find a market and leave a living remuneration to the producer; to the maintenance of friendly relations with all our neighbors and with distant nations; to the reestablishment of our commerce and share in the carrying trade upon the ocean; to the encouragement of such manufacturing industries as can be economically pursued in this country, to the end that the exports of home products and industries may pay for our imports—the only sure method of returning to and permanently maintaining a specie basis; to the elevation of labor; and, by a humane course, to bring the aborigines of the country under the benign influences of education and civilization. It is either this or war of extermination: wars of extermination, engaged in by people pursuing commerce and all industrial pursuits, are expensive even against the weakest people, and are demoralizing and wicked. Our superiority of strength and advantages of civilization should make us lenient toward the Indian. The wrong inflicted upon him should be taken into account and the balance placed to his credit. The moral view of the question should be considered and the question asked, Cannot the Indian be made a useful and productive member of society by proper teaching and treatment? If the effort is made in good faith, we will stand better before the civilized nations of the earth and in our own consciences for having made it.

All these things are not to be accomplished by one individual, but they will receive my support and such recommendations to Congress as will in my judgment best serve to carry them into effect. I beg your support and encouragement.

It has been, and is, my earnest desire to correct abuses that have grown up in the civil service of the country. To secure this reformation rules regulating methods of appointment and promotions were established and have been tried. My efforts for such reformation shall be continued to the best of my judgment. The spirit of the rules adopted will be maintained.

I acknowledge before this assemblage, representing, as it does, every section of our country, the obligation I am under to my countrymen for the great honor they have conferred on me by returning me to the highest office within their gift, and the further obligation resting on me to render to them the best services within my power. This I promise, looking forward with the greatest anxiety to the day when I shall be released from responsibilities that at times are almost overwhelming, and from which I have scarcely had a respite since the eventful firing upon Fort Sumter, in April 1861, to the present day. My services were then tendered and accepted under the first call for troops growing out of that event.

I did not ask for place or position, and was entirely without influence or the acquaintance of persons of influence, but was resolved to perform my part in a struggle threatening the very existence of the nation. I performed a conscientious duty, without asking promotion or command, and without a revengeful feeling toward any section or individual.

Notwithstanding this, throughout the war, and from my candidacy for my present office in 1868 to the close of the last presidential campaign, I have been the subject of abuse and slander scarcely ever equaled in political history, which today I feel that I can afford to disregard in view of your verdict, which I gratefully accept as my vindication.

Rutherford B. Hayes

· INAUGURAL ADDRESS ·

Monday, March 5, 1877

As the United States marked the centennial of its founding, Democrats had good reason to believe that 1876 would bring them victory for the first time in a generation. Republicans, winners of four straight presidential elections, had to answer for the scandals of President Grant's second term, the harsh recession that followed the panic of 1873, and continued tensions over Reconstruction.

The Democrats nominated the governor of New York, Samuel Tilden, who gained a national reputation as a reformer when he helped to expose the depredations of "Boss" William M. Tweed (a fellow Democrat) in New York City. Tilden's record in New York seemed perfectly suited for a party that sought to capitalize on national disgust over political boodle, patronage, and waste.

Demoralized Republicans turned to another governor, Rutherford B. Hayes of Ohio. Hayes had served in the Union army as a major and was wounded on four occasions. After serving in the House and as Ohio's governor, he retired from politics in the early 1870s, only to make a comeback as governor in 1875 and to be nominated for President just a few months later.

Hayes turned out to be a shrewd political strategist in an election year when the Republican Party needed all the help it could get. Republicans, Hayes believed, could exploit the lingering postwar resentments by reminding voters in the North and West which party had won the war, and which party had been defeatist if not treasonous.

The tactic became known in American political history as "waving the bloody shirt"; that is, reminding voters of Republican steadfastness during the war and Democratic disloyalty. Reform-minded Republicans had hoped to make inroads among white Democrats horrified by the brutal oppression of the South's newly freed blacks. But waving the bloody shirt would do nothing to advance the Republican Party's cause in the South. In the North, however, even in Tilden's home state of New York, the tactic resonated with voters inclined to associate national Democrats with being soft on rebellion. It also helped unite a party badly divided over Reconstruction, corruption, and personalities.

Just as shrewdly, Hayes promised to serve just a single term. The bulk of Grant's problems took place in his second term, so there was no mistaking Hayes's message. He clearly wished to separate himself from the outgoing President.

As votes were being counted on the night of Election Day, and transmitted by telegraph to the candidates' headquarters, it seemed as though Hayes's strategy almost worked—emphasis on "almost." The election was extremely close, but Tilden had the edge in the popular vote, winning 4.3 million votes to Hayes's 4 million. In the Electoral College, however, Tilden had 184 votes—a single vote short of a majority—while Hayes had 165. Twenty votes were disputed, in Oregon, South Carolina, Louisiana, and Florida. In a foreshadowing of the controversial election of 2000 between George W. Bush and Al Gore, accusations of voter fraud, cheating, and miscounting were hurled against election officials in the three southern states. Congress impaneled a special commission to study the results and decide how to award the disputed votes. The commission consisted of fifteen members, five from the House, five from the Senate, and five from the Supreme Court. The party in control of each house selected three members, the minority two, which led to an even split between Republicans and Democrats from Capitol Hill (Democrats ran the House, while Republicans ran the Senate). Of the five members of the Court, two were Democrats and two were Republican. The four justices were supposed to choose the fifth member from the Court. After their first choice, David Davis, was elected to the Senate, the justices were left with only Republicans from which to choose. This gave Republicans an unintended advantage.

In a series of party-line votes, Hayes was given all of the disputed electoral votes. The results were confirmed on March 2, just days before the new President was due to begin his term. During the backroom negotiations leading to Hayes's election, southern Democrats extracted Republican promises to withdraw federal troops from the South, share patronage with Democrats, and return the South to local rule, which meant a return to white supremacy.

Hayes was sworn in privately on March 3 in the White House because March 4, the traditional day for inaugural ceremonies, was a Sunday. The public ceremony took place on Monday, March 5, and it was something of an impromptu affair because of the uncertainty over who would be the new President. Thirty thousand people turned out to hear Hayes straddle a conciliatory middle ground on Reconstruction, emphasizing the return to local government in the South but asserting that local government ought to protect both races equally.

"The evils which afflict the southern states can only be removed or remedied by the united and harmonious efforts of both races," Hayes said, stating the obvious but elusive solution to the nation's still-unresolved issue of race. He then asked for "the cordial cooperation of all who cherish an interest in the welfare of the country, trusting that party ties and the prejudice of race will be freely surrendered in behalf of the great purpose to be accomplished."

Of course, that request was doomed from the moment Hayes uttered it. The federal retreat from the South and the formal end of Reconstruction empowered white supremacists, who terrorized and murdered blacks without fear of state punishment. Party ties and the prejudice of race would be defended, not surrendered, in the post-Reconstruction South.

With more sincerity, Hayes also pledged his administration to political reform, a less-controversial stance, although a welcome one after the high-level scandals of the Grant years. Hayes tried to strike a bipartisan note after one of the nation's closest elections, observing that "both the great political parties of the country . . . gave a prominent place to the subject of reform of our civil service." In a dramatic gesture designed to show his commitment to changing the nation's political culture, Hayes proposed "an amendment to the Constitution prescribing a term of six years for the presidential office and forbidding a reelection."

In a speech that had Grant's directness if not his brevity, Hayes managed one memorable line in a passage about partisanship and the national interest. The President of the United States, he noted, ought to remember that "he serves his party best who serves the country best."

Bibliographic Note
Ari Hoogenboom, *The Presidency of Rutherford B. Hayes*
(Lawrence, KS: University Press of Kansas, 1988).

We have assembled to repeat the public ceremonial, begun by Washington, observed by all my predecessors, and now a time-honored custom, which marks the commencement of a new term of the presidential office. Called to the duties of this great trust, I proceed, in compliance with usage, to announce some of the leading principles, on the subjects that now chiefly engage the public attention, by which it is my desire to be guided in the discharge of those duties. I shall not undertake to lay down irrevocably principles or measures of administration, but rather to speak of the motives which should animate us, and to suggest certain important ends to be attained in accordance with our institutions and essential to the welfare of our country.

At the outset of the discussions which preceded the recent presidential election it seemed to me fitting that I should fully make known my sentiments in regard to several of the important questions which then appeared to demand the consideration of the country. Following the example, and in part adopting the language, of one of my predecessors, I wish now, when every motive for misrepresentation has passed away, to repeat what was said before the election, trusting that my countrymen will candidly weigh and understand it, and that they will feel assured that the sentiments declared in accepting the nomination for the presidency will be the standard of my conduct in the path before me, charged, as I now am, with the grave and difficult task of carrying them out in the practical administration of the government so far as depends, under the Constitution and laws, on the chief executive of the nation.

The permanent pacification of the country upon such principles and by such measures as will secure the complete protection of all its citizens in the free enjoyment of all their constitutional rights is now the one subject in our public affairs which all thoughtful and patriotic citizens regard as of supreme importance.

Many of the calamitous efforts of the tremendous revolution which has passed over the southern states still remain. The immeasurable benefits which will surely follow, sooner or later, the hearty and generous acceptance of the legitimate results of that revolution have not yet been realized. Difficult and embarrassing questions meet us at the threshold of this subject. The people of those states are still impoverished, and the inestimable blessing of wise, honest, and peaceful local self-government is not fully enjoyed. Whatever difference of opinion may exist as to the cause of this condition of things, the fact is clear that in the progress of events the time has come when such government is the imperative necessity required by all the varied interests, public and private, of those states. But it must not be forgotten that only a local government which recognizes and maintains inviolate the rights of all is a true self-government.

With respect to the two distinct races whose peculiar relations to each other have brought upon us the deplorable complications and perplexities which exist in those states, it must be a government which guards the interests of both races carefully and equally. It must be a government which submits loyally and heartily to the Constitution and the laws—the laws of the nation and the laws of the states themselves—accepting and obeying faithfully the whole Constitution as it is.

Resting upon this sure and substantial foundation, the superstructure of beneficent local governments can be built up, and not otherwise. In furtherance of such obedience to the letter and the spirit of the Constitution, and in behalf of all that its attainment implies, all so-called party interests lose their apparent importance, and party lines may well be permitted to fade into insignificance. The question we have to consider for the immediate welfare of those states of the Union is the question of government or no government; of social order and all the peaceful industries and the happiness that belongs to it, or a return to barbarism. It is a question in which every citizen of the nation is deeply interested, and with respect to which we ought not to be, in a partisan sense, either Republicans or Democrats, but fellow citizens and fellow men, to whom the interests of a common country and a common humanity are dear.

The sweeping revolution of the entire labor system of a large portion of our country and the advance of four million people from a condition of servitude to that of citizenship, upon an equal footing with their former masters, could not occur without presenting problems of the gravest moment, to be dealt with by the emancipated race, by their former masters, and by the general government, the author of the act of emancipation. That it was a wise, just, and providential act, fraught with good for all concerned, is now generally conceded throughout the country. That a moral obligation rests upon the national government to employ its constitutional power and influence to establish the rights of the people it has emancipated, and to protect them in the enjoyment of those rights when they are infringed or assailed, is also generally admitted.

The evils which afflict the southern states can only be removed or remedied by the united and harmonious efforts of both races, actuated by motives of mutual sympathy and regard; and while in duty bound and fully determined to protect the rights of all by every constitutional means at the disposal of my administration, I am sincerely anxious to use every legitimate influence in favor of honest and efficient local self-government as the true resource of those states for the promotion of the contentment and prosperity of their citizens. In the effort I shall make to accomplish this purpose I ask the cordial cooperation of all who cherish an interest in the welfare of the country, trusting that party ties and the prejudice of race will be freely surrendered in behalf of the great purpose to be accomplished. In the important work of restoring the South it is not the political situation alone that merits attention. The material development of that section of the country has been arrested by the social and political revolution through which it has passed, and now needs and deserves the considerate care of the national government within the just limits prescribed by the Constitution and wise public economy.

But at the basis of all prosperity, for that as well as for every other part of the country, lies the improvement of the intellectual and moral condition of the people. Universal suffrage should rest upon universal education. To this end, liberal and permanent provision should be made for the support of free schools by the state governments, and, if need be, supplemented by legitimate aid from national authority.

Let me assure my countrymen of the southern states that it is my earnest desire to regard and promote their truest interest—the interests of the white and of the colored people both and equally—and to put forth my best efforts in behalf of a civil policy which will forever wipe out in our political affairs the color line and the distinction between North and South, to the end that we may have not merely a united North or a united South, but a united country.

I ask the attention of the public to the paramount necessity of reform in our civil service—a reform not merely as to certain abuses and practices of so-called official patronage which have come to have the sanction of usage in the several departments of our government, but a change in the system of appointment itself; a reform that shall be thorough, radical, and complete; a return to the principles and practices of the founders of the government. They neither expected nor desired from public officers any partisan service. They meant that public officers should owe their whole service to the government and to the people. They meant that the officer should be secure in his tenure as long as his personal character remained untarnished and the performance of his duties satisfactory. They held that appointments to office were not to be made nor expected merely as rewards for partisan services, nor merely on the nomination of members of Congress, as being entitled in any respect to the control of such appointments.

The fact that both the great political parties of the country, in declaring their principles prior to the election, gave a prominent place to the subject of reform of our civil service, recognizing and strongly urging its necessity, in terms almost identical in their specific import with those I have here employed, must be accepted as a conclusive argument in behalf of these measures. It must be regarded as the expression of the united voice and will of the whole country upon this subject, and both political parties are virtually pledged to give it their unreserved support.

The President of the United States of necessity owes his election to office to the suffrage and zealous labors of a political party, the members of which cherish with ardor and regard as of essential importance the principles of their party organization; but he should strive to be always mindful of the fact that he serves his party best who serves the country best.

In furtherance of the reform we seek, and in other important respects a change of great importance, I recommend an amendment to the Constitution prescribing a term of six years for the presidential office and forbidding a reelection.

With respect to the financial condition of the country, I shall not attempt an extended history of the embarrassment and prostration which we have suffered during the past three years. The depression in all our varied commercial and manufacturing interests throughout the country, which began in September 1873 still continues. It is very gratifying, however, to be able to say that there are indications all around us of a coming change to prosperous times.

Upon the currency question, intimately connected, as it is, with this topic, I may be permitted to repeat here the statement made in my letter of acceptance, that in my judgment the feeling of uncertainty inseparable from an irredeemable paper currency, with its fluctuation of values, is one of the greatest obstacles to a return to prosperous times. The only safe paper currency is one which rests upon a coin basis and is at all times and promptly convertible into coin.

I adhere to the views heretofore expressed by me in favor of congressional legislation in behalf of an early resumption of specie payments, and I am satisfied not only that this is wise, but that the interests, as well as the public sentiment, of the country imperatively demand it.

Passing from these remarks upon the condition of our own country to consider our relations with other lands, we are reminded by the international complications abroad, threatening the peace of Europe, that our traditional rule of noninterference in the affairs of foreign nations has proved of great value in past times and ought to be strictly observed.

The policy inaugurated by my honored predecessor, President Grant, of submitting to arbitration grave questions in dispute between ourselves and foreign powers points to a new, and incomparably the best, instrumentality for the preservation of peace, and will, as I believe, become a beneficent example of the course to be pursued in similar emergencies by other nations.

If, unhappily, questions of difference should at any time during the period of my administration arise between the United States and any foreign government, it will certainly be my disposition and my hope to aid in their settlement in the same peaceful and honorable way, thus securing to our country the great blessings of peace and mutual good offices with all the nations of the world.

Fellow citizens, we have reached the close of a political contest marked by the excitement which usually attends the contests between great political parties whose members espouse and advocate with earnest faith their respective creeds. The circumstances were, perhaps, in no respect extraordinary save in the closeness and the consequent uncertainty of the result.

For the first time in the history of the country it has been deemed best, in view of the peculiar circumstances of the case, that the objections and questions in dispute with reference to the counting of the electoral votes should be referred to the decision of a tribunal appointed for this purpose.

That tribunal—established by law for this sole purpose; its members, all of them, men of long-established reputation for integrity and intelligence, and, with the exception of those who are also members of the supreme judiciary, chosen equally from both political parties; its deliberations enlightened by the research and the arguments of able counsel—was entitled to the fullest confidence of the American people. Its decisions have been patiently waited for, and accepted as legally conclusive by the general judgment of the public. For the present, opinion will widely vary as to the wisdom of the several conclusions announced by that tribunal. This is to be anticipated in every instance where matters of dispute are made the subject of arbitration under the forms of law. Human judgment is never unerring, and is rarely regarded as otherwise than wrong by the unsuccessful party in the contest.

The fact that two great political parties have in this way settled a dispute in regard to which good men differ as to the facts and the law no less than as to the proper course to be pursued in solving the question in controversy is an occasion for general rejoicing.

Upon one point there is entire unanimity in public sentiment—that conflicting claims to the presidency must be amicably and peaceably adjusted, and that when so adjusted the general acquiescence of the nation ought surely to follow.

It has been reserved for a government of the people, where the right of suffrage is universal, to give to the world the first example in history of a great nation, in the midst of the struggle of opposing parties for power, hushing its party tumults to yield the issue of the contest to adjustment according to the forms of law.

Looking for the guidance of that Divine Hand by which the destinies of nations and individuals are shaped, I call upon you, senators, representatives, judges, fellow citizens, here and everywhere, to unite with me in an earnest effort to secure to our country the blessings, not only of material prosperity, but of justice, peace, and union—a union depending not upon the constraint of force, but upon the loving devotion of a free people; "and that all things may be so ordered and settled upon the best and surest foundations that peace and happiness, truth and justice, religion and piety, may be established among us for all generations."

James A. Garfield

Friday, March 4, 1881

With Reconstruction consigned to the dustbin, the nation focused its attention on other issues in 1880. Outside the South, the Civil War increasingly seemed to belong to another age as industry, changing demographics, and urbanization signaled a new chapter in U.S. history. The cities of the North and Midwest continued to receive tens of thousands of new immigrants, many of whom did not speak English. On the sprawling farms of the nation's heartland, farmers struggled to pay rent, mortgages, and disproportionately high taxes even as they subjected themselves to the fickle tastes of the market. Factory work was a disheartening and sometimes dehumanizing alternative to life behind a plow. Wages were low, jobs monotonous and dangerous, and a worker could expect to spend up to fourteen hours a day, seven days a week, on the job.

For the most part, the nation's political debates seemed disconnected with the realities of life on the isolated farms of rural America or the grimy cities of the Northeast and Midwest. The Republican Party was hampered by internal feuds that had little to do with the great issues of the day, while the Democratic Party was divided by regional interests—those of the urban North and the rural South—that seemed impossible to reconcile.

James A. Garfield, a native of Ohio, rose from poverty to become a lawyer and a member of his home state's senate. He was elected to the House of Representatives during the Civil War, becoming an important lawmaker and a top political operative for the Republican Party. His nomination for the presidency in 1880 took him by surprise—he supported a fellow Ohioan, John Sherman, brother of General William T. Sherman and the choice of outgoing President Rutherford B. Hayes.

In Garfield, the Republicans had themselves an experienced lawmaker and savvy politician who also happened to be a war hero from a politically important state. Republicans hoped his stature and experience would help unite a party divided into so-called Stalwarts, party regulars who had hoped to bring Grant back for another campaign in 1880, and Half-breeds, who opposed party machines and patronage. They were led by congressman James Blaine of Maine. Garfield was considered a Half-breed, and while the Stalwarts grudgingly went along with the nomination, they were not particularly enthusiastic. To placate them, the Republican convention nominated Chester Arthur, a Stalwart from New York, as the party's vice-presidential nominee.

To counter Garfield, the Democrats nominated another war hero, General Winfield Scott Hancock, who took part in the Battle of Gettysburg. Yet another Union general joined the race when the new Greenback Party nominated James Weaver, who fought at Shiloh, among other places. Weaver's presence in the race suggested that at least some Americans were more concerned with economic dislocation than they were with political intrigue. The Greenbackers supported the circulation of paper currency not backed by gold and advocated inflation as a means of lightening the load on debtors, particularly farmers.

In a three-way race, Garfield barely squeaked by in the popular vote. Only about seven thousand votes out of nine million cast separated Garfield and Hancock, although Garfield's total in the Electoral College was 214 compared with Hancock's 155. Hancock did manage to win all eleven states of the former Confederacy, the first time the Solid South emerged as a Democratic bloc. Weaver, the Greenback candidate, did not win any electoral votes, although he deprived Garfield of a majority.

Well-read and studious, Garfield assigned himself the task of reading all of his predecessors' inaugural addresses, hoping to draw inspiration for his own. Instead, after finishing his research, he wondered why he should even bother with a speech of his own. With a month to go before his swearing-in, he said he was "strangely disinclined to work on the inaugural." According to biographers Margaret Leech and Harry J. Brown, he thought about skipping the speech entirely.[1] The Constitution, after all, didn't require one. In the end, he decided to follow precedent.

Among the celebrants who gathered at the Capitol to hear Garfield's inaugural was the new President's eighty-year-old mother, Eliza Garfield.

---

1. Margaret Leech and Harry J. Brown, *The Garfield Orbit* (New York: Harper and Row, 1978), 222.

Widowed at a young age, with small children to support, she became the first mother to witness her son's inauguration as President.

Garfield's speech was expansive and ambitious. He insisted that there could be "no middle ground for the Negro race between slavery and equal citizenship" and that there could be "no permanent disfranchised peasantry in the United States." He asserted that the "emancipated race has already made remarkable progress," although he conceded that the "free enjoyment of equal suffrage is still in question." Denying blacks the right to vote, he said, was "more than an evil." It was, he said, "a crime" that "will destroy the government itself."

His most pointed remarks, however, offer a glimpse at a controversy in the West, where a new religious sect had established itself. Garfield conceded that the Constitution ensured religious freedom. But, he said, the "Mormon Church not only offends the moral sense of manhood by sanctioning polygamy, but prevents the administration of justice through ordinary instrumentalities of law." He proposed that Congress pass legislation banning "criminal practices . . . which destroy the family relations and endanger social order."

After Garfield finished, he and his wife and mother made the rounds of social engagements before the new President plunged ahead with the still-pressing issues of federal patronage.

On July 2, 1881, President Garfield was walking to a railroad station with several aides when he was shot by Charles Guiteau, a thirty-nine-year-old man who had hoped to win a job with the new administration. A bullet grazed the President's arm, another was lodged in his torso. He lay in bed for weeks while his medical team tried, in vain, to find the bullet in his torso. On August 11, he wrote to his mother: "Don't be disturbed by conflicting reports about my condition. It is true I am still weak, and on my back, but I am gaining every day."[2]

He died on September 19, just a few days shy of turning fifty years old. He was the second President to be murdered in fewer than twenty years. Recognizing, at last, the importance of reforming the nation's civil service and patronage systems, Congress passed the Pendleton Act in 1883, which established a merit system for certain jobs and banned political contributions from some federal officeholders.

---

2. Ibid., 317.

FELLOW CITIZENS:

We stand today upon an eminence which overlooks a hundred years of national life—a century crowded with perils, but crowned with the triumphs of liberty and law. Before continuing the onward march let us pause on this height for a moment to strengthen our faith and renew our hope by a glance at the pathway along which our people have traveled.

It is now three days more than a hundred years since the adoption of the first written constitution of the United States—the Articles of Confederation and Perpetual Union. The new Republic was then beset with danger on every hand. It had not conquered a place in the family of nations. The decisive battle of the war for independence, whose centennial anniversary will soon be gratefully celebrated at Yorktown, had not yet been fought. The colonists were struggling not only against the armies of a great nation, but against the settled opinions of mankind; for the world did not then believe that the supreme authority of government could be safely intrusted to the guardianship of the people themselves.

We cannot overestimate the fervent love of liberty, the intelligent courage, and the sum of common sense with which our fathers made the great experiment of self-government. When they found, after a short trial, that the confederacy of states was too weak to meet the necessities of a vigorous and expanding republic, they boldly set it aside, and in its stead established a national Union, founded directly upon the will of the people, endowed with full power of self-preservation and ample authority for the accomplishment of its great object.

Under this Constitution the boundaries of freedom have been enlarged, the foundations of order and peace have been strengthened, and the growth of our people in all the better elements of national life has indicated the wisdom of the founders and given new hope to their descendants. Under this Constitution our people long ago made themselves safe against danger from without and secured for their mariners and flag equality of rights on all the seas. Under this Constitution twenty-five states have been added to the Union, with constitutions and laws, framed and enforced by their own citizens, to secure the manifold blessings of local self-government.

The jurisdiction of this Constitution now covers an area fifty times greater than that of the original thirteen states and a population twenty times greater than that of 1780.

The supreme trial of the Constitution came at last under the tremendous pressure of civil war. We ourselves are witnesses that the Union emerged from the blood and fire of that conflict purified and made stronger for all the beneficent purposes of good government.

And now, at the close of this first century of growth, with the inspirations of its history in their hearts, our people have lately reviewed the condition of the nation, passed judgment upon the conduct and opinions of political parties, and have registered their will concerning the future administration of the government. To interpret and to execute that will in accordance with the Constitution is the paramount duty of the executive.

Even from this brief review it is manifest that the nation is resolutely facing to the front, resolved to employ its best energies in developing the great possibilities of the future. Sacredly preserving whatever has been gained to liberty and good government during the century, our people are determined to leave behind them all those bitter controversies concerning things which have been irrevocably settled, and the further discussion of which can only stir up strife and delay the onward march.

The supremacy of the nation and its laws should be no longer a subject of debate. That discussion, which for half a century threatened the existence of the Union, was closed at last in the high court of war by a decree from which there is no appeal—that the Constitution and the laws made in pursuance thereof are and shall continue to be the supreme law of the land, binding alike upon the states and the people. This decree does not disturb the autonomy of the states nor interfere with any of their necessary rights of local self-government, but it does fix and establish the permanent supremacy of the Union.

The will of the nation, speaking with the voice of battle and through the amended Constitution, has fulfilled the great promise of 1776 by proclaiming "liberty throughout the land to all the inhabitants thereof."

The elevation of the Negro race from slavery to the full rights of citizenship is the most important political change we have known since the adoption of the Constitution of 1787. No thoughtful man can fail to appreciate its beneficent effect upon our institutions and people. It has freed us from the perpetual danger of war and dissolution. It has added immensely to the moral and industrial forces of our people. It has liberated the master as well as the slave from a relation which wronged and enfeebled both. It has surrendered to their own guardianship the manhood of more than five million people, and has opened to each one of them a career of freedom and usefulness. It has given new inspiration to the power of self-help in both races by making labor more honorable to the one and more necessary to the other. The influence of this force will grow greater and bear richer fruit with the coming years.

No doubt this great change has caused serious disturbance to our southern communities. This is to be deplored, though it was perhaps unavoidable. But those who resisted the change should remember that under our institutions there was no middle ground for the Negro race between slavery and equal citizenship. There can be no permanent disfranchised peasantry in the United States. Freedom can never yield its fullness of blessings so long as the law or its administration places the smallest obstacle in the pathway of any virtuous citizen.

The emancipated race has already made remarkable progress. With unquestioning devotion to the Union, with a patience and gentleness not born of fear, they have "followed the light as God gave them to see the light." They are rapidly laying the material foundations of self-support, widening their circle of intelligence, and beginning to enjoy the blessings that gather around the homes of the industrious poor. They deserve the generous encouragement of all good men. So far as my authority can lawfully extend they shall enjoy the full and equal protection of the Constitution and the laws.

The free enjoyment of equal suffrage is still in question, and a frank statement of the issue may aid its solution. It is alleged that in many communities Negro citizens are practically denied the freedom of the ballot. In so far as the truth of this allegation is admitted, it is answered that in many places honest local government is impossible if the mass of uneducated Negroes are allowed to vote. These are grave allegations. So far as the latter is true, it is the only palliation that can be offered for opposing the freedom of the ballot. Bad local government is certainly a great evil, which ought to be prevented; but to violate the freedom and sanctities of the suffrage is more than an evil. It is a crime which, if persisted in, will destroy the government itself. Suicide is not a remedy. If in other lands it be high treason to compass the death of the king, it shall be counted no less a crime here to strangle our sovereign power and stifle its voice.

It has been said that unsettled questions have no pity for the repose of nations. It should be said with the utmost emphasis that this question of the suffrage will never give repose or safety to the states or to the nation until each, within its own jurisdiction, makes and keeps the ballot free and pure by the strong sanctions of the law.

But the danger which arises from ignorance in the voter cannot be denied. It covers a field far wider than that of Negro suffrage and the present condition of the race. It is a danger that lurks and hides in the sources and fountains of power in every state. We have no standard by which to measure the disaster that may be brought upon us by ignorance and vice in the citizens when joined to corruption and fraud in the suffrage.

The voters of the Union, who make and unmake constitutions, and upon whose will hang the destinies of our governments, can transmit their supreme authority to no successors save the coming generation of voters, who are the sole heirs of sovereign power. If that generation comes to its inheritance blinded by ignorance and corrupted by vice, the fall of the Republic will be certain and remediless.

The census has already sounded the alarm in the appalling figures which mark how dangerously high the tide of illiteracy has risen among our voters and their children.

To the South this question is of supreme importance. But the responsibility for the existence of slavery did not rest upon the South alone. The nation itself is responsible for the extension of the suffrage, and is under special obligations to aid in removing the illiteracy which it has added to the voting population. For the North and South alike there is but one remedy. All the constitutional power of the nation and of the states and all the volunteer forces of the people should be surrendered to meet this danger by the savory influence of universal education.

It is the high privilege and sacred duty of those now living to educate their successors and fit them, by intelligence and virtue, for the inheritance which awaits them.

In this beneficent work sections and races should be forgotten and partisanship should be unknown. Let our people find a new meaning in the divine oracle which declares that "a little child shall lead them," for our own little children will soon control the destinies of the Republic.

My countrymen, we do not now differ in our judgment concerning the controversies of past generations, and fifty years hence our children will not be divided in their opinions concerning our controversies. They will surely bless their fathers and their fathers' God that the Union was preserved, that slavery was overthrown, and that both races were made equal before the law. We may hasten or we may retard, but we cannot prevent, the final reconciliation. Is it not possible for us now to make a truce with time by anticipating and accepting its inevitable verdict?

Enterprises of the highest importance to our moral and material well-being unite us and offer ample employment of our best powers. Let all our people, leaving behind them the battlefields of dead issues, move forward and in their strength of liberty and the restored Union win the grander victories of peace.

The prosperity which now prevails is without parallel in our history. Fruitful seasons have done much to secure it, but they have not done all. The preservation of the public credit and the resumption of specie payments, so successfully attained by the administration of my predecessors, have enabled our people to secure the blessings which the seasons brought.

By the experience of commercial nations in all ages it has been found that gold and silver afford the only safe foundation for a monetary system. Confusion has recently been created by variations in the relative value of the two metals, but I confidently believe that arrangements can be made between the leading commercial nations which will secure the general use of both metals. Congress should provide that the compulsory coinage of silver now required by law may not disturb our monetary system by driving either metal out of circulation. If possible, such an adjustment should be made that the purchasing power of every coined dollar will be exactly equal to its debt-paying power in all the markets of the world.

The chief duty of the national government in connection with the currency of the country is to coin money and declare its value. Grave doubts have been entertained whether Congress is authorized by the Constitution to make any form of paper money legal tender. The present issue of United States notes has been sustained by the necessities of war; but such paper should depend for its value and currency upon its convenience in use and its prompt redemption in coin at the will of the holder, and not upon its compulsory circulation. These notes are not money, but promises to pay money. If the holders demand it, the promise should be kept.

The refunding of the national debt at a lower rate of interest should be accomplished without compelling the withdrawal of the national-bank notes, and thus disturbing the business of the country.

I venture to refer to the position I have occupied on financial questions during a long service in Congress, and to say that time and experience have strengthened the opinions I have so often expressed on these subjects.

The finances of the government shall suffer no detriment which it may be possible for my administration to prevent.

The interests of agriculture deserve more attention from the government than they have yet received. The farms of the United States afford homes and employment for more than one-half our people, and furnish much the largest part of all our exports. As the government lights our coasts for the protection of mariners and the benefit of commerce, so it should give to the tillers of the soil the best lights of practical science and experience.

Our manufacturers are rapidly making us industrially independent, and are opening to capital and labor new and profitable fields of employment. Their steady and healthy growth should still be matured. Our facilities for transportation should be promoted by the continued improvement of our harbors and great interior waterways and by the increase of our tonnage on the ocean.

The development of the world's commerce has led to an urgent demand for shortening the great sea voyage around Cape Horn by constructing ship canals or railways across the isthmus which unites the continents. Various plans to this end have been suggested and will need consideration, but none of them has been sufficiently matured to warrant the United States in extending pecuniary aid. The subject, however, is one which will immediately engage the attention of the government with a view to a thorough protection to American interests. We will urge no narrow policy nor seek peculiar or exclusive privileges in any commercial route; but, in the language of my predecessor, I believe it to be the right "and duty of the United States to assert and maintain such supervision and authority over any interoceanic canal across the isthmus that connects North and South America as will protect our national interest."

The Constitution guarantees absolute religious freedom. Congress is prohibited from making any law respecting an establishment of religion or prohibiting the free exercise thereof. The territories of the United States are subject to the direct legislative authority of Congress, and hence the general government is responsible for any violation of the Constitution in any of them. It is therefore a reproach to the government that in the most populous of the territories the constitutional guaranty is not enjoyed by the people and the authority of Congress is set at naught. The Mormon Church not only offends the moral sense of manhood by sanctioning polygamy, but prevents the administration of justice through ordinary instrumentalities of law.

In my judgment it is the duty of Congress, while respecting to the uttermost the conscientious convictions and religious scruples of every citizen, to prohibit within its jurisdiction all criminal practices, especially of that class which destroy the family relations and endanger social order. Nor can any ecclesiastical organization be safely permitted to usurp in the smallest degree the functions and powers of the national government.

The civil service can never be placed on a satisfactory basis until it is regulated by law. For the good of the service itself, for the protection of those who are intrusted with the appointing power against the waste of time and obstruction to the public business caused by the inordinate pressure for place, and for the protection of incumbents against intrigue and wrong, I shall at the proper time ask Congress to fix the tenure of the minor offices of the several executive departments and prescribe the grounds upon which removals shall be made during the terms for which incumbents have been appointed.

Finally, acting always within the authority and limitations of the Constitution, invading neither the rights of the states nor the reserved rights of the people, it will be the purpose of my administration to maintain the authority of the nation in all places within its jurisdiction; to enforce obedience to all the laws of the Union in the interests of the people; to demand rigid economy in all the expenditures of the government, and to require the honest and faithful service of all executive officers, remembering that the offices were created, not for the benefit of incumbents or their supporters, but for the service of the government.

And now, fellow citizens, I am about to assume the great trust which you have committed to my hands. I appeal to you for that earnest and thoughtful support which makes this government in fact, as it is in law, a government of the people.

I shall greatly rely upon the wisdom and patriotism of Congress and of those who may share with me the responsibilities and duties of administration, and, above all, upon our efforts to promote the welfare of this great people and their government I reverently invoke the support and blessings of Almighty God.

Grover Cleveland

Wednesday, March 4, 1885

Grover Cleveland broke the Republican Party's monopoly on the White House in 1884, becoming the first Democrat to win the presidency since James Buchanan in 1856. A political outsider, Cleveland seemed to be an antidote to the partisan bickering and corruption that had characterized the post–Civil War years and helped set the stage for the shocking murder of James A. Garfield in 1881.

Cleveland was a New York Democrat (although born in New Jersey) who forged a reputation for honesty as mayor of Buffalo and governor of New York, a position he held for just a year and a half before becoming President. But his reputation did not shield him from one of the best-remembered political scandals of the late nineteenth century.

Grover Cleveland was a forty-seven-year-old, never-married bachelor in 1884. He was not, however, childless. A decade earlier, he had fathered a son out of wedlock with a widow who lived in Buffalo. The child was placed in an orphanage and eventually adopted by another family. Cleveland helped the child's mother, Maria Halpin, start a business, and she eventually remarried.

Cleveland's opponents in 1884 made sure that voters understood the character of the man who proposed to lead the country. And so supporters of Republican nominee James Blaine tortured Cleveland and his backers with the memorable ditty "Ma, ma, where's my pa?" Democrats, however, gave as good as they got, coining a bit of verse of their own: "Blaine, Blaine, James A. Blaine, the continental liar from the state of Maine."

The nation certainly did not lack for more substantive issues to discuss. The growth of industry and of urban America continued at a fantastic pace. About one in three of the country's fifty-five million people now lived in cities. Waves of immigrants from eastern and southern Europe were beginning to change the character of those cities. Republicans continued to press for tariffs to protect American manufacturers and jobs, while Democrats emphasized civil service reform after nearly a quarter century of Republican control of presidential patronage. Historian Allan Nevins estimated that of the federal government's 126,000 employees, no fewer than 110,000 were patronage appointees.[1]

The divisions of post–Civil War politics were almost forgotten in 1884, no doubt because neither major party's presidential candidate was a Union army veteran. Cleveland, who was born in 1837, hired a substitute to take his place in the army, a common practice of draft evasion among men of means in the North. Blaine, who was thirty when the war broke out, was elected to Congress from Maine in 1862 and never put on a uniform.

The Republican Party remained bitterly divided, although both its factions agreed that Garfield's successor, Chester Arthur, should not be nominated for a term of his own. But the nomination of James Blaine, seemingly a pillar of the party's establishment, created an uproar among Republican reformers in New York and Massachusetts, who were dubbed the mugwumps, an anglicization of an Algonquian word for "great chief." They regarded Blaine as just another in a long line of corrupt Washington politicians, and so abandoned him for Cleveland, whom they regarded as more honest.

Cleveland had his own problems, having alienated New York's Tammany Hall machine and its Irish Catholic base of voters. Tammany was unhappy with Cleveland's refusal to meet its demands for patronage, and his support for free trade was interpreted as overly friendly to British economic interests—not a popular policy in the Irish-American community.

In the final week of the campaign, Blaine visited New York City, home to Tammany's wavering Democrats. During his visit, Blaine met with a delegation of Protestant ministers, one of whom, the Reverend Samuel Burchard, delivered a speech in which he equated the Democratic Party with "Rum, Romanism, and Rebellion." It was the "Romanism" part— a reference to the party's Roman Catholic base—that did the most damage, although with a fledgling temperance movement under way the "Rum" citation may have resonated with some voters. The pro-Confederacy accusation—"Rebellion"—was a perennial Republican charge against the Democrats, and still a powerful one.

---

1. Allan Nevins, *Grover Cleveland: A Study in Courage* (New York: Dodd, Mead, 1932), 199.

Cleveland's campaign got wind of the remarks and made sure they received widespread publicity. Blaine, caught off guard, took his time in distancing himself from Reverend Burchard. Irish Catholic Democrats who were thought to be leaning toward Blaine reconsidered.

On Election Day, the contest came down to Cleveland's home state. The result was uncertain for three days while New York officials tallied the vote. In the end, Governor Cleveland managed to win his state by just over a thousand votes. Nationally, the popular vote was 4.87 million for Cleveland and 4.85 million for Blaine. Had Blaine won New York, he would have won the election in the Electoral College. Cleveland finished with 219 electoral votes to Blaine's 182.

For the first time since Lincoln's election in 1860, the party in control of the White House changed hands when Grover Cleveland took the oath of office on March 5, 1885.

There was no mistaking the new President among the dignitaries gathered to witness the swearing-in on the East Portico of the Capitol. Cleveland was a large man, thanks to large appetites. When he rose to give his inaugural address, his audience noticed something besides his girth. He carried with him to the podium . . . nothing. He memorized his speech and recited it without notes. No President before or since has had the self-confidence and sheer audacity to attempt such a thing.

His speech offered a far more restricted view of federal power than James Garfield's. While the late President envisioned a federal role in education and in enforcing the right to vote, Cleveland embraced a more modest role in keeping with his Jeffersonian ideals and his nostalgic view of a rural America that was disappearing even as he spoke. "Public extravagance," he said, "begets extravagance among the people. We should never be ashamed of the simplicity and prudential economies which are best suited to the operation of a republican form of government and most compatible with the mission of the American people."

His speech set the tone for a conservative, business-oriented administration. For the most part, however, Cleveland's inaugural stands out not so much for what he said, but how he said it—from memory.

In the presence of this vast assemblage of my countrymen I am about to supplement and seal by the oath which I shall take the manifestation of the will of a great and free people. In the exercise of their power and right of self-government they have committed to one of their fellow citizens a supreme and sacred trust, and he here consecrates himself to their service.

This impressive ceremony adds little to the solemn sense of responsibility with which I contemplate the duty I owe to all the people of the land. Nothing can relieve me from anxiety lest by any act of mine their interests may suffer, and nothing is needed to strengthen my resolution to engage every faculty and effort in the promotion of their welfare.

Amid the din of party strife the people's choice was made, but its attendant circumstances have demonstrated anew the strength and safety of a government by the people. In each succeeding year it more clearly appears that our democratic principle needs no apology, and that in its fearless and faithful application is to be found the surest guaranty of good government.

But the best results in the operation of a government wherein every citizen has a share largely depend upon a proper limitation of purely partisan zeal and effort and a correct appreciation of the time when the heat of the partisan should be merged in the patriotism of the citizen.

Today the executive branch of the government is transferred to new keeping. But this is still the government of all the people, and it should be none the less an object of their affectionate solicitude. At this hour the animosities of political strife, the bitterness of partisan defeat, and the exultation of partisan triumph should be supplanted by an ungrudging acquiescence in the popular will and a sober, conscientious concern for the general weal. Moreover, if from this hour we cheerfully and honestly abandon all sectional prejudice and distrust, and determine, with manly confidence in one another, to work out harmoniously the achievements of our national destiny, we shall deserve to realize all the benefits which our happy form of government can bestow.

On this auspicious occasion we may well renew the pledge of our devotion to the Constitution, which, launched by the founders of the Republic and consecrated by their prayers and patriotic devotion, has for almost a century borne the hopes and the aspirations of a great people through prosperity and peace and through the shock of foreign conflicts and the perils of domestic strife and vicissitudes.

By the Father of His Country our Constitution was commended for adoption as "the result of a spirit of amity and mutual concession." In that same spirit it should be administered, in order to promote the lasting welfare of the country and to secure the full measure of its priceless benefits to us and to those who will succeed to the blessings of our national life. The large variety of diverse and competing interests subject to federal control, persistently seeking the recognition of their claims, need give us no fear that "the greatest good to the greatest number" will fail to be accomplished if in the halls of national legislation that spirit of amity and mutual concession shall prevail in which the Constitution had its birth. If this involves the surrender or postponement of private interests and the abandonment of local advantages, compensation will be found in the assurance that the common interest is subserved and the general welfare advanced.

In the discharge of my official duty I shall endeavor to be guided by a just and unstrained construction of the Constitution, a careful observance of the distinction between the powers granted to the federal government and those reserved to the states or to the people, and by a cautious appreciation of those functions which by the Constitution and laws have been especially assigned to the executive branch of the government.

But he who takes the oath today to preserve, protect, and defend the Constitution of the United States only assumes the solemn obligation which every patriotic citizen—on the farm, in the workshop, in the busy marts of trade, and everywhere—should share with him. The Constitution which prescribes his oath, my countrymen, is yours; the government you have chosen him to administer for a time is yours; the suffrage which executes the will of freemen is yours; the laws and the entire scheme of our civil rule, from the town meeting to the state capitals and the national capital, is yours. Your every voter, as surely as your chief magistrate, under the same high sanction, though in a different sphere, exercises a public trust. Nor is this all. Every citizen owes to the country a vigilant watch and close scrutiny of its public servants and a fair and reasonable estimate of their fidelity and usefulness. Thus is the people's will impressed upon the whole framework of our civil polity—municipal, state, and federal; and this is the price of our liberty and the inspiration of our faith in the Republic.

It is the duty of those serving the people in public place to closely limit public expenditures to the actual needs of the government economically administered, because this bounds the right of the government to exact tribute from the earnings of labor or the property of the citizen, and because public extravagance begets extravagance among the people. We should never be ashamed of the simplicity and prudential economies which are best suited to the operation of a republican form of government and most compatible with the mission of the American people. Those who are selected for a limited time to manage public affairs are still of the people, and may do much by their example to encourage, consistently with the dignity of their official functions, that plain way of life which among their fellow citizens aids integrity and promotes thrift and prosperity.

The genius of our institutions, the needs of our people in their home life, and the attention which is demanded for the settlement and development of the resources of our vast territory dictate the scrupulous avoidance of any departure from that foreign policy commended by the history, the traditions, and the prosperity of our Republic. It is the policy of independence, favored by our position and defended by our known love of justice and by our power. It is the policy of peace suitable to our interests. It is the policy of neutrality, rejecting any share in foreign broils and ambitions upon other continents and repelling their intrusion here. It is the policy of Monroe and of Washington and Jefferson—"Peace, commerce, and honest friendship with all nations; entangling alliance with none."

A due regard for the interests and prosperity of all the people demands that our finances shall be established upon such a sound and sensible basis as shall secure the safety and confidence of business interests and make the wage of labor sure and steady, and that our system of revenue shall be so adjusted as to relieve the people of unnecessary taxation, having a due regard to the interests of capital invested and workingmen employed in American industries, and preventing the accumulation of a surplus in the Treasury to tempt extravagance and waste.

Care for the property of the nation and for the needs of future settlers requires that the public domain should be protected from purloining schemes and unlawful occupation.

The conscience of the people demands that the Indians within our boundaries shall be fairly and honestly treated as wards of the government and their education and civilization promoted with a view to their ultimate citizenship, and that polygamy in the territories, destructive of the family relation and offensive to the moral sense of the civilized world, shall be repressed.

The laws should be rigidly enforced which prohibit the immigration of a servile class to compete with American labor, with no intention of acquiring citizenship, and bringing with them and retaining habits and customs repugnant to our civilization.

The people demand reform in the administration of the government and the application of business principles to public affairs. As a means to this end, civil-service reform should be in good faith enforced. Our citizens have the right to protection from the incompetency of public employees who hold their places solely as the reward of partisan service, and from the corrupting influence of those who promise and the vicious methods of those who expect such rewards; and those who worthily seek public employment have the right to insist that merit and competency shall be recognized instead of party subserviency or the surrender of honest political belief.

In the administration of a government pledged to do equal and exact justice to all men there should be no pretext for anxiety touching the protection of the freedmen in their rights or their security in the enjoyment of their privileges under the Constitution and its amendments. All discussion as to their fitness for the place accorded to them as American citizens is idle and unprofitable except as it suggests the necessity for their improvement. The fact that they are citizens entitles them to all the rights due to that relation and charges them with all its duties, obligations, and responsibilities.

These topics and the constant and ever-varying wants of an active and enterprising population may well receive the attention and the patriotic endeavor of all who make and execute the federal law. Our duties are practical and call for industrious application, an intelligent perception of the claims of public office, and, above all, a firm determination, by united action, to secure to all the people of the land the full benefits of the best form of government ever vouchsafed to man. And let us not trust to human effort alone, but humbly acknowledging the power and goodness of Almighty God, who presides over the destiny of nations, and who has at all times been revealed in our country's history, let us invoke His aid and His blessings upon our labors.

Benjamin Harrison

Monday, March 4, 1889

For the second time in a dozen years, the winning candidate in the 1888 presidential election did not receive the most popular votes. But, like Rutherford B. Hayes in 1876, Benjamin Harrison collected enough votes in the Electoral College to win the presidency, denying incumbent Grover Cleveland another term.

Harrison, a fifty-five-year-old Civil War veteran and U.S. senator from Indiana, was the grandson of President William Henry Harrison—the only grandchild of a President to be elected to the White House. He captured 233 votes in the Electoral College to Cleveland's 168, but the incumbent won 5.4 million votes, about 48.6 percent of those cast, to Harrison's 5.4 million, or 47.8 percent cast. Incredibly, given how presidential politics usually works, Cleveland lost his home state of New York and with it 36 electoral votes. Had he managed to hold the state he had formerly governed, Cleveland would have held on in the Electoral College.

Harrison was a surprise choice of the Republican Party, which had been expected to support James Blaine again. But Blaine declined to run, and Harrison emerged as a compromise candidate. His eventual victory can be attributed to issues small and large. Cleveland was a fiscally conservative, business-friendly Democrat who opposed high tariffs on imported goods. That position served him well in the South, which was dependent on imported manufactured goods, but lost him support in the North, where business leaders demanded protection against foreign competition. Harrison and his fellow Republicans took advantage of Cleveland's weakness, crafting a pro-tariff campaign that targeted the large industrial states of the Northeast and the Great Lakes.

Still, Cleveland might have held off Harrison were it not for yet another mini-scandal that touched the increasingly important nerve of ethnic politics. Late in the campaign, a Republican operative from California played a brilliant but largely forgotten political dirty trick on the incumbent President. The Republican, George Osgoodby, assumed the role of a fictitious British-born American citizen and wrote a letter to the British ambassador to the United States, Sir Lionel Sackville-West, asking him for electoral advice. Sackville-West should have ignored the letter; instead, he wrote back and said that Cleveland was preferable to Harrison because he was more sympathetic to British interests in the United States.

Inevitably, the letter made its way into the public press, and it created an uproar in Cleveland's home state with its sizeable and politically active Irish-American community. The swing of New York's Irish Americans, usually reliable Democrats, to Harrison may have cost Cleveland the presidency. He lost New York by a single percentage point.

Harrison took the oath of office on a dreary, rain-drenched Monday in Washington, D.C. Outgoing President Cleveland held an umbrella over his successor's head after Harrison took off his hat to recite the oath of office. The uncooperative weather was somewhat reminiscent of the cold day in 1841 when the new President's grandfather was sworn into office. And the comparison didn't stop there. Like his grandfather, Benjamin Harrison had a good deal to say in his maiden speech as President. At nearly forty-five hundred words, Benjamin Harrison's inaugural was the longest since James Polk's, which was longer than Harrison's by about three hundred words. The final draft was finished the night before the ceremony.

Harrison's fellow Republicans no doubt were overjoyed to resume what they may have regarded as their rightful place at the head of the federal government. But their excitement had to be dampened, literally, as their standard-bearer made his way through his long speech. He put his hat back on before embarking on the speech, while the thousands gathered to hear him took shelter under umbrellas.

The speech is wordy because Harrison sought to cover as many specifics as possible, thus sacrificing soaring rhetoric (or empty bombast) for a laundry list of his priorities. He reaffirmed his support for protective tariffs, urged equal treatment for blacks, pledged support for civil-service reform, and celebrated the imminent statehood of the Dakotas, Montana, and Washington. As a huge wave of immigration from southern and eastern Europe began to arrive—swelling the nation's population to sixty million and leading to the creation of a new processing center on Ellis Island in New York Harbor in 1892—Harrison called for changes in naturalization laws so that "inquiry into the character and good disposition of persons applying for citizenship" could be "more careful and searching." He explained, "We accept the man as a citizen without any knowledge of his fitness, and he assumes the duties of citizenship without any knowledge as to what they are . . . We should not cease to be hospitable to immigration, but we should cease to be careless as to the character of it. There are men of all races, even the best, whose coming is necessarily a burden upon our public revenues or a threat to social order. These should be identified and excluded." One such mass exclusion already had taken place with passage of the Chinese Exclusion Act of 1882.

Despite Harrison's intent to include as many specific ideas as he could, his speech was not completely devoid of eloquence. "Let us exalt patriotism and moderate our party contentions," he said. "Let those who would die for the flag on the field of battle give a better proof of their patriotism and a higher glory to their country by promoting fraternity and justice." The pacific tone was striking from a man who had indeed served his flag as a general in the Union army. Harrison sought to remind his listeners that there were many ways to serve one's country, and perhaps the best way might not be as a soldier in a time of war. Perhaps there were higher aspirations than mere military might.

Harrison's day in the sun, speaking only metaphorically, ended with pageantry that included music by the Marine Corps band, conducted by John Philip Sousa. After the pomp and ceremony, Harrison was confronted with the task of filling jobs in accordance with the wishes of his party, a task he found distasteful. After his presidency, he wrote of the press of job seekers during his early days in the White House. "It is a rare piece of good fortune during the early months of an administration if the President gets one wholly uninterrupted hour at his desk each day," he noted.

As Harrison settled into this busy routine, Grover Cleveland returned to New York. He believed Washington, and Harrison, had not seen the last of him.

FELLOW CITIZENS:

There is no constitutional or legal requirement that the President shall take the oath of office in the presence of the people, but there is so manifest an appropriateness in the public induction to office of the chief executive officer of the nation that from the beginning of the government the people, to whose service the official oath consecrates the officer, have been called to witness the solemn ceremonial. The oath taken in the presence of the people becomes a mutual covenant. The officer covenants to serve the whole body of the people by a faithful execution of the laws, so that they may be the unfailing defense and security of those who respect and observe them, and that neither wealth, station, nor the power of combinations shall be able to evade their just penalties or to wrest them from a beneficent public purpose to serve the ends of cruelty or selfishness.

My promise is spoken; yours unspoken, but not the less real and solemn. The people of every state have here their representatives. Surely I do not misinterpret the spirit of the occasion when I assume that the whole body of the people covenant with me and with each other today to support and defend the Constitution and the Union of the states, to yield willing obedience to all the laws and each to every other citizen his equal civil and political rights. Entering thus solemnly into covenant with each other, we may reverently invoke and confidently expect the favor and help of Almighty God—that He will give to me wisdom, strength, and fidelity, and to our people a spirit of fraternity and a love of righteousness and peace.

This occasion derives peculiar interest from the fact that the presidential term which begins this day is the twenty-sixth under our Constitution. The first inauguration of President Washington took place in New York, where Congress was then sitting, on the 30th day of April 1789, having been deferred by reason of delays attending the organization of the Congress and the canvass of the electoral vote. Our people have already worthily observed the centennials of the Declaration of Independence, of the battle of Yorktown, and of the adoption of the Constitution, and will shortly celebrate in New York the institution of the second great department of our constitutional scheme of government. When the centennial of the institution of the judicial department, by the organization of the Supreme Court, shall have been suitably observed, as I trust it will be, our nation will have fully entered its second century.

I will not attempt to note the marvelous and in great part happy contrasts between our country as it steps over the threshold into its second century of organized existence under the Constitution and that weak but wisely ordered young nation that looked undauntedly down the first century, when all its years stretched out before it.

Our people will not fail at this time to recall the incidents which accompanied the institution of government under the Constitution, or to find inspiration and guidance in the teachings and example of Washington and his great associates, and hope and courage in the contrast which thirty-eight populous and prosperous states offer to the thirteen states, weak in everything except courage and the love of liberty, that then fringed our Atlantic seaboard.

The territory of Dakota has now a population greater than any of the original states (except Virginia) and greater than the aggregate of five of the smaller states in 1790. The center of population when our national capital was located was east of Baltimore, and it was argued by many well-informed persons that it would move eastward rather than westward; yet in 1880 it was found to be near Cincinnati, and the new census about to be taken will show another stride to the westward. That which was the body has come to be only the rich fringe of the nation's robe. But our growth has not been limited to territory, population, and aggregate wealth, marvelous as it has been in each of those directions. The masses of our people are better fed, clothed, and housed than their fathers were. The facilities for popular education have been vastly enlarged and more generally diffused.

The virtues of courage and patriotism have given recent proof of their continued presence and increasing power in the hearts and over the lives of our people. The influences of religion have been multiplied and strengthened. The sweet offices of charity have greatly increased. The virtue of temperance is held in higher estimation. We have not attained an ideal condition. Not all of our people are happy and prosperous; not all of them are virtuous and law-abiding. But on the whole the opportunities offered to the individual to secure the comforts of life are better than are found elsewhere and largely better than they were here one hundred years ago.

The surrender of a large measure of sovereignty to the general government, effected by the adoption of the Constitution, was not accomplished until the suggestions of reason were strongly reenforced by the more imperative voice of experience. The divergent interests of peace speedily demanded a "more perfect union." The merchant, the shipmaster, and the manufacturer discovered and disclosed to our statesmen and to the people that commercial emancipation must be added to the political freedom which had been so bravely won. The commercial policy of the mother country had not relaxed any of its hard and oppressive features. To hold in check the development of our commercial marine, to prevent or retard the establishment and growth of manufactures in the states, and so to secure the American market for their shops and the carrying trade for their ships, was the policy of European statesmen, and was pursued with the most selfish vigor.

Petitions poured in upon Congress urging the imposition of discriminating duties that should encourage the production of needed things at home. The patriotism of the people, which no longer found a field of exercise in war, was energetically directed to the duty of equipping the young Republic for the defense of its independence by making its people self-dependent. Societies for the promotion of home manufactures and for encouraging the use of domestics in the dress of the people were organized in many of the states. The revival at the end of the century of the same patriotic interest in the preservation and development of domestic industries and the defense of our working people against injurious foreign competition is an incident worthy of attention. It is not a departure but a return that we have witnessed. The protective policy had then its opponents. The argument was made, as now, that its benefits inured to particular classes or sections.

If the question became in any sense or at any time sectional, it was only because slavery existed in some of the states. But for this there was no reason why the cotton-producing states should not have led or walked abreast with the New England states in the production of cotton fabrics. There was this reason only why the states that divide with Pennsylvania the mineral treasures of the great southeastern and central mountain ranges should have been so tardy in bringing to the smelting furnace and to the mill the coal and iron from their near opposing hillsides. Mill fires were lighted at the funeral pile of slavery. The emancipation proclamation was heard in the depths of the earth as well as in the sky; men were made free, and material things became our better servants.

The sectional element has happily been eliminated from the tariff discussion. We have no longer states that are necessarily only planting states. None are excluded from achieving that diversification of pursuits among the people which brings wealth and contentment. The cotton plantation will not be less valuable when the product is spun in the country town by operatives whose necessities call for diversified crops and create a home demand for garden and agricultural products. Every new mine, furnace, and factory is an extension of the productive capacity of the state more real and valuable than added territory.

Shall the prejudices and paralysis of slavery continue to hang upon the skirts of progress? How long will those who rejoice that slavery no longer exists cherish or tolerate the incapacities it put upon their communities? I look hopefully to the continuance of our protective system and to the consequent development of manufacturing and mining enterprises in the states hitherto wholly given to agriculture as a potent influence in the perfect unification of our people. The men who have invested their capital in these enterprises, the farmers who have felt the benefit of their neighborhood, and the men who work in shop or field will not fail to find and to defend a community of interest.

Is it not quite possible that the farmers and the promoters of the great mining and manufacturing enterprises which have recently been established in the South may yet find that the free ballot of the workingman, without distinction of race, is needed for their defense as well as for his own? I do not doubt that if those men in the South who now accept the tariff views of Clay and the constitutional expositions of Webster would courageously avow and defend their real convictions they would not find it difficult, by friendly instruction and cooperation, to make the black man their efficient and safe ally, not only in establishing correct principles in our national administration, but in preserving for their local communities the benefits of social order and economical and honest government. At least until the good offices of kindness and education have been fairly tried the contrary conclusion can not be plausibly urged.

I have altogether rejected the suggestion of a special executive policy for any section of our country. It is the duty of the executive to administer and enforce in the methods and by the instrumentalities pointed out and provided by the Constitution all the laws enacted by Congress. These laws are general and their administration should be uniform and equal. As a citizen may not elect what laws he will obey, neither may the executive eject which he will enforce. The duty to obey and to execute embraces the Constitution in its entirety and the whole code of laws enacted under it. The evil example of permitting individuals, corporations, or communities to nullify the laws because they cross some selfish or local interest or prejudices is full of danger, not only to the nation at large, but much more to those who use this pernicious expedient to escape their just obligations or to obtain an unjust advantage over others. They will presently themselves be compelled to appeal to the law for protection, and those who would use the law as a defense must not deny that use of it to others.

If our great corporations would more scrupulously observe their legal limitations and duties, they would have less cause to complain of the unlawful limitations of their rights or of violent interference with their operations. The community that by concert, open or secret, among its citizens

denies to a portion of its members their plain rights under the law has sev-
ered the only safe bond of social order and prosperity. The evil works from
a bad center both ways. It demoralizes those who practice it and destroys
the faith of those who suffer by it in the efficiency of the law as a safe protector. The man in whose breast that faith has been darkened is naturally
the subject of dangerous and uncanny suggestions. Those who use unlawful methods, if moved by no higher motive than the selfishness that
prompted them, may well stop and inquire what is to be the end of this.

An unlawful expedient cannot become a permanent condition of government. If the educated and influential classes in a community either
practice or connive at the systematic violation of laws that seem to them to
cross their convenience, what can they expect when the lesson that convenience or a supposed class interest is a sufficient cause for lawlessness has
been well learned by the ignorant classes? A community where law is the
rule of conduct and where courts, not mobs, execute its penalties is the
only attractive field for business investments and honest labor.

Our naturalization laws should be so amended as to make the inquiry
into the character and good disposition of persons applying for citizenship more careful and searching. Our existing laws have been in their administration an unimpressive and often an unintelligible form. We accept
the man as a citizen without any knowledge of his fitness, and he assumes
the duties of citizenship without any knowledge as to what they are. The
privileges of American citizenship are so great and its duties so grave that
we may well insist upon a good knowledge of every person applying for
citizenship and a good knowledge by him of our institutions. We should
not cease to be hospitable to immigration, but we should cease to be careless as to the character of it. There are men of all races, even the best,
whose coming is necessarily a burden upon our public revenues or a threat
to social order. These should be identified and excluded.

We have happily maintained a policy of avoiding all interference with
European affairs. We have been only interested spectators of their contentions in diplomacy and in war, ready to use our friendly offices to promote
peace, but never obtruding our advice and never attempting unfairly to
coin the distresses of other powers into commercial advantage to ourselves. We have a just right to expect that our European policy will be the
American policy of European courts.

It is so manifestly incompatible with those precautions for our peace
and safety which all the great powers habitually observe and enforce in
matters affecting them that a shorter waterway between our eastern and

western seaboards should be dominated by any European government that we may confidently expect that such a purpose will not be entertained by any friendly power.

We shall in the future, as in the past, use every endeavor to maintain and enlarge our friendly relations with all the great powers, but they will not expect us to look kindly upon any project that would leave us subject to the dangers of a hostile observation or environment. We have not sought to dominate or to absorb any of our weaker neighbors, but rather to aid and encourage them to establish free and stable governments resting upon the consent of their own people. We have a clear right to expect, therefore, that no European government will seek to establish colonial dependencies upon the territory of these independent American states. That which a sense of justice restrains us from seeking they may be reasonably expected willingly to forego.

It must not be assumed, however, that our interests are so exclusively American that our entire inattention to any events that may transpire elsewhere can be taken for granted. Our citizens domiciled for purposes of trade in all countries and in many of the islands of the sea demand and will have our adequate care in their personal and commercial rights. The necessities of our navy require convenient coaling stations and dock and harbor privileges. These and other trading privileges we will feel free to obtain only by means that do not in any degree partake of coercion, however feeble the government from which we ask such concessions. But having fairly obtained them by methods and for purposes entirely consistent with the most friendly disposition toward all other powers, our consent will be necessary to any modification or impairment of the concession.

We shall neither fail to respect the flag of any friendly nation or the just rights of its citizens, nor to exact the like treatment for our own. Calmness, justice, and consideration should characterize our diplomacy. The offices of an intelligent diplomacy or of friendly arbitration in proper cases should be adequate to the peaceful adjustment of all international difficulties. By such methods we will make our contribution to the world's peace, which no nation values more highly, and avoid the opprobrium which must fall upon the nation that ruthlessly breaks it.

The duty devolved by law upon the President to nominate and, by and with the advice and consent of the Senate, to appoint all public officers whose appointment is not otherwise provided for in the Constitution or by act of Congress has become very burdensome and its wise and efficient discharge full of difficulty. The civil list is so large that a personal knowledge of any large number of the applicants is impossible. The President must rely upon the representations of others, and these are often made inconsiderately and without any just sense of responsibility. I have a right, I think, to insist that those who volunteer or are invited to give advice as to appointments shall exercise consideration and fidelity. A high sense of duty and an ambition to improve the service should characterize all public officers.

There are many ways in which the convenience and comfort of those who have business with our public offices may be promoted by a thoughtful and obliging officer, and I shall expect those whom I may appoint to justify their selection by a conspicuous efficiency in the discharge of their duties. Honorable party service will certainly not be esteemed by me a disqualification for public office, but it will in no case be allowed to serve as a shield of official negligence, incompetence, or delinquency. It is entirely creditable to seek public office by proper methods and with proper motives, and all applicants will be treated with consideration; but I shall need, and the heads of departments will need, time for inquiry and deliberation. Persistent importunity will not, therefore, be the best support of an application for office. Heads of departments, bureaus, and all other public officers having any duty connected therewith will be expected to enforce the civil-service law fully and without evasion. Beyond this obvious duty I hope to do something more to advance the reform of the civil service. The ideal, or even my own ideal, I shall probably not attain. Retrospect will be a safer basis of judgment than promises. We shall not, however, I am sure, be able to put our civil service upon a nonpartisan basis until we have secured an incumbency that fair-minded men of the opposition will approve for impartiality and integrity. As the number of such in the civil list is increased removals from office will diminish.

While a Treasury surplus is not the greatest evil, it is a serious evil. Our revenue should be ample to meet the ordinary annual demands upon our Treasury, with a sufficient margin for those extraordinary but scarcely less imperative demands which arise now and then. Expenditure should always be made with economy and only upon public necessity. Wastefulness, profligacy, or favoritism in public expenditures is criminal. But there is nothing in the condition of our country or of our people to suggest that anything presently necessary to the public prosperity, security, or honor should be unduly postponed.

It will be the duty of Congress wisely to forecast and estimate these extraordinary demands, and, having added them to our ordinary expenditures, to so adjust our revenue laws that no considerable annual surplus will remain. We will fortunately be able to apply to the redemption of the public debt any small and unforeseen excess of revenue. This is better than to reduce our income below our necessary expenditures, with the resulting choice between another change of our revenue laws and an increase of the public debt. It is quite possible, I am sure, to effect the necessary reduction in our revenues without breaking down our protective tariff or seriously injuring any domestic industry.

The construction of a sufficient number of modern warships and of their necessary armament should progress as rapidly as is consistent with care and perfection in plans and workmanship. The spirit, courage, and skill of our naval officers and seamen have many times in our history given to weak ships and inefficient guns a rating greatly beyond that of the naval list. That they will again do so upon occasion I do not doubt; but they ought not, by premeditation or neglect, to be left to the risks and exigencies of an unequal combat. We should encourage the establishment of American steamship lines. The exchanges of commerce demand stated, reliable, and rapid means of communication, and until these are provided the development of our trade with the states lying south of us is impossible.

Our pension laws should give more adequate and discriminating relief to the Union soldiers and sailors and to their widows and orphans. Such occasions as this should remind us that we owe everything to their valor and sacrifice.

It is a subject of congratulation that there is a near prospect of the admission into the Union of the Dakotas and Montana and Washington territories. This act of justice has been unreasonably delayed in the case of some of them. The people who have settled these territories are intelligent, enterprising, and patriotic, and the accession of these new states will add strength to the nation. It is due to the settlers in the territories who have availed themselves of the invitations of our land laws to make homes upon the public domain that their titles should be speedily adjusted and their honest entries confirmed by patent.

It is very gratifying to observe the general interest now being manifested in the reform of our election laws. Those who have been for years calling attention to the pressing necessity of throwing about the ballot box and about the elector further safeguards, in order that our elections might not only be free and pure, but might clearly appear to be so, will welcome the accession of any who did not so soon discover the need of reform. The national Congress has not as yet taken control of elections in that case over which the Constitution gives it jurisdiction, but has accepted and adopted the election laws of the several states, provided penalties for their violation and a method of supervision. Only the inefficiency of the state laws or an unfair partisan administration of them could suggest a departure from this policy.

It was clearly, however, in the contemplation of the framers of the Constitution that such an exigency might arise, and provision was wisely made for it. The freedom of the ballot is a condition of our national life, and no power vested in Congress or in the executive to secure or perpetuate it should remain unused upon occasion. The people of all the congressional districts have an equal interest that the election in each shall truly express the views and wishes of a majority of the qualified electors residing within it. The results of such elections are not local, and the insistence of electors residing in other districts that they shall be pure and free does not savor at all of impertinence.

If in any of the states the public security is thought to be threatened by ignorance among the electors, the obvious remedy is education. The sympathy and help of our people will not be withheld from any community struggling with special embarrassments or difficulties connected with the suffrage if the remedies proposed proceed upon lawful lines and are promoted by just and honorable methods. How shall those who practice election frauds recover that respect for the sanctity of the ballot which is the first condition and obligation of good citizenship? The man who has come to regard the ballot box as a juggler's hat has renounced his allegiance.

Let us exalt patriotism and moderate our party contentions. Let those who would die for the flag on the field of battle give a better proof of their patriotism and a higher glory to their country by promoting fraternity and justice. A party success that is achieved by unfair methods or by practices that partake of revolution is hurtful and evanescent even from a party standpoint. We should hold our differing opinions in mutual respect, and, having submitted them to the arbitrament of the ballot, should accept an adverse judgment with the same respect that we would have demanded of our opponents if the decision had been in our favor.

No other people have a government more worthy of their respect and love or a land so magnificent in extent, so pleasant to look upon, and so full of generous suggestion to enterprise and labor. God has placed upon our head a diadem and has laid at our feet power and wealth beyond definition or calculation. But we must not forget that we take these gifts upon the condition that justice and mercy shall hold the reins of power and that the upward avenues of hope shall be free to all the people.

I do not mistrust the future. Dangers have been in frequent ambush along our path, but we have uncovered and vanquished them all. Passion has swept some of our communities, but only to give us a new demonstration that the great body of our people are stable, patriotic, and law-abiding. No political party can long pursue advantage at the expense of public honor or by rude and indecent methods without protest and fatal disaffection in its own body. The peaceful agencies of commerce are more fully revealing the necessary unity of all our communities, and the increasing intercourse of our people is promoting mutual respect. We shall find unalloyed pleasure in the revelation which our next census will make of the swift development of the great resources of some of the states. Each state will bring its generous contribution to the great aggregate of the nation's increase. And when the harvests from the fields, the cattle from the hills, and the ores of the earth shall have been weighed, counted, and valued, we will turn from them all to crown with the highest honor the state that has most promoted education, virtue, justice, and patriotism among its people.

Grover Cleveland

· SECOND INAUGURAL ADDRESS ·

March 4, 1893

rover Cleveland accepted his defeat in 1888 with uncommon grace and more than a little relief. He told a supporter that he didn't regret his loss to Benjamin Harrison. "It is better to be defeated battling for an honest principle than to win by a cowardly subterfuge," he wrote. "We were defeated, it is true, but the principles of tariff reform will surely win in the end."[1]

Upon leaving the White House, Cleveland and his young bride, Frances Folsom, whom he married in the White House in 1886, might have seemed candidates for a quiet, prosperous life of domestic bliss. But if the Clevelands were indeed looking forward to a vacation, they assumed it would be short-lived. Not long before she left the White House, Mrs. Cleveland reportedly told a member of the staff to "take good care" of the "furniture and ornaments. I want to find everything just as it is now when we come back again. We are coming back just four years from today."[2]

She was right. In a rematch of the 1888 election, Grover Cleveland became the first and only President to serve nonconsecutive terms as he ousted the man who had ousted him, Benjamin Harrison. He also became the first candidate to win his party's nomination three consecutive times, a record that would be only be surpassed when Franklin Roosevelt was nominated for a fourth term in 1944.

With the assistance of his supporters in business, Cleveland brought together the Democratic Party's disparate factions under the banner of low tariffs. It was not the most exciting issue, although quite important at the time, but it turned out to be the source of his redemption in 1892. For

---

1. Nevins, *Cleveland: A Study,* 439.

2. Ibid., 448.

Benjamin Harrison had been true to his word. He signed what became known as the McKinley Tariff, named for its chief sponsor, Representative William McKinley of Ohio, which raised prices on imported textiles and food such as wheat and some dairy products. The tariffs were supposed to produce prosperity by encouraging homegrown industry; instead, they led to a run-up in the cost of living.

Another point of contention during the Harrison years was a controversy over the coinage of silver. Farmers and other supporters of cheap money concluded that unlimited coinage of silver, rather than gold, would lead to the inflation they supported. Owners of silver mines in the West, needless to say, wholeheartedly supported the demands for silver currency. In 1890, Congress responded by passing the Sherman Silver Purchase Act, which empowered the Treasury to purchase 4.5 million ounces of silver per month.

Further evidence of growing discontent came in the form of a new third party, the Populist Party, which tried to articulate the grievances of farmers in the Midwest and working people in the East. The party nominated James Weaver, a U.S. representative from Iowa who was elected on the Greenback platform of cheap money and hostility to the railroad industry. Weaver won a million votes and twenty-two electoral votes, most of them from Kansas, Colorado, Idaho, and Nevada. (Several states split their electoral votes, giving Weaver more than the combined total of the four states he won.) Weaver's success foreshadowed a backlash against the laissez-faire politics of the post–Civil War generation.

Cleveland won 5.5 million popular votes and 277 electoral votes, while Harrison had 5.1 million in the popular vote and 145 electoral votes. Weaver's presence in the race helped to deny Cleveland a majority of the popular vote, and he finished with only about 46 percent of the total.

In his second inaugural, delivered on a cold, wet afternoon, Cleveland asserted that the voters had "condemned the injustice of maintaining protection for protection's sake," a rather sweeping statement from a man who was a minority President. He returned to familiar themes, condemning "the waste of public money" and demanding a return to a "sound and stable currency." More surprisingly, however, the generally pro-business President recognized that "combinations of business interests . . . frequently constitute conspiracies against the interests of the people" and should be "restrained by Federal power."

There was little of note about affairs beyond U.S. borders in Cleveland's speech, although there would have been much to discuss. In early 1893, during Harrison's last weeks in office, the United States supported the overthrow of Queen Liliuokalani of Hawaii, who was perceived to be hostile to a small but growing American presence in the islands. Harrison supported the outright annexation of Hawaii, but Cleveland opposed the idea and would soon put a stop to an annexation treaty.

And even as Cleveland addressed his damp, cold audience, a storm of another sort was brewing, this one much closer to home than sunny Hawaii. In May, only two months into Cleveland's term, stock prices collapsed, plunging the nation into a depression comparable to that of the Great Depression of the 1930s. Unemployment would reach 20 percent; thousands of businesses, railroads, and banks shut down; the price of farm goods fell; and creditors foreclosed on farms.

As the economic crisis unfolded, the President was diagnosed with cancer of the jaw. Doctors removed a portion of Cleveland's upper jaw in late June, although they did so without leaving visible scars. Weeks after the operation, Cleveland was outfitted with a prosthetic jaw. The public knew nothing of the President's illness, and he recovered well enough that rumors of ill health were quickly dismissed. He never referred to his illness or his permanent disability.

His stoicism was remarkable, but the next four years brought unending pain of all sorts. The economy resisted remedies, and the poor and unemployed became impatient. In the spring of 1894, a man named Jacob Coxey led hundreds of protestors to the Capitol to demand cheaper money and increased spending on public works projects, a revolutionary notion at a time of limited government interference in the economy.

Grover Cleveland, the only nineteenth-century Democrat to break the Republican Party's post–Civil War hold on the White House, was rejected by his party in 1896. By then, the forces of protest could not be ignored.

My Fellow citizens:

In obedience of the mandate of my countrymen I am about to dedicate myself to their service under the sanction of a solemn oath. Deeply moved by the expression of confidence and personal attachment which has called me to this service, I am sure my gratitude can make no better return than the pledge I now give before God and these witnesses of unreserved and complete devotion to the interests and welfare of those who have honored me.

I deem it fitting on this occasion, while indicating the opinion I hold concerning public questions of present importance, to also briefly refer to the existence of certain conditions and tendencies among our people which seem to menace the integrity and usefulness of their government.

While every American citizen must contemplate with the utmost pride and enthusiasm the growth and expansion of our country, the sufficiency of our institutions to stand against the rudest shocks of violence, the wonderful thrift and enterprise of our people, and the demonstrated superiority of our free government, it behooves us to constantly watch for every symptom of insidious infirmity that threatens our national vigor.

The strong man who in the confidence of sturdy health courts the sternest activities of life and rejoices in the hardihood of constant labor may still have lurking near his vitals the unheeded disease that dooms him to sudden collapse.

It cannot be doubted that our stupendous achievements as a people and our country's robust strength have given rise to heedlessness of those laws governing our national health which we can no more evade than human life can escape the laws of God and nature.

Manifestly nothing is more vital to our supremacy as a nation and to the beneficent purposes of our government than a sound and stable currency. Its exposure to degradation should at once arouse to activity the most enlightened statesmanship, and the danger of depreciation in the purchasing power of the wages paid to toil should furnish the strongest incentive to prompt and conservative precaution.

In dealing with our present embarrassing situation as related to this subject we will be wise if we temper our confidence and faith in our national strength and resources with the frank concession that even these will not permit us to defy with impunity the inexorable laws of finance and trade. At the same time, in our efforts to adjust differences of opinion we should be free from intolerance or passion, and our judgments should be unmoved by alluring phrases and unvexed by selfish interests.

I am confident that such an approach to the subject will result in prudent and effective remedial legislation. In the meantime, so far as the executive branch of the government can intervene, none of the powers with which it is invested will be withheld when their exercise is deemed necessary to maintain our national credit or avert financial disaster.

Closely related to the exaggerated confidence in our country's greatness which tends to a disregard of the rules of national safety, another danger confronts us not less serious. I refer to the prevalence of a popular disposition to expect from the operation of the government especial and direct individual advantages.

The verdict of our voters which condemned the injustice of maintaining protection for protection's sake enjoins upon the people's servants the duty of exposing and destroying the brood of kindred evils which are the unwholesome progeny of paternalism. This is the bane of republican institutions and the constant peril of our government by the people. It degrades to the purposes of wily craft the plan of rule our fathers established and bequeathed to us as an object of our love and veneration. It perverts the patriotic sentiments of our countrymen and tempts them to pitiful calculation of the sordid gain to be derived from their government's maintenance. It undermines the self-reliance of our people and substitutes in its place dependence upon governmental favoritism. It stifles the spirit of true Americanism and stupefies every ennobling trait of American citizenship.

The lessons of paternalism ought to be unlearned and the better lesson taught that while the people should patriotically and cheerfully support their government its functions do not include the support of the people.

The acceptance of this principle leads to a refusal of bounties and subsidies, which burden the labor and thrift of a portion of our citizens to aid ill-advised or languishing enterprises in which they have no concern. It leads also to a challenge of wild and reckless pension expenditure, which overleaps the bounds of grateful recognition of patriotic service and prostitutes to vicious uses the people's prompt and generous impulse to aid those disabled in their country's defense.

Every thoughtful American must realize the importance of checking at its beginning any tendency in public or private station to regard frugality and economy as virtues which we may safely outgrow. The toleration of this idea results in the waste of the people's money by their chosen servants and encourages prodigality and extravagance in the home life of our countrymen.

Under our scheme of government the waste of public money is a crime against the citizen, and the contempt of our people for economy and frugality in their personal affairs deplorably saps the strength and sturdiness of our national character.

It is a plain dictate of honesty and good government that public expenditures should be limited by public necessity, and that this should be measured by the rules of strict economy; and it is equally clear that frugality among the people is the best guaranty of a contented and strong support of free institutions.

One mode of the misappropriation of public funds is avoided when appointments to office, instead of being the rewards of partisan activity, are awarded to those whose efficiency promises a fair return of work for the compensation paid to them. To secure the fitness and competency of appointees to office and remove from political action the demoralizing madness for spoils, civil-service reform has found a place in our public policy and laws. The benefits already gained through this instrumentality and the further usefulness it promises entitle it to the hearty support and encouragement of all who desire to see our public service well performed or who hope for the elevation of political sentiment and the purification of political methods.

The existence of immense aggregations of kindred enterprises and combinations of business interests formed for the purpose of limiting production and fixing prices is inconsistent with the fair field which ought to be open to every independent activity. Legitimate strife in business should not be superseded by an enforced concession to the demands of combinations that have the power to destroy, nor should the people to be served lose the benefit of cheapness which usually results from wholesome competition. These aggregations and combinations frequently constitute conspiracies against the interests of the people, and in all their phases they are unnatural and opposed to our American sense of fairness. To the extent that they can be reached and restrained by federal power the general government should relieve our citizens from their interference and exactions.

Loyalty to the principles upon which our government rests positively demands that the equality before the law which it guarantees to every citizen should be justly and in good faith conceded in all parts of the land. The enjoyment of this right follows the badge of citizenship wherever found, and, unimpaired by race or color, it appeals for recognition to American manliness and fairness.

Our relations with the Indians located within our border impose upon us responsibilities we cannot escape. Humanity and consistency require us to treat them with forbearance and in our dealings with them to honestly and considerately regard their rights and interests. Every effort should be made to lead them, through the paths of civilization and education, to self-supporting and independent citizenship. In the meantime, as the nation's wards, they should be promptly defended against the cupidity of designing men and shielded from every influence or temptation that retards their advancement.

The people of the United States have decreed that on this day the control of their government in its legislative and executive branches shall be given to a political party pledged in the most positive terms to the accomplishment of tariff reform. They have thus determined in favor of a more just and equitable system of federal taxation. The agents they have chosen to carry out their purposes are bound by their promises not less than by the command of their masters to devote themselves unremittingly to this service.

While there should be no surrender of principle, our task must be undertaken wisely and without heedless vindictiveness. Our mission is not punishment, but the rectification of wrong. If in lifting burdens from the daily life of our people we reduce inordinate and unequal advantages too long enjoyed, this is but a necessary incident of our return to right and justice. If we exact from unwilling minds acquiescence in the theory of an honest distribution of the fund of the governmental beneficence treasured up for all, we but insist upon a principle which underlies our free institutions. When we tear aside the delusions and misconceptions which have blinded our countrymen to their condition under vicious tariff laws, we but show them how far they have been led away from the paths of contentment and prosperity. When we proclaim that the necessity for revenue to support the government furnishes the only justification for taxing the people, we announce a truth so plain that its denial would seem to indicate the extent to which judgment may be influenced by familiarity with perversions of the taxing power. And when we seek to reinstate the self-confidence and business enterprise of our citizens by discrediting an abject dependence upon governmental favor, we strive to stimulate those elements of American character which support the hope of American achievement.

Anxiety for the redemption of the pledges which my party has made and solicitude for the complete justification of the trust the people have reposed in us constrain me to remind those with whom I am to cooperate that we can succeed in doing the work which has been especially set before us only by the most sincere, harmonious, and disinterested effort. Even if insuperable obstacles and opposition prevent the consummation of our task, we shall hardly be excused; and if failure can be traced to our fault or neglect we may be sure the people will hold us to a swift and exacting accountability.

The oath I now take to preserve, protect, and defend the Constitution of the United States not only impressively defines the great responsibility I assume, but suggests obedience to constitutional commands as the rule by which my official conduct must be guided. I shall to the best of my ability and within my sphere of duty preserve the Constitution by loyally protecting every grant of federal power it contains, by defending all its restraints when attacked by impatience and restlessness, and by enforcing its limitations and reservations in favor of the states and the people.

Fully impressed with the gravity of the duties that confront me and mindful of my weakness, I should be appalled if it were my lot to bear unaided the responsibilities which await me. I am, however, saved from discouragement when I remember that I shall have the support and the counsel and cooperation of wise and patriotic men who will stand at my side in cabinet places or will represent the people in their legislative halls.

I find also much comfort in remembering that my countrymen are just and generous and in the assurance that they will not condemn those who by sincere devotion to their service deserve their forbearance and approval.

Above all, I know there is a Supreme Being who rules the affairs of men and whose goodness and mercy have always followed the American people, and I know He will not turn from us now if we humbly and reverently seek His powerful aid.

William McKinley

Thursday, March 4, 1897

The election of 1896 tore apart the Gilded Age political alignment that had framed politics since the end of the Civil War. While, in the end, voters reverted to type by electing William McKinley, a solid Republican and Union army veteran, losing candidate William Jennings Bryan symbolized the power of discontent in the nation's agrarian heartland.

The nation was ripe for sweeping change in 1896. The depression following the stock market crash of 1893 was persistent and demoralizing. Despite a slight upturn in 1895, unemployment hovered near 20 percent at a time when government did not offer unemployment insurance and other social welfare programs. The depression's lingering effects were felt not only at home but abroad as well, as diplomats and business leaders began to look overseas to new markets to help the economy recover and to shield it from future shocks.

The election turned on the coinage of gold and silver. In essence, economic populists advocated free coinage of silver, which would encourage currency inflation and so reduce the value of debt. This view was popular in the West and South. More conservative politicians, aligned with eastern business and financial interests, opposed the coinage of silver and instead insisted that gold and only gold ought to support the nation's currency. Incumbent President Grover Cleveland blamed the pro-silver advocates for worsening the nation's economic plight, successfully urging Congress to repeal the Sherman Silver Purchase Act.

Two years into Cleveland's second presidency, it was clear that the President and his policies had fallen into spectacular disrepute. The midterm elections produced an historic rout of the Democratic Party, which lost 113 seats in the House of Representatives.

At their convention in Chicago in 1896, the Democratic Party repudi-

ated Cleveland with a platform that advocated the coinage of silver as well
as gold, and declared its opposition to the gold standard and to Cleveland's
use of bond issues to cover debt, a practice that, his party now asserted,
led to "enormous profits" for the "banking syndicates." In Chicago, the
party that Grover Cleveland dominated for a dozen years transformed
itself, but it lacked a leader. Until, that is, the young man from Nebraska
rose to speak to the convention's twenty thousand delegates, party activists, and journalists.

William Jennings Bryan was an attorney, a failed candidate for the
U.S. Senate, an orator of uncommon ability, and a voice of prairie discontent. With a booming voice, he said the party had to choose between the
interests of "the idle holders of idle capital" and "the struggling masses."
Demagogic, passionate, and memorable, the speech ended with the line
for which it is best remembered: "We will answer their demand for a
gold standard by saying to them: you shall not press down upon the brow
of labor this crown of thorns; you shall not crucify mankind upon a cross
of gold."

The convention floor erupted. Bryan became an instant contender for
nomination by the party that had rejected its standard-bearer. He was
nominated the following day on its fifth ballot, becoming, at age thirty-
six, the youngest major-party candidate for President ever. Bryan then received the nomination of the Populist Party, which had won twenty-two
electoral votes in 1892.

Republicans, who convened before the Democrats, declared their continued support for the gold standard and for protective tariffs. They chose
McKinley, the governor of Ohio, to lead the party into the general election. McKinley was dependable, solid, and mature at age fifty-three. He
would offer an appealing contrast to voters frightened by the young rabble-
rouser who led the Democratic column.

Bryan, in keeping with his populist message and image, refused to play
the role of remote presidential candidate. He campaigned extensively in
the farm belt, delivering hundreds of speeches to tens of thousands.
McKinley, by contrast, remained at home in Ohio, where he greeted
would-be supporters with handshakes and a standard speech. His campaign manager, Mark Hanna, raised millions of dollars to fund pamphlets,
newspaper advertisements, and other tools of mass politics.

Hanna's strategy succeeded. McKinley, marketed as "the advance agent
of prosperity," won 7 million votes and 271 electoral votes to Bryan's

6.4 million and 176. The difference in the popular vote was the largest since Grant's reelection in 1872.

William McKinley's inauguration on March 4, 1897, marked the beginning of a new administration, but it would also be the end of an era. Thirty-two years after the end of the Civil War, the generation that fought it was about to pass the torch of political leadership to its children. The great figures of the war were dead. The United States was home to 72 million people in 1897, most of them with no adult memories of the war and its immediate aftermath. While most Americans still lived in rural areas, 22 million lived in large towns or cities. Places like New York, on its way to 3.4 million residents in 1900, and Chicago, 1.6 million, were becoming not just cities but metropolitan capitals. Race relations throughout the country were dealt a blow during the campaign with the Supreme Court's ruling in *Plessy v. Ferguson*. The justices upheld legal segregation, as long as separate facilities for African Americans were equal to those of whites— although separate facilities such as schools and other public places rarely were, in fact, equal.

The speech offered little to suggest that the nation had undergone a transforming election, although it did acknowledge the country's difficulties in circuitous, passive language. "The country is suffering from industrial disturbances from which speedy relief must be had," McKinley said. He held that sound money, not inflated currency, was the answer to the nation's ills.

McKinley, a cannier politician than he is given credit for, was careful to acknowledge the passions and protests to which Bryan had given voice. "The depression of the past four years," he said, "has fallen with especial severity upon the great body of toilers in the country, and upon none more than the holders of small farms." To assist the toilers, he proposed helping the producers of wealth and jobs because, he argued, "Legislation helpful to producers is beneficial to all."

Foreign affairs, which would come to dominate McKinley's administration, merited only a passing, albeit significant, reference to the importance of supporting arbitration as a means of settling international disputes. The United States had acted as a mediator in several South and Central American boundary disputes beginning in 1876, and McKinley's nod to the arbitration movement was more than just boilerplate rhetoric.

Soon enough, however, the idea of arbitration as a means to avoid war would become decidedly unfashionable. William McKinley was not eager to go to war with Spain over Cuba, but he eventually did so. The Spanish-

American War was the stage on which the United States became a world power.

FELLOW CITIZENS:

In obedience to the will of the people, and in their presence, by the authority vested in me by this oath, I assume the arduous and responsible duties of President of the United States, relying upon the support of my countrymen and invoking the guidance of Almighty God. Our faith teaches that there is no safer reliance than upon the God of our fathers, who has so singularly favored the American people in every national trial, and who will not forsake us so long as we obey His commandments and walk humbly in His footsteps.

The responsibilities of the high trust to which I have been called—always of grave importance—are augmented by the prevailing business conditions entailing idleness upon willing labor and loss to useful enterprises. The country is suffering from industrial disturbances from which speedy relief must be had. Our financial system needs some revision; our money is all good now, but its value must not further be threatened. It should all be put upon an enduring basis, not subject to easy attack, nor its stability to doubt or dispute. Our currency should continue under the supervision of the government. The several forms of our paper money offer, in my judgment, a constant embarrassment to the government and a safe balance in the Treasury. Therefore I believe it necessary to devise a system which, without diminishing the circulating medium or offering a premium for its contraction, will present a remedy for those arrangements which, temporary in their nature, might well in the years of our prosperity have been displaced by wiser provisions. With adequate revenue secured, but not until then, we can enter upon such changes in our fiscal laws as will, while insuring safety and volume to our money, no longer impose upon the government the necessity of maintaining so large a gold reserve, with its attendant and inevitable temptations to speculation. Most of our financial laws are the outgrowth of experience and trial, and should not be amended without investigation and demonstration of the wisdom of the proposed changes. We must be both "sure we are right" and "make haste slowly." If, therefore, Congress, in its wisdom, shall deem it expedient to create a commission to take under early consideration the revision of our coinage, banking, and currency laws, and give them that exhaustive, careful, and dispassionate examination that their importance demands, I shall cordially concur in such action. If such power is vested in the President, it is my purpose to appoint a commission of prominent, well-informed citizens of different parties, who will command public confidence, both on account of their ability and special fitness for the work. Business experience and public training may thus be combined, and the patriotic zeal of the friends of the country be so directed that such a report will be made as to receive the support of all parties, and our finances cease to be the subject of mere partisan contention. The experiment is, at all events, worth a trial, and, in my opinion, it can but prove beneficial to the entire country.

The question of international bimetallism will have early and earnest attention. It will be my constant endeavor to secure it by cooperation with the other great commercial powers of the world. Until that condition is realized when the parity between our gold and silver money springs from and is supported by the relative value of the two metals, the value of the silver already coined and of that which may hereafter be coined, must be kept constantly at par with gold by every resource at our command. The credit of the government, the integrity of its currency, and the inviolability of its obligations must be preserved. This was the commanding verdict of the people, and it will not be unheeded.

Economy is demanded in every branch of the government at all times, but especially in periods, like the present, of depression in business and distress among the people. The severest economy must be observed in all public expenditures, and extravagance stopped wherever it is found, and prevented wherever in the future it may be developed. If the revenues are to remain as now, the only relief that can come must be from decreased expenditures. But the present must not become the permanent condition of the government. It has been our uniform practice to retire, not increase, our outstanding obligations, and this policy must again be resumed and vigorously enforced. Our revenues should always be large enough to meet with ease and promptness not only our current needs and the principal and interest of the public debt, but to make proper and liberal provision for that most deserving body of public creditors, the soldiers and sailors and the widows and orphans who are the pensioners of the United States.

The government should not be permitted to run behind or increase its debt in times like the present. Suitably to provide against this is the mandate of duty—the certain and easy remedy for most of our financial difficulties. A deficiency is inevitable so long as the expenditures of the government exceed its receipts. It can only be met by loans or an increased revenue. While a large annual surplus of revenue may invite waste and extravagance, inadequate revenue creates distrust and undermines public and private credit. Neither should be encouraged. Between more loans and more revenue there ought to be but one opinion. We should have more revenue, and that without delay, hindrance, or postponement. A surplus in the Treasury created by loans is not a permanent or safe reliance. It will suffice while it lasts, but it cannot last long while the outlays of the government are greater than its receipts, as has been the case during the past two years. Nor must it be forgotten that however much such loans may temporarily relieve the situation, the government is still indebted for the amount of the surplus thus accrued, which it must ultimately pay, while its ability to pay is not strengthened, but weakened by a continued deficit. Loans are imperative in great emergencies to preserve the government or its credit, but a failure to supply needed revenue in time of peace for the maintenance of either has no justification.

The best way for the government to maintain its credit is to pay as it goes—not by resorting to loans, but by keeping out of debt—through an adequate income secured by a system of taxation, external or internal, or both. It is the settled policy of the government, pursued from the beginning and practiced by all parties and administrations, to raise the bulk of our revenue from taxes upon foreign productions entering the United States for sale and consumption, and avoiding, for the most part, every form of direct taxation, except in time of war. The country is clearly opposed to any needless additions to the subject of internal taxation, and is committed by its latest popular utterance to the system of tariff taxation. There can be no misunderstanding, either, about the principle upon which this tariff taxation shall be levied. Nothing has ever been made plainer at a general election than that the controlling principle in the raising of revenue from duties on imports is zealous care for American interests and American labor. The people have declared that such legislation should be had as will give ample protection and encouragement to the industries and the development of our country. It is, therefore, earnestly hoped and expected that Congress will, at the earliest practicable moment, enact revenue legislation that shall be fair, reasonable, conservative, and just, and which, while supplying sufficient revenue for public purposes, will still be signally beneficial and helpful to every section and every enterprise of the people. To this policy we are all, of whatever party, firmly bound by the voice of the people—a power vastly more potential than the expression of any political platform. The paramount duty of Congress is to stop deficiencies by the restoration of that protective legislation which has always been the firmest prop of the Treasury. The passage of such a law or laws would strengthen the credit of the government both at home and abroad, and go far toward stopping the drain upon the gold reserve held for the redemption of our currency, which has been heavy and well-nigh constant for several years.

In the revision of the tariff especial attention should be given to the re-enactment and extension of the reciprocity principle of the law of 1890, under which so great a stimulus was given to our foreign trade in new and advantageous markets for our surplus agricultural and manufactured products. The brief trial given this legislation amply justifies a further experiment and additional discretionary power in the making of commercial treaties, the end in view always to be the opening up of new markets for the products of our country, by granting concessions to the products of other lands that we need and cannot produce ourselves, and which do not involve any loss of labor to our own people, but tend to increase their employment.

The depression of the past four years has fallen with especial severity upon the great body of toilers of the country, and upon none more than the holders of small farms. Agriculture has languished and labor suffered. The revival of manufacturing will be a relief to both. No portion of our population is more devoted to the institution of free government nor more loyal in their support, while none bears more cheerfully or fully its proper share in the maintenance of the government or is better entitled to its wise and liberal care and protection. Legislation helpful to producers is beneficial to all. The depressed condition of industry on the farm and in the mine and factory has lessened the ability of the people to meet the demands upon them, and they rightfully expect that not only a system of revenue shall be established that will secure the largest income with the least burden, but that every means will be taken to decrease, rather than increase, our public expenditures. Business conditions are not the most promising. It will take time to restore the prosperity of former years. If we cannot promptly attain it, we can resolutely turn our faces in that direction and aid its return by friendly legislation. However troublesome the situation may appear, Congress will not, I am sure, be found lacking in disposition or ability to relieve it as far as legislation can do so. The restoration of confidence and the revival of business, which men of all parties so much desire, depend more largely upon the prompt, energetic, and intelligent action of Congress than upon any other single agency affecting the situation.

It is inspiring, too, to remember that no great emergency in the 108 years of our eventful national life has ever arisen that has not been met with wisdom and courage by the American people, with fidelity to their best interests and highest destiny, and to the honor of the American name. These years of glorious history have exalted mankind and advanced the cause of freedom throughout the world, and immeasurably strengthened the precious free institutions which we enjoy. The people love and will sustain these institutions. The great essential to our happiness and prosperity is that we adhere to the principles upon which the government was established and insist upon their faithful observance. Equality of rights must prevail, and our laws be always and everywhere respected and obeyed. We may have failed in the discharge of our full duty as citizens of the great Republic, but it is consoling and encouraging to realize that free speech, a free press, free thought, free schools, the free and unmolested right of religious liberty and worship, and free and fair elections are dearer and more universally enjoyed today than ever before. These guaranties must be sacredly preserved and wisely strengthened. The constituted authorities must be cheerfully and vigorously upheld. Lynchings must not be tolerated in a great and civilized country like the United States; courts, not mobs, must execute the penalties of the law. The preservation of public order, the right of discussion, the integrity of courts, and the orderly administration of justice must continue forever the rock of safety upon which our government securely rests.

One of the lessons taught by the late election, which all can rejoice in, is that the citizens of the United States are both law-respecting and law-abiding people, not easily swerved from the path of patriotism and honor. This is in entire accord with the genius of our institutions, and but emphasizes the advantages of inculcating even a greater love for law and order in the future. Immunity should be granted to none who violate the laws, whether individuals, corporations, or communities; and as the Constitution imposes upon the President the duty of both its own execution, and of the statutes enacted in pursuance of its provisions, I shall endeavor carefully to carry them into effect. The declaration of the party now restored to power has been in the past that of "opposition to all combinations of capital organized in trusts, or otherwise, to control arbitrarily the condition of trade among our citizens," and it has supported "such legislation as will prevent the execution of all schemes to oppress the people by undue charges on their supplies, or by unjust rates for the transportation of their products to the market." This purpose will be steadily pursued, both by the enforcement of the laws now in existence and the recommendation and support of such new statutes as may be necessary to carry it into effect.

Our naturalization and immigration laws should be further improved to the constant promotion of a safer, a better, and a higher citizenship. A grave peril to the Republic would be a citizenship too ignorant to understand or too vicious to appreciate the great value and beneficence of our institutions and laws, and against all who come here to make war upon them our gates must be promptly and tightly closed. Nor must we be unmindful of the need of improvement among our own citizens, but with the zeal of our forefathers encourage the spread of knowledge and free education. Illiteracy must be banished from the land if we shall attain that high destiny as the foremost of the enlightened nations of the world which, under Providence, we ought to achieve.

Reforms in the civil service must go on; but the changes should be real and genuine, not perfunctory, or prompted by a zeal in behalf of any party simply because it happens to be in power. As a member of Congress I voted and spoke in favor of the present law, and I shall attempt its enforcement in the spirit in which it was enacted. The purpose in view was to secure the most efficient service of the best men who would accept appointment under the government, retaining faithful and devoted public servants in office, but shielding none, under the authority of any rule or custom, who are inefficient, incompetent, or unworthy. The best interests of the country demand this, and the people heartily approve the law wherever and whenever it has been thus administrated.

Congress should give prompt attention to the restoration of our American merchant marine, once the pride of the seas in all the great ocean highways of commerce. To my mind, few more important subjects so imperatively demand its intelligent consideration. The United States has progressed with marvelous rapidity in every field of enterprise and endeavor until we have become foremost in nearly all the great lines of inland trade, commerce, and industry. Yet, while this is true, our American merchant marine has been steadily declining until it is now lower, both in the percentage of tonnage and the number of vessels employed, than it was prior to the Civil War. Commendable progress has been made of late years in the upbuilding of the American navy, but we must supplement these efforts by providing as a proper consort for it a merchant marine amply sufficient for our own carrying trade to foreign countries. The question is one that appeals both to our business necessities and the patriotic aspirations of a great people.

It has been the policy of the United States since the foundation of the government to cultivate relations of peace and amity with all the nations of the world, and this accords with my conception of our duty now. We have cherished the policy of noninterference with affairs of foreign governments wisely inaugurated by Washington, keeping ourselves free from entanglement, either as allies or foes, content to leave undisturbed with them the settlement of their own domestic concerns. It will be our aim to pursue a firm and dignified foreign policy, which shall be just, impartial, ever watchful of our national honor, and always insisting upon the enforcement of the lawful rights of American citizens everywhere. Our diplomacy should seek nothing more and accept nothing less than is due us. We want no wars of conquest; we must avoid the temptation of territorial aggression. War should never be entered upon until every agency of peace has failed; peace is preferable to war in almost every contingency. Arbitration is the true method of settlement of international as well as local or individual differences. It was recognized as the best means of adjustment of differences between employers and employees by the Forty-ninth Congress, in 1886, and its application was extended to our diplomatic relations by the unanimous concurrence of the Senate and House of the Fifty-first Congress in 1890. The latter resolution was accepted as the basis of negotiations with us by the British House of Commons in 1893, and upon our invitation a treaty of arbitration between the United States and Great Britain was signed at Washington and transmitted to the Senate for its ratification in January last. Since this treaty is clearly the result of our own initiative; since it has been recognized as the leading feature of our foreign policy throughout our entire national history—the adjustment of difficulties by judicial methods rather than force of arms—and since it presents to the world the glorious example of reason and peace, not passion and war, controlling the relations between two of the greatest nations in the world, an example certain to be followed by others, I respectfully urge the early action of the Senate thereon, not merely as a matter of policy, but as a duty to mankind. The importance and moral influence of the ratification of such a treaty can hardly be overestimated in the cause of advancing civilization. It may well engage the best thought of the statesmen and people of every country, and I cannot but consider it fortunate that it was reserved to the United States to have the leadership in so grand a work.

It has been the uniform practice of each President to avoid, as far as possible, the convening of Congress in extraordinary session. It is an example which, under ordinary circumstances and in the absence of a public necessity, is to be commended. But a failure to convene the representatives of the people in Congress in extra session when it involves neglect of a public duty places the responsibility of such neglect upon the executive himself. The condition of the public Treasury, as has been indicated, demands the immediate consideration of Congress. It alone has the power to provide revenues for the government. Not to convene it under such circumstances I can view in no other sense than the neglect of a plain duty. I do not sympathize with the sentiment that Congress in session is dangerous to our general business interests. Its members are the agents of the people, and their presence at the seat of government in the execution of the sovereign should not operate as an injury, but a benefit. There could be no better time to put the government upon a sound financial and economic basis than now. The people have only recently voted that this should be done, and nothing is more binding upon the agents of their will than the obligation of immediate action. It has always seemed to me that the postponement of the meeting of Congress until more than a year after it has been chosen deprived Congress too often of the inspiration of the popular will and the country of the corresponding benefits. It is evident, therefore, that to postpone action in the presence of so great a necessity would be unwise on the part of the executive because unjust to the interests of the people. Our action now will be freer from mere partisan consideration than if the question of tariff revision was postponed until the regular session of Congress. We are nearly two years from a congressional election, and politics cannot so greatly distract us as if such contest was immediately pending. We can approach the problem calmly and patriotically, without fearing its effect upon an early election.

Our fellow citizens who may disagree with us upon the character of this legislation prefer to have the question settled now, even against their preconceived views, and perhaps settled so reasonably, as I trust and believe it will be, as to insure great permanence, than to have further uncertainty menacing the vast and varied business interests of the United States. Again, whatever action Congress may take will be given a fair opportunity for trial before the people are called to pass judgment upon it, and this I consider a great essential to the rightful and lasting settlement of the question. In view of these considerations, I shall deem it my duty as President to convene Congress in extraordinary session on Monday, the 15th day of March 1897.

In conclusion, I congratulate the country upon the fraternal spirit of the people and the manifestations of good will everywhere so apparent. The recent election not only most fortunately demonstrated the obliteration of sectional or geographical lines, but to some extent also the prejudices which for years have distracted our councils and marred our true greatness as a nation. The triumph of the people, whose verdict is carried into effect today, is not the triumph of one section, nor wholly of one party, but of all sections and all the people. The North and the South no longer divide on the old lines, but upon principles and policies; and in this fact surely every lover of the country can find cause for true felicitation.

Let us rejoice in and cultivate this spirit; it is ennobling and will be both a gain and a blessing to our beloved country. It will be my constant aim to do nothing, and permit nothing to be done, that will arrest or disturb this growing sentiment of unity and cooperation, this revival of esteem and affiliation which now animates so many thousands in both the old antagonistic sections, but I shall cheerfully do everything possible to promote and increase it. Let me again repeat the words of the oath administered by the chief justice which, in their respective spheres, so far as applicable, I would have all my countrymen observe: "I will faithfully execute the office of President of the United States, and will, to the best of my ability, preserve, protect, and defend the Constitution of the United States." This is the obligation I have reverently taken before the Lord Most High. To keep it will be my single purpose, my constant prayer; and I shall confidently rely upon the forbearance and assistance of all the people in the discharge of my solemn responsibilities.

William McKinley

Monday, March 4, 1901

William McKinley's second inaugural, delivered during a driving rainstorm that left his listeners wet and miserable, captured the country's triumphant mood at the dawn of the twentieth century. War with Spain had been quick, easy, and victorious, allowing the United States to take its place on the world stage along with the European powers and Japan. The economy at home, meanwhile, was fully recovered from the depression of the 1890s. On all counts, foreign and domestic, the solid leadership of William McKinley—as opposed to the frightening elixirs offered by William Jennings Bryan—seemed the cure for America's turn-of-the-century ailments.

McKinley had beaten Bryan once again in November 1900, although this time around the President had a new running mate. Vice president Garret Hobart of New Jersey had died in office in 1899, leaving the nation without a vice president. The vacancy offered an opportunity for New York's Republican bosses who already were weary of their reform-minded new governor, Theodore Roosevelt. They persuaded McKinley and his chief political adviser, Mark Hanna, to put young Roosevelt on the ticket as vice president, where he would be out of their way. McKinley obliged.

Roosevelt did most of the campaigning in 1900 while McKinley stuck with his archaic back-porch strategy. Bryan campaigned with characteristic energy, denouncing McKinley's drift toward imperial rule overseas. Few seemed similarly concerned about a U.S. presence in the Philippines, which came into America's possession after the Spanish-American War, or in Cuba where the war started. McKinley won 292 electoral votes to Bryan's 155 and took the popular vote with 7.2 million votes, compared with Bryan's 6.3 million.

McKinley addressed a nation whose fortunes and outlook had changed dramatically since his first inaugural, a point he was eager to make. His opening paragraph, in fact, took note of the nation's "anxiety" in 1897, contrasting it to the nation's mood on that rainy afternoon, when "we have a surplus instead of a deficit" and "every avenue of production is crowded with activity, labor is well employed, and American products find good markets at home and abroad."

The country's seventy-six million people indeed were enjoying better times thanks to the economic recovery of McKinley's first term. National income would nearly double over the next two decades, from $36.5 billion in 1900 to more than $60 billion in 1920. While most industrial workers put in long days and workweeks (more than ten hours a day, six days a week), they also were developing leisure and consumption habits that were the building blocks of a new mass culture.

But events abroad, rather than at home, defined McKinley's presidency and his second inaugural address. His was a very modern inaugural, the speech of a leader of a world power. He was less concerned with paying long-winded tribute to the wonders of the Constitution—a staple of so many nineteenth-century inaugurals—than he was with preparing the country for global leadership.

Victory against Spain in 1898, he said, "imposed upon us obligations from which we cannot escape and from which it would be dishonorable to seek escape." Those "obligations" included an overseas empire with the annexation of Hawaii, a colonial relationship with the Philippine Islands, a military presence in Cuba, and a commercial presence in China. The United States, born of a revolt against colonial rule, became an imperial power thanks to its victory over a fading colonial power, Spain.

McKinley saw this expansion as a logical continuation of Manifest Destiny and the pioneering spirit of the American people. Referring to the country's new possessions, he said, "The Republic has marched on and on, and its step has exalted freedom and humanity. We are undergoing the same ordeal as did our predecessors nearly a century ago. We are following the course they blazed."

Although McKinley had seemed reluctant to go to war with Spain, a fact he acknowledged in this speech, he now embraced the growing idea that America was obliged to bring its civilizing mission overseas. Social Darwinism was behind much of this rhetoric, along with notions that Rudyard Kipling described as the "white man's burden." The United States would march on and on, McKinley said, bringing with it "freedom and humanity."

FELLOW CITIZENS

Not everybody was as sanguine about this march as McKinley was. Bryan had argued during the campaign that the United States had no business getting into the business of empire building. And in the Philippines themselves, U.S. troops were fighting a bloody war against Filipino rebels who did not support American rule. McKinley acknowledged the conflict in the last part of his inaugural, asserting that U.S. troops "are not waging war against the inhabitants of the Philippine Islands. A portion of them are making war against the United States." McKinley vowed that Filipinos loyal to U.S. rule "shall not be abandoned." The war continued until 1902.

McKinley's speech was the last inaugural given by a veteran of the Civil War and was the first one in decades to focus so intently on foreign affairs. Since the 1850s, the trauma of impending civil war, the war itself, and its aftermath had dominated presidential inaugurals. With McKinley's speech, however, the war moved off stage, even if certain divisions did not. While McKinley insisted that the sectionalism had "disappeared" and that differences could "no longer be traced by the war maps of 1861," sharp divisions certainly did exist. The Solid South continued to reject Republican presidential candidates. About 85 percent of the nation's African Americans lived in the South, while some 80 percent of the country's burgeoning foreign-born population lived in the North. The scars of the war and of sectionalism still were evident, despite McKinley's assertions.

Nevertheless, prosperity at home and triumph abroad were the themes of McKinley's inaugural. But the nation's buoyant spirits would be dealt a blow six months after this speech, when McKinley was shot while attending the Pan-American Exposition in Buffalo, New York. His assailant was an anarchist named Leon Czolgosz, and his Old World last name and his ideology encapsulated the nation's repressed fears of immigrants and the ideas they brought with them across the Atlantic Ocean.

McKinley died on September 14, 1901, after a week of suffering. He was the third President to be gunned down in fewer than forty years.

Theodore Roosevelt, the man New York Republicans believed they had buried in obscurity, was now the youngest President in U.S. history.

Bibliographic Note
Lewis L. Gould, *The Presidency of William McKinley*
(Lawrence, KS: University Press of Kansas, 1980).

My fellow citizens:

When we assembled here on the fourth of March 1897 there was great anxiety with regard to our currency and credit. None exists now. Then our Treasury receipts were inadequate to meet the current obligations of the government. Now they are sufficient for all public needs, and we have a surplus instead of a deficit. Then I felt constrained to convene the Congress in extraordinary session to devise revenues to pay the ordinary expenses of the government. Now I have the satisfaction to announce that the Congress just closed has reduced taxation in the sum of $41 million. Then there was deep solicitude because of the long depression in our manufacturing, mining, agricultural, and mercantile industries and the consequent distress of our laboring population. Now every avenue of production is crowded with activity, labor is well employed, and American products find good markets at home and abroad.

Our diversified productions, however, are increasing in such unprecedented volume as to admonish us of the necessity of still further enlarging our foreign markets by broader commercial relations. For this purpose reciprocal trade arrangements with other nations should in liberal spirit be carefully cultivated and promoted.

The national verdict of 1896 has for the most part been executed. Whatever remains unfulfilled is a continuing obligation resting with undiminished force upon the executive and the Congress. But fortunate as our condition is, its permanence can only be assured by sound business methods and strict economy in national administration and legislation. We should not permit our great prosperity to lead us to reckless ventures in business or profligacy in public expenditures. While the Congress determines the objects and the sum of appropriations, the officials of the executive departments are responsible for honest and faithful disbursement, and it should be their constant care to avoid waste and extravagance.

Honesty, capacity, and industry are nowhere more indispensable than in public employment. These should be fundamental requisites to original appointment and the surest guaranties against removal.

Four years ago we stood on the brink of war without the people knowing it and without any preparation or effort at preparation for the impending peril. I did all that in honor could be done to avert the war, but without avail. It became inevitable; and the Congress at its first regular session, without party division, provided money in anticipation of the crisis and in preparation to meet it. It came. The result was signally favorable to American arms and in the highest degree honorable to the government. It imposed upon us obligations from which we cannot escape and from which it would be dishonorable to seek escape. We are now at peace with the world, and it is my fervent prayer that if differences arise between us and other powers they may be settled by peaceful arbitration and that hereafter we may be spared the horrors of war.

Intrusted by the people for a second time with the office of President, I enter upon its administration appreciating the great responsibilities which attach to this renewed honor and commission, promising unreserved devotion on my part to their faithful discharge and reverently invoking for my guidance the direction and favor of Almighty God. I should shrink from the duties this day assumed if I did not feel that in their performance I should have the cooperation of the wise and patriotic men of all parties. It encourages me for the great task which I now undertake to believe that those who voluntarily committed to me the trust imposed upon the chief executive of the Republic will give to me generous support in my duties to "preserve, protect, and defend the Constitution of the United States" and to "care that the laws be faithfully executed." The national purpose is indicated through a national election. It is the constitutional method of ascertaining the public will. When once it is registered it is a law to us all, and faithful observance should follow its decrees.

Strong hearts and helpful hands are needed, and, fortunately, we have them in every part of our beloved country. We are reunited. Sectionalism has disappeared. Division on public questions can no longer be traced by the war maps of 1861. These old differences less and less disturb the judgment. Existing problems demand the thought and quicken the conscience of the country, and the responsibility for their presence, as well as for their righteous settlement, rests upon us all—no more upon me than upon you. There are some national questions in the solution of which patriotism should exclude partisanship. Magnifying their difficulties will not take them off our hands nor facilitate their adjustment. Distrust of the capacity, integrity, and high purposes of the American people will not be an inspiring theme for future political contests. Dark pictures and gloomy forebodings are worse than useless. These only becloud, they do not help to point the way of safety and honor. "Hope maketh not ashamed." The prophets of evil were not the builders of the Republic, nor in its crises since have they saved or served it. The faith of the fathers was a mighty force in its creation, and the faith of their descendants has wrought its progress and furnished its defenders. They are obstructionists who despair, and who would destroy confidence in the ability of our people to solve wisely and for civilization the mighty problems resting upon them. The American people, intrenched in freedom at home, take their love for it with them wherever they go, and they reject as mistaken and unworthy the doctrine that we lose our own liberties by securing the enduring foundations of liberty to others. Our institutions will not deteriorate by extension, and our sense of justice will not abate under tropic suns in distant seas. As heretofore, so hereafter will the nation demonstrate its fitness to administer any new estate which events devolve upon it, and in the fear of God will "take occasion by the hand and make the bounds of freedom wider yet." If there are those among us who would make our way more difficult, we must not be disheartened, but the more earnestly dedicate ourselves to the task upon which we have rightly entered. The path of progress is seldom smooth. New things are often found hard to do. Our fathers found them so. We find them so. They are inconvenient. They cost us something. But are we not made better for the effort and sacrifice, and are not those we serve lifted up and blessed?

We will be consoled, to, with the fact that opposition has confronted every onward movement of the Republic from its opening hour until now, but without success. The Republic has marched on and on, and its step has exalted freedom and humanity. We are undergoing the same ordeal as did our predecessors nearly a century ago. We are following the course they blazed. They triumphed. Will their successors falter and plead organic impotency in the nation? Surely after 125 years of achievement for mankind we will not now surrender our equality with other powers on matters fundamental and essential to nationality. With no such purpose was the nation created. In no such spirit has it developed its full and independent sovereignty. We adhere to the principle of equality among ourselves, and by no act of ours will we assign to ourselves a subordinate rank in the family of nations.

My fellow citizens, the public events of the past four years have gone into history. They are too near to justify recital. Some of them were unforeseen; many of them momentous and far-reaching in their consequences to ourselves and our relations with the rest of the world. The part which the United States bore so honorably in the thrilling scenes in China, while new to American life, has been in harmony with its true spirit and best traditions, and in dealing with the results its policy will be that of moderation and fairness.

We face at this moment a most important question: that of the future relations of the United States and Cuba. With our near neighbors we must remain close friends. The declaration of the purposes of this government in the resolution of April 20, 1898, must be made good. Ever since the evacuation of the island by the army of Spain, the executive, with all practicable speed, has been assisting its people in the successive steps necessary to the establishment of a free and independent government prepared to assume and perform the obligations of international law which now rest upon the United States under the Treaty of Paris. The convention elected by the people to frame a constitution is approaching the completion of its labors. The transfer of American control to the new government is of such great importance, involving an obligation resulting from our intervention and the treaty of peace, that I am glad to be advised by the recent act of Congress of the policy which the legislative branch of the government deems essential to the best interests of Cuba and the United States. The principles which led to our intervention require that the fundamental law upon which the new government rests should be adapted to secure a government capable of performing the duties and discharging the functions of a separate nation, of observing its international obligations of protecting life and property, insuring order, safety, and liberty, and conforming to the established and historical policy of the United States in its relation to Cuba.

The peace which we are pledged to leave to the Cuban people must carry with it the guaranties of permanence. We became sponsors for the pacification of the island, and we remain accountable to the Cubans, no less than to our own country and people, for the reconstruction of Cuba as a free commonwealth on abiding foundations of right, justice, liberty, and assured order. Our enfranchisement of the people will not be completed until free Cuba shall "be a reality, not a name; a perfect entity, not a hasty experiment bearing within itself the elements of failure."

While the treaty of peace with Spain was ratified on the 6th of February 1899, and ratifications were exchanged nearly two years ago, the Congress has indicated no form of government for the Philippine Islands. It has, however, provided an army to enable the executive to suppress insurrection, restore peace, give security to the inhabitants, and establish the authority of the United States throughout the archipelago. It has authorized the organization of native troops as auxiliary to the regular force. It has been advised from time to time of the acts of the military and naval officers in the islands, of my action in appointing civil commissions, of the instructions with which they were charged, of their duties and powers, of their recommendations, and of their several acts under executive commission, together with the very complete general information they have submitted. These reports fully set forth the conditions, past and present, in the islands, and the instructions clearly show the principles which will guide the executive until the Congress shall, as it is required to do by the treaty, determine "the civil rights and political status of the native inhabitants." The Congress having added the sanction of its authority to the powers already possessed and exercised by the executive under the Constitution, thereby leaving with the executive the responsibility for the government of the Philippines, I shall continue the efforts already begun until order shall be restored throughout the islands, and as fast as conditions permit will establish local governments, in the formation of which the full cooperation of the people has been already invited, and when established will encourage the people to administer them. The settled purpose, long ago proclaimed, to afford the inhabitants of the islands self-government as fast as they were ready for it will be pursued with earnestness and fidelity. Already something has been accomplished in this direction. The government's representatives, civil and military, are doing faithful and noble work in their mission of emancipation and merit the approval and support of their countrymen. The most liberal terms of amnesty have already been communicated to the insurgents, and the way is still open for those who have raised their arms against the government for honorable submission to its authority. Our countrymen should not be deceived. We are not waging war against the inhabitants of the Philippine Islands. A portion of them are making war against the United States. By far the greater part of the inhabitants recognize American sovereignty and welcome it as a guaranty of order and of security for life, property, liberty, freedom of conscience, and the pursuit of happiness. To them full protection will be given. They shall not be abandoned. We will not leave the destiny of the loyal millions [of] the islands to the disloyal thousands who are in rebellion against the United States. Order under civil institutions will come as soon as those who now break the peace shall keep it. Force will not be needed or used when those who make war against us shall make it no more. May it end without further bloodshed, and there be ushered in the reign of peace to be made permanent by a government of liberty under law!

Theodore Roosevelt

· INAUGURAL ADDRESS ·

Saturday, March 4, 1905

$\mathscr{H}$ istorian Edmund Morris, describing the blustery conditions in Washington when Theodore Roosevelt took the oath of office in 1905, noted that as the President spoke flags "whipped and cracked," women and men constantly reached to secure their hats, caps from young men in the service academies spun through the air, and a "whirl of pigeons" circled the Capitol dome.

It was, Morris observed, a scene "of constant movement, as if Roosevelt's energy had animated the entire body politic."[1]

Both the President and the nation had energy to spare. The United States and its President were young, confident, and eager, ready for a place on the world stage alongside the older, more established powers destined to give way to the American colossus as the new century progressed.

At the age of forty-five, Theodore Roosevelt personified the nation's spirit as few presidents have before or since. He was no longer a youth, but neither had he settled into sedentary, stolid middle age. He was an advocate of what he called "the strenuous life," and he governed himself, and the nation, accordingly.

He succeeded the slain William McKinley in September 1901 and came to embody the energetic spirit of a new age, the Progressive Era, ushered in by urban reformers who stood for political and economic change, who believed in scientific solutions to the nation's woes, and who were anxious about race and the supposed diluting effects of new Americans on the nation's traditional Anglo-Saxon Protestant stock.

Seeing himself, as so many Progressives did, as a mediator between an idle upper class and a mass of poor, urban, industrial workers from other

---

1. Edmund Morris, *Theodore Rex* (New York: Random House, 2001), 276.

nations or descended from recent immigrants, Roosevelt advocated reforms that sought to dampen class conflict and to regulate the excesses of capital. He intervened in a dispute between mine owners and workers in late 1902, resulting in an agreement that kept labor peace in the mines for more than a decade. His administration successfully sued the Northern Securities Company, which controlled several railroads, for violating antitrust laws. And, avid outdoorsman that he was, he supported measures to conserve the nation's resources and its physical landmarks, especially in the West.

In finishing McKinley's term, Roosevelt put his own stamp on the modern presidency, embracing a politically strenuous life of active engagement, public persuasion, and innovative policy making. Not since the administration of Andrew Jackson had the nation witnessed such a whirlwind of policy and politics. "I declined to adopt the view that what was imperatively necessary for the nation could not be done by the President unless he could find some specific authorization to do it,"[2] Roosevelt wrote in his autobiography years later.

Roosevelt applied that expansive view of executive power not just to domestic issues but to foreign affairs as well. Under Roosevelt, America's presence in the world became more defined and more aggressive. In 1903, the United States and Colombia signed the Hay-Herrán Treaty, which gave the United States authority to build a canal in Panama, then a province of Colombia. But the treaty was rejected in the Colombian Senate. Panamanians, long discontented with Colombian rule, rebelled, and Roosevelt sent a warship to the area that ensured the rebellion's success. Washington and newly independent Panama quickly signed a treaty to allow construction of the canal, and work got under way in 1904.

The U.S. role in Central America signaled the beginning of an increased U.S. presence in that region, in the Caribbean, and in South America. Roosevelt issued an amendment to the Monroe Doctrine—meant to keep Old World powers out of the Americas—which asserted that the United States could intervene in the Americas to keep order.

Americans enjoyed the spectacle of this young, activist President who seemed to embody the virtues they ascribed to the nation as a whole. He was so popular in 1904 that the opposition Democrats could find nobody of stature to challenge him, settling on a judge named Alton B. Parker,

---

2. Wayne Andrew, ed., *The Autobiography of Theodore Roosevelt*, abridged (New York: Scribners, 1958), 197.

who, like Roosevelt, came from New York. Judge Parker disappeared from history after Roosevelt thrashed him on Election Day, winning 7.6 million popular votes to Parker's 5 million, and taking 336 electoral votes to Parker's 140. He became the first president to win a term in his own right after succeeding a fallen president.

In addressing the nation on that windy Washington afternoon in 1905, Roosevelt offered not the slightest hint of his agenda for the next four years. Part of his reticence was strategic, as his biographer Morris has noted.[3] Behind the scenes, there was a move to have him mediate a war between Russia and Japan. Roosevelt omitted any references to the conflict.

Rather than an outline of the future, or even a celebration of the past, the speech was a sermon, fit for a bully pulpit, littered with ideas Roosevelt had articulated throughout his young but amazingly productive life. "Much has been given us, and much will rightfully be expected from us," he said as he called the nation to "duties to others and duties to ourselves."

His ebullient spirit, absolute confidence, and enthusiasm for leadership are evident in nearly every sentence of this short speech. Though students of Roosevelt might see it as a hodgepodge of the President's greatest hits—his citation of "the manlier and hardier virtues" is the most obvious example—the speech does reveal its speaker's deepest held beliefs. He asserted, in a foreshadowing of Cold War rhetoric a half century later, that self-government around the world depended on the success of the "experiment" in the United States. That notion of American leadership and American exceptionalism neatly encapsulated Roosevelt's world view.

The strenuous life was not just about exercise and physical activity. It was an intellectual process, too, as Roosevelt suggested in his speech. Americans knew that self-government was "difficult," but they had little choice but to overcome those difficulties, to display "practical intelligence . . . courage . . . hardihood, and endurance, and above all the power of devotion to a lofty ideal."

A deceptive speech, seemingly replete with generalities that, upon closer inspection, offered a glimpse of its speaker's soul, Roosevelt's inaugural marked the beginning of a new era in the American presidency. From 1905 on, the President's first speech would be addressed not just to the nation but to the world.

---

3. Morris, *Theodore Rex,* 377.

No people on earth have more cause to be thankful than ours, and this is said reverently, in no spirit of boastfulness in our own strength, but with gratitude to the Giver of Good who has blessed us with the conditions which have enabled us to achieve so large a measure of well-being and of happiness. To us as a people it has been granted to lay the foundations of our national life in a new continent. We are the heirs of the ages, and yet we have had to pay few of the penalties which in old countries are exacted by the dead hand of a bygone civilization. We have not been obliged to fight for our existence against any alien race; and yet our life has called for the vigor and effort without which the manlier and hardier virtues wither away. Under such conditions it would be our own fault if we failed; and the success which we have had in the past, the success which we confidently believe the future will bring, should cause in us no feeling of vainglory, but rather a deep and abiding realization of all which life has offered us; a full acknowledgment of the responsibility which is ours; and a fixed determination to show that under a free government a mighty people can thrive best, alike as regards the things of the body and the things of the soul.

Much has been given us, and much will rightfully be expected from us. We have duties to others and duties to ourselves; and we can shirk neither. We have become a great nation, forced by the fact of its greatness into relations with the other nations of the earth, and we must behave as beseems a people with such responsibilities. Toward all other nations, large and small, our attitude must be one of cordial and sincere friendship. We must show not only in our words, but in our deeds, that we are earnestly desirous of securing their good will by acting toward them in a spirit of just and generous recognition of all their rights. But justice and generosity in a nation, as in an individual, count most when shown not by the weak but by the strong. While ever careful to refrain from wrongdoing others, we must be no less insistent that we are not wronged ourselves. We wish peace, but we wish the peace of justice, the peace of righteousness. We wish it because we think it is right and not because we are afraid. No weak nation that acts manfully and justly should ever have cause to fear us, and no strong power should ever be able to single us out as a subject for insolent aggression.

Our relations with the other powers of the world are important; but still more important are our relations among ourselves. Such growth in wealth, in population, and in power as this nation has seen during the century and a quarter of its national life is inevitably accompanied by a like growth in the problems which are ever before every nation that rises to greatness. Power invariably means both responsibility and danger. Our forefathers faced certain perils which we have outgrown. We now face other perils, the very existence of which it was impossible that they should foresee. Modern life is both complex and intense, and the tremendous changes wrought by the extraordinary industrial development of the last half century are felt in every fiber of our social and political being. Never before have men tried so vast and formidable an experiment as that of administering the affairs of a continent under the forms of a Democratic republic. The conditions which have told for our marvelous material well-being, which have developed to a very high degree our energy, self-reliance, and individual initiative, have also brought the care and anxiety inseparable from the accumulation of great wealth in industrial centers. Upon the success of our experiment much depends, not only as regards our own welfare, but as regards the welfare of mankind. If we fail, the cause of free self-government throughout the world will rock to its foundations, and therefore our responsibility is heavy, to ourselves, to the world as it is to-day, and to the generations yet unborn. There is no good reason why we should fear the future, but there is every reason why we should face it seriously, neither hiding from ourselves the gravity of the problems before us nor fearing to approach these problems with the unbending, unflinching purpose to solve them aright.

Yet, after all, though the problems are new, though the tasks set before us differ from the tasks set before our fathers who founded and preserved this Republic, the spirit in which these tasks must be undertaken and these problems faced, if our duty is to be well done, remains essentially unchanged. We know that self-government is difficult. We know that no people needs such high traits of character as that people which seeks to govern its affairs aright through the freely expressed will of the freemen who compose it. But we have faith that we shall not prove false to the memories of the men of the mighty past. They did their work, they left us the splendid heritage we now enjoy. We in our turn have an assured confidence that we shall be able to leave this heritage unwasted and enlarged to our children and our children's children. To do so we must show, not merely in great crises, but in the everyday affairs of life, the qualities of practical intelligence, of courage, of hardihood, and endurance, and above all the power of devotion to a lofty ideal, which made great the men who founded this Republic in the days of Washington, which made great the men who preserved this Republic in the days of Abraham Lincoln.

William Howard Taft

Thursday, March 4, 1909

A late-winter blizzard covered Washington on the morning of March 4, 1909, forcing the inaugural ceremonies to be moved indoors, into the Senate chamber. As the new President's listeners tried to follow an inaugural address of more than five thousand words, they no doubt offered silent thanks that they were not seated outside in the cold and snow.

William Howard Taft was the handpicked heir to outgoing President Theodore Roosevelt, although the two men would have seemed to be very different. Taft's size—he was more than three hundred pounds on March 4, and destined to get larger—suggested a somewhat less-than-strenuous life, and his politics were not as instinctively progressive as Roosevelt's. Though pledged to continue the policies of his popular predecessor, Taft had a more restricted view of executive power, particularly on issues involving private property. The differences between Roosevelt and Taft would lead to a disastrous split in the Republican Party by 1912, but on Inauguration Day 1909 the two men were pleased with each other and with the direction of the nation.

On that snowy day in Washington, the nation was growing at a startling rate. The population would increase by 20 percent by the end of the decade, to ninety-one million. A million immigrants a year were pouring into the country, most of them settling into ethnic enclaves in the nation's cities. Urban life and industrial work were replacing the village and farm of nineteenth-century America.

The election of 1908 was approximately midway into the period known to historians as the Progressive Era, when journalists, political activists, and politicians agitated for change in the laissez-faire, individualistic consensus of the Gilded Age. Crusading reporters exposed horrifying conditions in the nation's cities and factories while exposing corruption in political clubhouses and corporate boardrooms. Women were demanding full voting rights. A nascent labor movement demanded the right to organize. And African American activists like Ida B. Wells campaigned against the scourge of lynchings in the South.

Theodore Roosevelt personified the era's determination to change not just the nation's politics, but its civic culture as well. Roosevelt scorned the go-it-alone dogma of the Gilded Age, saying that democracy and individualism could coexist only in a "simple and poor society." Giant corporations so dwarfed individual citizens, he wrote, that people had no choice but "to combine in their turn" for "self-defense."[1] By 1908, no serious presidential candidate could suggest that the nation had taken a step in the wrong direction.

The election of 1908 featured, as the main contestants, Roosevelt's designated heir, Taft, and the ubiquitous William Jennings Bryan, attempting his third national campaign in twelve years. The Democrat tried to push the limits of progressive change, assailing privilege and big business and suggesting that Taft and the Republicans represented both. But Taft, whose appearance surely suggested satisfaction with the status quo, successfully articulated the Roosevelt program of trust-busting, conservation, support for labor's right to unionize, and backing an income tax. Taft, then, was seen as a responsible reformer; Bryan, a potentially dangerous radical at a time when middle-class Progressives feared a confrontation between labor and capital, rich and poor. Taft won the election with ease, taking 7.6 million popular votes (about 51 percent) and 321 electoral votes to Bryan's 6.4 million (about 43 percent) and 162, respectively.

Inauguration Day offered Taft the opportunity to demonstrate that while he might have been Roosevelt's choice, he was not Roosevelt's echo. The sheer length of his inaugural address and its unadorned prose surely reminded official Washington, and the nation, that the age of the bully pulpit was over. Taft moved briskly from trust-busting to the need to revise heavily protective tariffs to the importance of scientific experiments in the Department of Agriculture, as if reading from a checklist. Some phrases, however, still leap out, not because of their artful construction but because of what they say about the times and the presidency. For example, Taft framed the triumph of progressive thought in asserting that the "scope of modern government in what it can and ought to accomplish

---

1. Andrew, ed., *Autobiography of Theodore Roosevelt*, abridged, 257–58.

for its people has been widened far beyond the principles laid down by the old 'laissez faire' school of political writers, and this widening has met popular approval."

By far the most interesting and moving part of the speech, and indeed its longest sustained treatment of any subject, was Taft's direct appeal to the solidly Democratic South and his plea for racial reconciliation in the former Confederacy. The native of Ohio spent several weeks in the South after his election and before the inauguration, which may have played a role in his decision to highlight the region and speak directly to its residents.

Taft spent a thousand words—more than Roosevelt's entire inaugural—on the South and on race. He asked for an "increase in the tolerance of political views of all kinds" in the South. He noted that the Fifteenth Amendment, which assured blacks of the right to vote, "has not been generally observed" but "it ought to be observed." He felt obliged to note that "Personally, I have not the slightest race prejudice or feeling," and he sympathized with the victims of racial prejudice. But what to do about it? He urged the "appointment to office" of blacks as "appreciation of their progress," but he also said no such appointments should be made where and when "race feeling is so widespread . . . as to interfere with the ease and facility with which the local government business can be done." The Fifteenth Amendment ought to be enforced, but he insisted that the federal government would not "interfere with the regulation by southern states of their domestic affairs."

Taft, it was said, was a very good dancer despite his girth. On the issues of the South and race, he tried to avoid stepping on toes, but in doing so, simply seemed to be moving backward.

Bibliographic Note
Paolo E. Coletta, *The Presidency of William Howard Taft*
(Lawrence, KS: University Press of Kansas, 1973).

MY FELLOW CITIZENS:

Anyone who has taken the oath I have just taken must feel a heavy weight of responsibility. If not, he has no conception of the powers and duties of the office upon which he is about to enter, or he is lacking in a proper sense of the obligation which the oath imposes.

The office of an inaugural address is to give a summary outline of the main policies of the new administration, so far as they can be anticipated. I have had the honor to be one of the advisers of my distinguished predecessor, and, as such, to hold up his hands in the reforms he has initiated. I should be untrue to myself, to my promises, and to the declarations of the party platform upon which I was elected to office, if I did not make the maintenance and enforcement of those reforms a most important feature of my administration. They were directed to the suppression of the lawlessness and abuses of power of the great combinations of capital invested in railroads and in industrial enterprises carrying on interstate commerce. The steps which my predecessor took and the legislation passed on his recommendation have accomplished much, have caused a general halt in the vicious policies which created popular alarm, and have brought about in the business affected a much higher regard for existing law.

To render the reforms lasting, however, and to secure at the same time freedom from alarm on the part of those pursuing proper and progressive business methods, further legislative and executive action are needed. Relief of the railroads from certain restrictions of the antitrust law have been urged by my predecessor and will be urged by me. On the other hand, the administration is pledged to legislation looking to a proper federal supervision and restriction to prevent excessive issues of bonds and stock by companies owning and operating interstate commerce railroads.

Then, too, a reorganization of the Department of Justice, of the Bureau of Corporations in the Department of Commerce and Labor, and of the Interstate Commerce Commission, looking to effective cooperation of these agencies, is needed to secure a more rapid and certain enforcement of the laws affecting interstate railroads and industrial combinations.

I hope to be able to submit at the first regular session of the incoming Congress, in December next, definite suggestions in respect to the needed amendments to the antitrust and the interstate commerce law and the changes required in the executive departments concerned in their enforcement.

It is believed that with the changes to be recommended American business can be assured of that measure of stability and certainty in respect to those things that may be done and those that are prohibited which is essential to the life and growth of all business. Such a plan must include the right of the people to avail themselves of those methods of combining capital and effort deemed necessary to reach the highest degree of economic efficiency, at the same time differentiating between combinations based upon legitimate economic reasons and those formed with the intent of creating monopolies and artificially controlling prices.

The work of formulating into practical shape such changes is creative work of the highest order, and requires all the deliberation possible in the interval. I believe that the amendments to be proposed are just as necessary in the protection of legitimate business as in the clinching of the reforms which properly bear the name of my predecessor.

A matter of most pressing importance is the revision of the tariff. In accordance with the promises of the platform upon which I was elected, I shall call Congress into extra session to meet on the 15th day of March, in order that consideration may be at once given to a bill revising the Dingley Act. This should secure an adequate revenue and adjust the duties in such a manner as to afford to labor and to all industries in this country, whether of the farm, mine, or factory, protection by tariff equal to the difference between the cost of production abroad and the cost of production here, and have a provision which shall put into force, upon executive determination of certain facts, a higher or maximum tariff against those countries whose trade policy toward us equitably requires such discrimination. It is thought that there has been such a change in conditions since the enactment of the Dingley Act, drafted on a similarly protective principle, that the measure of the tariff above stated will permit the reduction of rates in certain schedules and will require the advancement of few, if any.

The proposal to revise the tariff made in such an authoritative way as to lead the business community to count upon it necessarily halts all those branches of business directly affected; and as these are most important, it disturbs the whole business of the country. It is imperatively necessary, therefore, that a tariff bill be drawn in good faith in accordance with promises made before the election by the party in power, and as promptly passed as due consideration will permit. It is not that the tariff is more important in the long run than the perfecting of the reforms in respect to antitrust legislation and interstate commerce regulation, but the need for action when the revision of the tariff has been determined upon is more immediate to avoid embarrassment of business. To secure the needed speed in the passage of the tariff bill, it would seem wise to attempt no other legislation at the extra session. I venture this as a suggestion only, for the course to be taken by Congress, upon the call of the executive, is wholly within its discretion.

In the mailing of a tariff bill the prime motive is taxation and the securing thereby of a revenue. Due largely to the business depression which followed the financial panic of 1907, the revenue from customs and other sources has decreased to such an extent that the expenditures for the current fiscal year will exceed the receipts by $100 million. It is imperative that such a deficit shall not continue, and the framers of the tariff bill must, of course, have in mind the total revenues likely to be produced by it and so arrange the duties as to secure an adequate income. Should it be impossible to do so by import duties, new kinds of taxation must be adopted, and among these I recommend a graduated inheritance tax as correct in principle and as certain and easy of collection.

The obligation on the part of those responsible for the expenditures made to carry on the government, to be as economical as possible, and to make the burden of taxation as light as possible, is plain, and should be affirmed in every declaration of government policy. This is especially true when we are face to face with a heavy deficit. But when the desire to win the popular approval leads to the cutting off of expenditures really needed to make the government effective and to enable it to accomplish its proper objects, the result is as much to be condemned as the waste of government funds in unnecessary expenditure. The scope of a modern government in what it can and ought to accomplish for its people has been widened far beyond the principles laid down by the old "laissez-faire" school of political writers, and this widening has met popular approval.

In the Department of Agriculture the use of scientific experiments on a large scale and the spread of information derived from them for the improvement of general agriculture must go on.

The importance of supervising business of great railways and industrial combinations and the necessary investigation and prosecution of unlawful business methods are another necessary tax upon government which did not exist half a century ago.

The putting into force of laws which shall secure the conservation of our resources, so far as they may be within the jurisdiction of the federal government, including the most important work of saving and restoring our forests and the great improvement of waterways, are all proper government functions which must involve large expenditure if properly performed. While some of them, like the reclamation of arid lands, are made to pay for themselves, others are of such an indirect benefit that this cannot be expected of them. A permanent improvement, like the Panama Canal, should be treated as a distinct enterprise, and should be paid for by the proceeds of bonds, the issue of which will distribute its cost between the present and future generations in accordance with the benefits derived. It may well be submitted to the serious consideration of Congress whether the deepening and control of the channel of a great river system, like that of the Ohio or of the Mississippi, when definite and practical plans for the enterprise have been approved and determined upon, should not be provided for in the same way.

Then, too, there are expenditures of government absolutely necessary if our country is to maintain its proper place among the nations of the world, and is to exercise its proper influence in defense of its own trade interests in the maintenance of traditional American policy against the colonization of European monarchies in this hemisphere, and in the promotion of peace and international morality. I refer to the cost of maintaining a proper army, a proper navy, and suitable fortifications upon the mainland of the United States and in its dependencies.

We should have an army so organized and so officered as to be capable in time of emergency, in cooperation with the national militia and under the provisions of a proper national volunteer law, rapidly to expand into a force sufficient to resist all probable invasion from abroad and to furnish a respectable expeditionary force if necessary in the maintenance of our traditional American policy which bears the name of President Monroe.

Our fortifications are yet in a state of only partial completeness, and the number of men to man them is insufficient. In a few years however, the usual annual appropriations for our coast defenses, both on the mainland and in the dependencies, will make them sufficient to resist all direct attack, and by that time we may hope that the men to man them will be provided as a necessary adjunct. The distance of our shores from Europe and Asia of course reduces the necessity for maintaining under arms a great army, but it does not take away the requirement of mere prudence— that we should have an army sufficiently large and so constituted as to form a nucleus out of which a suitable force can quickly grow.

What has been said of the army may be affirmed in even a more emphatic way of the navy. A modern navy cannot be improvised. It must be built and in existence when the emergency arises which calls for its use and operation. My distinguished predecessor has in many speeches and messages set out with great force and striking language the necessity for maintaining a strong navy commensurate with the coastline, the governmental resources, and the foreign trade of our nation; and I wish to reiterate all the reasons which he has presented in favor of the policy of maintaining a strong navy as the best conservator of our peace with other nations, and the best means of securing respect for the assertion of our rights, the defense of our interests, and the exercise of our influence in international matters.

Our international policy is always to promote peace. We shall enter into any war with a full consciousness of the awful consequences that it always entails, whether successful or not, and we, of course, shall make every effort consistent with national honor and the highest national interest to avoid a resort to arms. We favor every instrumentality, like that of the Hague Tribunal and arbitration treaties made with a view to its use in all international controversies, in order to maintain peace and to avoid war. But we should be blind to existing conditions and should allow ourselves to become foolish idealists if we did not realize that, with all the nations of the world armed and prepared for war, we must be ourselves in a similar condition, in order to prevent other nations from taking advantage of us and of our inability to defend our interests and assert our rights with a strong hand.

In the international controversies that are likely to arise in the Orient growing out of the question of the open door and other issues the United States can maintain her interests intact and can secure respect for her just demands. She will not be able to do so, however, if it is understood that she never intends to back up her assertion of right and her defense of her interest by anything but mere verbal protest and diplomatic note. For these reasons the expenses of the army and navy and of coast defenses should always be considered as something which the government must pay for, and they should not be cut off through mere consideration of economy. Our government is able to afford a suitable army and a suitable navy. It may maintain them without the slightest danger to the Republic or the cause of free institutions, and fear of additional taxation ought not to change a proper policy in this regard.

The policy of the United States in the Spanish war and since has given it a position of influence among the nations that it never had before, and should be constantly exerted to securing to its bona fide citizens, whether native or naturalized, respect for them as such in foreign countries. We should make every effort to prevent humiliating and degrading prohibition against any of our citizens wishing temporarily to sojourn in foreign countries because of race or religion.

The admission of Asiatic immigrants who cannot be amalgamated with our population has been made the subject either of prohibitory clauses in our treaties and statutes or of strict administrative regulation secured by diplomatic negotiation. I sincerely hope that we may continue to minimize the evils likely to arise from such immigration without unnecessary friction and by mutual concessions between self-respecting governments. Meantime we must take every precaution to prevent, or failing that, to punish outbursts of race feeling among our people against foreigners of whatever nationality who have by our grant a treaty right to pursue lawful business here and to be protected against lawless assault or injury.

This leads me to point out a serious defect in the present federal jurisdiction, which ought to be remedied at once. Having assured to other countries by treaty the protection of our laws for such of their subjects or citizens as we permit to come within our jurisdiction, we now leave to a state or a city, not under the control of the federal government, the duty of performing our international obligations in this respect. By proper legislation we may, and ought to, place in the hands of the federal executive the means of enforcing the treaty rights of such aliens in the courts of the federal government. It puts our government in a pusillanimous position to make definite engagements to protect aliens and then to excuse the failure to perform those engagements by an explanation that the duty to keep them is in states or cities, not within our control. If we would promise we must put ourselves in a position to perform our promise. We cannot permit the possible failure of justice, due to local prejudice in any state or municipal government, to expose us to the risk of a war which might be avoided if federal jurisdiction was asserted by suitable legislation by Congress and carried out by proper proceedings instituted by the executive in the courts of the national government.

One of the reforms to be carried out during the incoming administration is a change of our monetary and banking laws, so as to secure greater elasticity in the forms of currency available for trade and to prevent the limitations of law from operating to increase the embarrassment of a financial panic. The monetary commission, lately appointed, is giving full consideration to existing conditions and to all proposed remedies, and will doubtless suggest one that will meet the requirements of business and of public interest.

We may hope that the report will embody neither the narrow view of those who believe that the sole purpose of the new system should be to secure a large return on banking capital or of those who would have greater expansion of currency with little regard to provisions for its immediate redemption or ultimate security. There is no subject of economic discussion so intricate and so likely to evoke differing views and dogmatic statements as this one.

The commission, in studying the general influence of currency on business and of business on currency, have wisely extended their investigations in European banking and monetary methods. The information that they have derived from such experts as they have found abroad will undoubtedly be found helpful in the solution of the difficult problem they have in hand.

The incoming Congress should promptly fulfill the promise of the Republican platform and pass a proper postal savings bank bill. It will not be unwise or excessive paternalism. The promise to repay by the government will furnish an inducement to savings deposits which private enterprise cannot supply and at such a low rate of interest as not to withdraw custom from existing banks. It will substantially increase the funds available for investment as capital in useful enterprises. It will furnish absolute security which makes the proposed scheme of government guaranty of deposits so alluring, without its pernicious results.

I sincerely hope that the incoming Congress will be alive, as it should be, to the importance of our foreign trade and of encouraging it in every way feasible. The possibility of increasing this trade in the Orient, in the Philippines, and in South America are known to everyone who has given the matter attention. The direct effect of free trade between this country and the Philippines will be marked upon our sales of cottons, agricultural machinery, and other manufactures. The necessity of the establishment of direct lines of steamers between North and South America has been brought to the attention of Congress by my predecessor and by Mr. Root before and after his noteworthy visit to that continent, and I sincerely hope that Congress may be induced to see the wisdom of a tentative effort to establish such lines by the use of mail subsidies.

The importance of the part which the Departments of Agriculture and of Commerce and Labor may play in ridding the markets of Europe of prohibitions and discriminations against the importation of our products is fully understood, and it is hoped that the use of the maximum and minimum feature of our tariff law to be soon passed will be effective to remove many of those restrictions.

The Panama Canal will have a most important bearing upon the trade between the eastern and far western sections of our country, and will greatly increase the facilities for transportation between the eastern and the western seaboard, and may possibly revolutionize the transcontinental rates with respect to bulky merchandise. It will also have a most beneficial effect to increase the trade between the eastern seaboard of the United States and the western coast of South America, and, indeed, with some of the important ports on the east coast of South America reached by rail from the west coast.

The work on the canal is making most satisfactory progress. The type of the canal as a lock canal was fixed by Congress after a full consideration of the conflicting reports of the majority and minority of the consulting board, and after the recommendation of the War Department and the executive upon those reports. Recent suggestion that something had occurred on the isthmus to make the lock type of the canal less feasible than it was supposed to be when the reports were made and the policy determined on led to a visit to the isthmus of a board of competent engineers to examine the Gatun dam and locks, which are the key of the lock type. The report of that board shows nothing has occurred in the nature of newly revealed evidence which should change the views once formed in the original discussion. The construction will go on under a most effective organization controlled by Colonel Goethals and his fellow army engineers associated with him, and will certainly be completed early in the next administration, if not before.

Some type of canal must be constructed. The lock type has been selected. We are all in favor of having it built as promptly as possible. We must not now, therefore, keep up a fire in the rear of the agents whom we have authorized to do our work on the isthmus. We must hold up their hands, and speaking for the incoming administration I wish to say that I propose to devote all the energy possible and under my control to the pushing of this work on the plans which have been adopted, and to stand behind the men who are doing faithful, hard work to bring about the early completion of this, the greatest constructive enterprise of modern times.

The governments of our dependencies in Puerto Rico and the Philippines are progressing as favorably as could be desired. The prosperity of Puerto Rico continues unabated. The business conditions in the Philippines are not all that we could wish them to be, but with the passage of the new tariff bill permitting free trade between the United States and the archipelago, with such limitations on sugar and tobacco as shall prevent injury to domestic interests in those products, we can count on an improvement in business conditions in the Philippines and the development of a mutually profitable trade between this country and the islands. Meantime our government in each dependency is upholding the traditions of civil liberty and increasing popular control which might be expected under American auspices. The work which we are doing there redounds to our credit as a nation.

I look forward with hope to increasing the already good feeling between the South and the other sections of the country. My chief purpose is not to effect a change in the electoral vote of the southern states. That is a secondary consideration. What I look forward to is an increase in the tolerance of political views of all kinds and their advocacy throughout the South, and the existence of a respectable political opposition in every state; even more than this, to an increased feeling on the part of all the people in the South that this government is their government, and that its officers in their states are their officers.

The consideration of this question cannot, however, be complete and full without reference to the Negro race, its progress and its present condition. The Thirteenth Amendment secured them freedom; the Fourteenth Amendment due process of law, protection of property, and the pursuit of happiness; and the Fifteenth Amendment attempted to secure the Negro against any deprivation of the privilege to vote because he was a Negro. The Thirteenth and Fourteenth amendments have been generally enforced and have secured the objects for which they are intended. While the Fifteenth Amendment has not been generally observed in the past, it ought to be observed, and the tendency of southern legislation today is toward the enactment of electoral qualifications which shall square with that amendment. Of course, the mere adoption of a constitutional law is only one step in the right direction. It must be fairly and justly enforced as well. In time both will come. Hence it is clear to all that the domination of an ignorant, irresponsible element can be prevented by constitutional laws which shall exclude from voting both Negroes and whites not having education or other qualifications thought to be necessary for a proper electorate. The danger of the control of an ignorant electorate has therefore passed. With this change, the interest which many of the southern white citizens take in the welfare of the Negroes has increased. The colored men must base their hope on the results of their own industry, self-restraint, thrift, and business success, as well as upon the aid and comfort and sympathy which they may receive from their white neighbors of the South.

There was a time when Northerners who sympathized with the Negro in his necessary struggle for better conditions sought to give him the suffrage as a protection to enforce its exercise against the prevailing sentiment of the South. The movement proved to be a failure. What remains is the Fifteenth Amendment to the Constitution and the right to have statutes of states specifying qualifications for electors subjected to the test of compliance with that amendment. This is a great protection to the Negro.

It never will be repealed, and it never ought to be repealed. If it had not passed, it might be difficult now to adopt it; but with it in our fundamental law, the policy of southern legislation must and will tend to obey it, and so long as the statutes of the states meet the test of this amendment and are not otherwise in conflict with the Constitution and laws of the United States, it is not the disposition or within the province of the federal government to interfere with the regulation by southern states of their domestic affairs. There is in the South a stronger feeling than ever among the intelligent, well-to-do, and influential element in favor of the industrial education of the Negro and the encouragement of the race to make themselves useful members of the community. The progress which the Negro has made in the last fifty years, from slavery, when its statistics are reviewed, is marvelous, and it furnishes every reason to hope that in the next twenty-five years a still greater improvement in his condition as a productive member of society, on the farm, and in the shop, and in other occupations may come.

The Negroes are now Americans. Their ancestors came here years ago against their will, and this is their only country and their only flag. They have shown themselves anxious to live for it and to die for it. Encountering the race feeling against them, subjected at times to cruel injustice growing out of it, they may well have our profound sympathy and aid in the struggle they are making. We are charged with the sacred duty of making their path as smooth and easy as we can. Any recognition of their distinguished men, any appointment to office from among their number, is properly taken as an encouragement and an appreciation of their progress, and this just policy should be pursued when suitable occasion offers.

But it may well admit of doubt whether, in the case of any race, an appointment of one of their number to a local office in a community in which the race feeling is so widespread and acute as to interfere with the ease and facility with which the local government business can be done by the appointee is of sufficient benefit by way of encouragement to the race to outweigh the recurrence and increase of race feeling which such an appointment is likely to engender. Therefore the executive, in recognizing the Negro race by appointments, must exercise a careful discretion not thereby to do it more harm than good. On the other hand, we must be careful not to encourage the mere pretense of race feeling manufactured in the interest of individual political ambition.

Personally, I have not the slightest race prejudice or feeling, and recognition of its existence only awakens in my heart a deeper sympathy for those who have to bear it or suffer from it, and I question the wisdom of a policy which is likely to increase it. Meantime, if nothing is done to prevent it, a better feeling between the Negroes and the whites in the South will continue to grow, and more and more of the white people will come to realize that the future of the South is to be much benefited by the industrial and intellectual progress of the Negro. The exercise of political franchises by those of this race who are intelligent and well-to-do will be acquiesced in, and the right to vote will be withheld only from the ignorant and irresponsible of both races.

There is one other matter to which I shall refer. It was made the subject of great controversy during the election and calls for at least a passing reference now. My distinguished predecessor has given much attention to the cause of labor, with whose struggle for better things he has shown the sincerest sympathy. At his instance Congress has passed the bill fixing the liability of interstate carriers to their employees for injury sustained in the course of employment, abolishing the rule of fellow servant and the common-law rule as to contributory negligence, and substituting therefor the so-called rule of comparative negligence. It has also passed a law fixing the compensation of government employees for injuries sustained in the employ of the government through the negligence of the superior. It has also passed a model child-labor law for the District of Columbia. In previous administrations an arbitration law for interstate commerce railroads and their employees, and laws for the application of safety devices to save the lives and limbs of employees of interstate railroads had been passed. Additional legislation of this kind was passed by the outgoing Congress.

I wish to say that insofar as I can I hope to promote the enactment of further legislation of this character. I am strongly convinced that the government should make itself as responsible to employees injured in its employ as an interstate-railway corporation is made responsible by federal law to its employees; and I shall be glad, whenever any additional reasonable safety device can be invented to reduce the loss of life and limb among railway employees, to urge Congress to require its adoption by interstate railways.

Another labor question has arisen which has awakened the most excited discussion. That is in respect to the power of the federal courts to issue injunctions in industrial disputes. As to that, my convictions are fixed. Take away from the courts, if it could be taken away, the power to issue injunctions in labor disputes, and it would create a privileged class among the laborers and save the lawless among their number from a most needful remedy available to all men for the protection of their business against lawless invasion. The proposition that business is not a property or pecuniary right which can be protected by equitable injunction is utterly without foundation in precedent or reason. The proposition is usually linked with one to make the secondary boycott lawful. Such a proposition is at variance with the American instinct, and will find no support, in my judgment, when submitted to the American people. The secondary boycott is an instrument of tyranny, and ought not to be made legitimate.

The issue of a temporary restraining order without notice has in several instances been abused by its inconsiderate exercise, and to remedy this the platform upon which I was elected recommends the formulation in a statute of the conditions under which such a temporary restraining order ought to issue. A statute can and ought to be framed to embody the best modern practice, and can bring the subject so closely to the attention of the court as to make abuses of the process unlikely in the future. The American people, if I understand them, insist that the authority of the courts shall be sustained, and are opposed to any change in the procedure by which the powers of a court may be weakened and the fearless and effective administration of justice be interfered with.

Having thus reviewed the questions likely to recur during my administration, and having expressed in a summary way the position which I expect to take in recommendations to Congress and in my conduct as an executive, I invoke the considerate sympathy and support of my fellow citizens and the aid of the Almighty God in the discharge of my responsible duties.

Woodrow Wilson

· FIRST INAUGURAL ADDRESS ·

Tuesday, March 4, 1913

The election of 1912 featured a former President, Theodore Roosevelt, the incumbent President, William Howard Taft, and a future President, Woodrow Wilson. All three candidates could and did claim to represent the Progressive Era's faith in activist government, although it was Roosevelt who actually ran under the banner of the Progressive Party.

While the 1912 election produced a President, Wilson, who would have a profound effect on America's role as a world power, the story of the campaign really is the story of Roosevelt and Taft, mentor and protégé. Taft, of course, was Roosevelt's designated heir in 1908, but by 1910 the former President had grown impatient with what he viewed as Taft's pro-business, conservative style of governance. Liberal Republicans such as Robert La Follette of Wisconsin were similarly hostile to Taft as 1912 approached, making a challenge to the President's renomination likely. The presidential campaign that year featured a new wrinkle—the first widespread use of state primary elections to determine the allocation of delegates to national conventions. Most state delegations remained in the control of party bosses, but in North Dakota, New York, Wisconsin, Illinois, California, and several other states, party rank and file had a hand in influencing the nomination process. Roosevelt formally entered the race in February, as did La Follette.

The ensuing battle was epic in its drama, pathos, and personal tragedy. Both Roosevelt and Taft felt betrayed by the other. Friends and fellow Republicans were forced to choose sides in the party's civil war. When the Republicans gathered for their convention in Chicago, Taft had the upper hand, especially after one of his allies, Elihu Root, was elected the convention's temporary chairman. Roosevelt's forces realized they were outnumbered, so they walked out of the convention, met separately, and nominated

Roosevelt as their candidate. The Republicans who remained nominated Taft over La Follette. The Progressive Party nominated Roosevelt and gained a new nickname when its candidate announced that he was as fit as a bull moose.

Into this mess stepped Thomas Woodrow Wilson, a man new to the practice of politics though no stranger to political theory and history. He was a lawyer with a PhD in political science and history from Johns Hopkins University, an accomplished author and an academic spokesman for the Progressive movement's agenda of reform. In 1912, Wilson, a one-term governor of New Jersey, benefited from the Democratic Party's dearth of presidential timber and its absolute desperation. Having won the presidency only twice since the Civil War, the Democrats were not about to turn again to three-time loser William Jennings Bryan. Wilson, who had traveled widely during his short tenure in New Jersey, was a known quantity among the party's leaders, if not among the rank and file. He was the new face the Democrats needed after a generation of losing.

As Wilson, Roosevelt, and, to a lesser extent, a demoralized Taft scoured the country for votes, the ideas and agenda of the Progressive movement set the terms of debate. In the decade leading to 1912, horrified Americans read lurid accounts of the nation's factories, food plants, and political clubhouses through the journalism of Lincoln Steffens, Ray S. Baker, and Ida M. Tarbell, among others. A fire in the Triangle Shirtwaist Factory in New York in 1911 killed nearly 150 young women, and a subsequent investigation shed light on dangerous working conditions and the nation's shocking reliance on child labor—about four million children younger than fifteen years old were in the workforce in 1912.

Advocates for women's suffrage, for the prohibition of alcohol consumption, and for the rights of labor unions demanded that their voices be heard amid the clatter of politics as usual. Change, then, was very much the theme of the 1912 campaign, further dooming the incumbent, Taft, who was seen as a symbol of the status quo.

With the Republicans fatally and bitterly divided, Woodrow Wilson won 435 votes in the Electoral College at a time when 266 were required to win. Roosevelt placed second in the electoral vote with 88, while the incumbent, Taft, managed only 8. Still, Wilson's overwhelming electoral vote total did not translate into a majority of the popular vote. With the total split among four candidates—the Socialist Eugene V. Debs ran one of his stronger races—Wilson managed just about 42 percent of the votes cast, with about 6.2 million.

In addressing the nation and an audience of about a hundred thousand for the first time as President, Wilson emphasized the need for change. And part of that change would be in the way in which the President spoke to his fellow citizens. In a marked departure from most of his predecessors, Wilson took the occasion to outline not the country's strengths, but its flaws as he saw them. He believed the nation had "squandered a great part of what we might have used," that Americans had not considered "the human cost, the cost of lives snuffed out" by industrial and economic growth. "The great government we loved has too often been made use of for private and selfish purposes, and those who used it had forgotten the people." That, he said, was about to change. "The scales of heedlessness have fallen from our eyes."

Wilson's inaugural contained an outline of his goals, chief among them an end to a "tariff which cuts us off from our proper part in the commerce of the world." He emphasized the need for conservation, banking reforms, and more laws to protect workers and consumers.

Rhetorically, however, the heart of the speech is Wilson's mini-sermon about a government "too often debauched and made an instrument of evil." But who had so debauched the government? Was it the man seated on the platform with him, Taft? Was it members of Congress, with whom he would have to work closely to pass his program? Was it the bosses who opposed him in New Jersey and at the Democratic National Convention?

He was discreet about such details. But there was no mistaking his grim determination to bring about change. "This is not a day of triumph," he said in closing, "it is a day of dedication. Here muster, not the forces of party, but the forces of humanity."

And Wilson surely saw himself marching at the head of humanity's column.

Bibliographic Note
Francis L. Broderick, *Progressivism at Risk: Electing a President in 1912*
(New York: Greenwood Press, 1989).

Kendrick A. Clements, *The Presidency of Woodrow Wilson*
(Lawrence, KS: University Press of Kansas, 1992).

THERE HAS BEEN A CHANGE OF GOVERNMENT. It began two years ago, when the House of Representatives became Democratic by a decisive majority. It has now been completed. The Senate about to assemble will also be Democratic. The offices of President and vice president have been put into the hands of Democrats. What does the change mean? That is the question that is uppermost in our minds today. That is the question I am going to try to answer, in order, if I may, to interpret the occasion.

It means much more than the mere success of a party. The success of a party means little except when the nation is using that party for a large and definite purpose. No one can mistake the purpose for which the nation now seeks to use the Democratic Party. It seeks to use it to interpret a change in its own plans and point of view. Some old things with which we had grown familiar, and which had begun to creep into the very habit of our thought and of our lives, have altered their aspect as we have latterly looked critically upon them, with fresh, awakened eyes; have dropped their disguises and shown themselves alien and sinister. Some new things, as we look frankly upon them, willing to comprehend their real character, have come to assume the aspect of things long believed in and familiar, stuff of our own convictions. We have been refreshed by a new insight into our own life.

We see that in many things that life is very great. It is incomparably great in its material aspects, in its body of wealth, in the diversity and sweep of its energy, in the industries which have been conceived and built up by the genius of individual men and the limitless enterprise of groups of men. It is great, also, very great, in its moral force. Nowhere else in the world have noble men and women exhibited in more striking forms the beauty and the energy of sympathy and helpfulness and counsel in their efforts to rectify wrong, alleviate suffering, and set the weak in the way of strength and hope. We have built up, moreover, a great system of government, which has stood through a long age as in many respects a model for those who seek to set liberty upon foundations that will endure against fortuitous change, against storm and accident. Our life contains every great thing, and contains it in rich abundance.

But the evil has come with the good, and much fine gold has been corroded. With riches has come inexcusable waste. We have squandered a great part of what we might have used, and have not stopped to conserve the exceeding bounty of nature, without which our genius for enterprise would have been worthless and impotent, scorning to be careful, shamefully prodigal as well as admirably efficient. We have been proud of our industrial achievements, but we have not hitherto stopped thoughtfully enough to count the human cost, the cost of lives snuffed out, of energies overtaxed and broken, the fearful physical and spiritual cost to the men and women and children upon whom the dead weight and burden of it all has fallen pitilessly the years through. The groans and agony of it all had not yet reached our ears, the solemn, moving undertone of our life, coming up out of the mines and factories, and out of every home where the struggle had its intimate and familiar seat. With the great government went many deep secret things which we too long delayed to look into and scrutinize with candid, fearless eyes. The great government we loved has too often been made use of for private and selfish purposes, and those who used it had forgotten the people.

At last a vision has been vouchsafed us of our life as a whole. We see the bad with the good, the debased and decadent with the sound and vital. With this vision we approach new affairs. Our duty is to cleanse, to reconsider, to restore, to correct the evil without impairing the good, to purify and humanize every process of our common life without weakening or sentimentalizing it. There has been something crude and heartless and unfeeling in our haste to succeed and be great. Our thought has been "Let every man look out for himself, let every generation look out for itself," while we reared giant machinery which made it impossible that any but those who stood at the levers of control should have a chance to look out for themselves. We had not forgotten our morals. We remembered well enough that we had set up a policy which was meant to serve the humblest as well as the most powerful, with an eye single to the standards of justice and fair play, and remembered it with pride. But we were very heedless and in a hurry to be great.

We have come now to the sober second thought. The scales of heedlessness have fallen from our eyes. We have made up our minds to square every process of our national life again with the standards we so proudly set up at the beginning and have always carried at our hearts. Our work is a work of restoration.

We have itemized with some degree of particularity the things that ought to be altered and here are some of the chief items: A tariff which cuts us off from our proper part in the commerce of the world, violates the just principles of taxation, and makes the government a facile instrument in the hand of private interests; a banking and currency system based upon the necessity of the government to sell its bonds fifty years ago and perfectly adapted to concentrating cash and restricting credits; an industrial system which, take it on all its sides, financial as well as administrative, holds capital in leading strings, restricts the liberties and limits the opportunities of labor, and exploits without renewing or conserving the natural resources of the country; a body of agricultural activities never yet given the efficiency of great business undertakings or served as it should be through the instrumentality of science taken directly to the farm, or afforded the facilities of credit best suited to its practical needs; watercourses undeveloped, waste places unreclaimed, forests untended, fast disappearing without plan or prospect of renewal, unregarded waste heaps at every mine. We have studied as perhaps no other nation has the most effective means of production, but we have not studied cost or economy as we should either as organizers of industry, as statesmen, or as individuals.

Nor have we studied and perfected the means by which government may be put at the service of humanity, in safeguarding the health of the nation, the health of its men and its women and its children, as well as their rights in the struggle for existence. This is no sentimental duty. The firm basis of government is justice, not pity. These are matters of justice. There can be no equality or opportunity, the first essential of justice in the body politic, if men and women and children be not shielded in their lives, their very vitality, from the consequences of great industrial and social processes which they cannot alter, control, or singly cope with. Society must see to it that it does not itself crush or weaken or damage its own constituent parts. The first duty of law is to keep sound the society it serves. Sanitary laws, pure food laws, and laws determining conditions of labor which individuals are powerless to determine for themselves are intimate parts of the very business of justice and legal efficiency.

These are some of the things we ought to do, and not leave the others undone, the old-fashioned, never-to-be-neglected, fundamental safeguarding of property and of individual right. This is the high enterprise of the new day: To lift everything that concerns our life as a nation to the light that shines from the hearthfire of every man's conscience and vision of the right. It is inconceivable that we should do this as partisans; it is inconceivable we should do it in ignorance of the facts as they are or in blind haste. We shall restore, not destroy. We shall deal with our economic system as it is and as it may be modified, not as it might be if we had a clean sheet of paper to write upon; and step by step we shall make it what it should be, in the spirit of those who question their own wisdom and seek counsel and knowledge, not shallow self-satisfaction or the excitement of excursions whither they cannot tell. Justice, and only justice, shall always be our motto.

And yet it will be no cool process of mere science. The nation has been deeply stirred, stirred by a solemn passion, stirred by the knowledge of wrong, of ideals lost, of government too often debauched and made an instrument of evil. The feelings with which we face this new age of right and opportunity sweep across our heartstrings like some air out of God's own presence, where justice and mercy are reconciled and the judge and the brother are one. We know our task to be no mere task of politics but a task which shall search us through and through, whether we be able to understand our time and the need of our people, whether we be indeed their spokesmen and interpreters, whether we have the pure heart to comprehend and the rectified will to choose our high course of action.

This is not a day of triumph; it is a day of dedication. Here muster, not the forces of party, but the forces of humanity. Men's hearts wait upon us; men's lives hang in the balance; men's hopes call upon us to say what we will do. Who shall live up to the great trust? Who dares fail to try? I summon all honest men, all patriotic, all forward-looking men, to my side. God helping me, I will not fail them, if they will but counsel and sustain me!

Woodrow Wilson

Monday, March 5, 1917

On the night of March 3, a Saturday, President Woodrow Wilson worked late into the night. Reelected by a narrow margin in November, he was scheduled to take the oath of office as President in a private ceremony the following day. But Europe was in flames, and the United States was attempting, vainly, to keep the fire of war from spreading. Wilson was in no mood to celebrate. The public ceremonies, in any case, would not take place until Monday, in keeping with the practice of not holding inaugurations on Sunday.

The matter at hand on March 3 was proving to be bitter and frustrating for the President. After several German attacks on American shipping in the Atlantic Ocean, Wilson had asked Congress to approve a plan to arm merchant vessels. The House of Representatives agreed, their enthusiasm fueled by the President's release of an intercepted German telegram proposing an alliance between Imperial Germany and Mexico. The document, transmitted by German foreign secretary Arthur Zimmerman to his country's ambassador to Mexico, inflamed the American public against the Germans, in part because it promised to return to Mexico lands lost to the United States in the Mexican War of 1846.

The Senate, however, was not so easily convinced that merchant ships required guns. Senators Robert La Follette of Wisconsin, William Borah of Idaho, and others mounted a filibuster against the plan. The proposal died an agonizing death on the Senate floor on March 3, the last day of the expiring Congress. Wilson, monitoring events in the White House that night, was enraged. He wrote and released a statement whose tone was absolutely authentic and undoubtedly counterproductive. "A little group of willful men, representing no opinion but their own, have rendered the great government of the United States helpless and contemptible,"[1] Wilson said. Those words, especially the first phrase, made headlines and overshadowed the words he would speak on March 5.

Seldom has a reelected President suffered so bitter a loss on the eve of a second term. But Wilson's victory in November had been extremely close: he defeated Charles Evans Hughes, the first and only sitting Supreme Court justice to run for President on a major party line, by just 23 votes in the Electoral College, 277 to 254. The election hinged on California's 13 votes—Wilson won the state by fewer than four thousand votes.

The war in Europe dominated the campaign, and, of course, would dominate Wilson's second administration after Congress declared war on Germany and its allies on April 6, 1917. Wilson's campaign slogan reminded voters that "He kept us out of war." He had, indeed, despite provocations and threats from the Old World, and even after attacks on U.S. shipping.

But as Wilson prepared for his second term, it was evident not only to him but to the country that war was not far away, that unlike European wars in the nineteenth century this one coincided with the spread of American interests around the globe, and this one involved weaponry—the submarine—that could reach American shores.

"We are provincials no longer," Wilson told a somber crowd on March 5 after retaking the oath of office, this time in its traditional setting on the Capitol's East Portico. A quiet crowd of about fifty thousand witnessed the ceremony, although most could not hear a word Wilson said. The world was changing even as Wilson spoke, but the President's inaugural address still lacked even primitive amplification.

Wilson's speech was solemn and almost resigned in tone, a speech appropriate for a nation at war rather than at peace but without the defiance and rhetorical flourishes of wartime addresses. Instead, Wilson began setting the moral framework for war, should it come, with a recitation of principles "we shall stand for, whether in war or in peace." They included equality of nations, freedom of the seas (which Germany, of course, had disrupted), a limit on armaments, and a rejection of Old World balance-of-power arrangements. These principles, he said, "spring up native amongst us."

In this speech and in others, Wilson saw himself as the spokesman for values shared by all his fellow citizens. He asserted that he did not have to

---

1. August Heckscher, *Woodrow Wilson* (New York: Scribner, 1991) 431.

list the principles Americans stood for because they were "part and parcel of your own thinking." It invariably came as a surprise, it seems, when Wilson discovered that not everybody shared the same values after all, whether they were a "little group of willful men" in the Senate, or political radicals in the immediate postwar years.

For decades, the United States had been able to stand aside while the Old World marched from one battlefield to the next, from one war to another. It avoided entangling alliances, it had no interest in choosing sides. But now, with the United States a world power with interests spanning the globe, the catastrophic contest in Europe was "impossible to avoid." That did not necessarily mean war, Wilson said, but simply that the war had "affected the life of the whole world . . . We are the blood of all the nations that are at war. The currents of our thoughts as well as the currents of our trade run quick at all seasons back and forth between us and them." Isolationism, he said in so many words, was no longer an option, and that would be so regardless of the war's outcome.

Gone from this speech was the idea that Woodrow Wilson would or could continue to keep the United States out of war. War, in fact, had become a possibility that Wilson acknowledged in asserting that the nation "may even be drawn on, by circumstances, not by our own purpose or desire, to a more active assertion of our rights as we see them and a more immediate association with the great struggle itself." His speech may have been hard to hear, but there was no mistaking what those words meant.

On April 2, 1917, Wilson gave a far more momentous speech as he asked a special session of Congress to declare war on Germany and the Central Powers. American intervention ensured the French and British of victory and seemed to mark America's entry into the power arrangements of the Old World. Wilson hoped that an American presence among the great powers would help to purify them, to lift them up toward some admirable, righteous goal. He believed Americans shared that vision. But he was wrong.

My fellow citizens:

The four years which have elapsed since last I stood in this place have been crowded with counsel and action of the most vital interest and consequence. Perhaps no equal period in our history has been so fruitful of important reforms in our economic and industrial life or so full of significant changes in the spirit and purpose of our political action. We have sought very thoughtfully to set our house in order, correct the grosser errors and abuses of our industrial life, liberate and quicken the processes of our national genius and energy, and lift our politics to a broader view of the people's essential interests.

It is a record of singular variety and singular distinction. But I shall not attempt to review it. It speaks for itself and will be of increasing influence as the years go by. This is not the time for retrospect. It is time rather to speak our thoughts and purposes concerning the present and the immediate future.

Although we have centered counsel and action with such unusual concentration and success upon the great problems of domestic legislation to which we addressed ourselves four years ago, other matters have more and more forced themselves upon our attention—matters lying outside our own life as a nation and over which we had no control, but which, despite our wish to keep free of them, have drawn us more and more irresistibly into their own current and influence.

It has been impossible to avoid them. They have affected the life of the whole world. They have shaken men everywhere with a passion and an apprehension they never knew before. It has been hard to preserve calm counsel while the thought of our own people swayed this way and that under their influence. We are a composite and cosmopolitan people. We are of the blood of all the nations that are at war. The currents of our thoughts as well as the currents of our trade run quick at all seasons back and forth between us and them. The war inevitably set its mark from the first alike upon our minds, our industries, our commerce, our politics, and our social action. To be indifferent to it, or independent of it, was out of the question.

And yet all the while we have been conscious that we were not part of it. In that consciousness, despite many divisions, we have drawn closer together. We have been deeply wronged upon the seas, but we have not wished to wrong or injure in return; have retained throughout the consciousness of standing in some sort apart, intent upon an interest that transcended the immediate issues of the war itself.

As some of the injuries done us have become intolerable we have still been clear that we wished nothing for ourselves that we were not ready to demand for all mankind—fair dealing, justice, the freedom to live and to be at ease against organized wrong.

It is in this spirit and with this thought that we have grown more and more aware, more and more certain that the part we wished to play was the part of those who mean to vindicate and fortify peace. We have been obliged to arm ourselves to make good our claim to a certain minimum of right and of freedom of action. We stand firm in armed neutrality since it seems that in no other way we can demonstrate what it is we insist upon and cannot forget. We may even be drawn on, by circumstances, not by our own purpose or desire, to a more active assertion of our rights as we see them and a more immediate association with the great struggle itself. But nothing will alter our thought or our purpose. They are too clear to be obscured. They are too deeply rooted in the principles of our national life to be altered. We desire neither conquest nor advantage. We wish nothing that can be had only at the cost of another people. We always professed unselfish purpose and we covet the opportunity to prove our professions are sincere.

There are many things still to be done at home, to clarify our own politics and add new vitality to the industrial processes of our own life, and we shall do them as time and opportunity serve, but we realize that the greatest things that remain to be done must be done with the whole world for a stage and in cooperation with the wide and universal forces of mankind, and we are making our spirits ready for those things.

We are provincials no longer. The tragic events of the thirty months of vital turmoil through which we have just passed have made us citizens of the world. There can be no turning back. Our own fortunes as a nation are involved whether we would have it so or not.

And yet we are not the less Americans on that account. We shall be the more American if we but remain true to the principles in which we have been bred. They are not the principles of a province or of a single continent. We have known and boasted all along that they were the principles of a liberated mankind. These, therefore, are the things we shall stand for, whether in war or in peace:

That all nations are equally interested in the peace of the world and in the political stability of free peoples, and equally responsible for their maintenance; that the essential principle of peace is the actual equality of nations in all matters of right or privilege; that peace cannot securely or justly rest upon an armed balance of power; that governments derive all their just powers from the consent of the governed and that no other powers should be supported by the common thought, purpose, or power of the family of nations; that the seas should be equally free and safe for the use of all peoples, under rules set up by common agreement and consent, and that, so far as practicable, they should be accessible to all upon equal terms; that national armaments shall be limited to the necessities of national order and domestic safety; that the community of interest and of power upon which peace must henceforth depend imposes upon each nation the duty of seeing to it that all influences proceeding from its own citizens meant to encourage or assist revolution in other states should be sternly and effectually suppressed and prevented.

I need not argue these principles to you, my fellow countrymen; they are your own part and parcel of your own thinking and your own motives in affairs. They spring up native amongst us. Upon this as a platform of purpose and of action we can stand together. And it is imperative that we should stand together. We are being forged into a new unity amidst the fires that now blaze throughout the world. In their ardent heat we shall, in God's Providence, let us hope, be purged of faction and division, purified of the errant humors of party and of private interest, and shall stand forth in the days to come with a new dignity of national pride and spirit. Let each man see to it that the dedication is in his own heart, the high purpose of the nation in his own mind, ruler of his own will and desire.

I stand here and have taken the high and solemn oath to which you have been audience because the people of the United States have chosen me for this august delegation of power and have by their gracious judgment named me their leader in affairs.

I know now what the task means. I realize to the full the responsibility which it involves. I pray God I may be given the wisdom and the prudence to do my duty in the true spirit of this great people. I am their servant and can succeed only as they sustain and guide me by their confidence and their counsel. The thing I shall count upon, the thing without which neither counsel nor action will avail, is the unity of America—an America united in feeling, in purpose, and in its vision of duty, of opportunity, and of service.

We are to beware of all men who would turn the tasks and the necessities of the nation to their own private profit or use them for the building up of private power.

United alike in the conception of our duty and in the high resolve to perform it in the face of all men, let us dedicate ourselves to the great task to which we must now set our hand. For myself I beg your tolerance, your countenance, and your united aid.

The shadows that now lie dark upon our path will soon be dispelled, and we shall walk with the light all about us if we be but true to ourselves— to ourselves as we have wished to be known in the counsels of the world and in the thought of all those who love liberty and justice and the right exalted.

Warren G. Harding

INAUGURAL ADDRESS

Friday, March 4, 1921

Woodrow Wilson was ahead of his time when he declared that the United States had entered into a new age, an age in which its people were "no longer provincials," an age where the responsibilities of a great power could no longer be avoided. The clarion call America heard and responded to in 1920 was not a call to leadership but a return to what Warren G. Harding called "normalcy."

For Harding, his fellow Republicans, and millions of voters, "normalcy" meant more than a return to peacetime. It meant an end to the disruptions and chaos that had been unleashed after the war. Some changes clearly could not be reversed: in 1920 women voted for the first time. But perhaps, by returning to "normalcy," the passions of political radicals, the agitation of labor leaders, and the ever-increasing influence of the nation's cities could be tamed.

Harding, a senator and newspaper publisher from Ohio, was the compromise candidate of a Republican Party eager to challenge the party of a suddenly unpopular Woodrow Wilson. As a candidate, Harding had an impressive bearing, a face made for a sculptor's chisel, and a pleasing style. With a record devoid of substance and thus unable to inspire criticism, he was a perfect candidate for a party wary of revisiting the divisions that had handed Wilson his victory in 1912. Or so it seemed. Before his nomination, party leaders asked him if there were any scandals in his private life that might haunt him during the campaign. None, he replied, although it later turned out that he was less than a faithful husband.

As American troops sailed home in triumph in late 1918 and early 1919, the nation seemed anything but triumphant. The cost of living for millions of American workers rose dramatically during the war: food nearly doubled in price, clothing more than doubled.[1] Some four million workers went on strike in 1919. Authorities feared that the Bolshevik revolution in Russia would be exported to the United States, leading Attorney General A. Mitchell Palmer to authorize raids on suspected radicals that swooped up ten thousand people in 1920.

The world of statecraft was undergoing a similar breakdown. President Wilson returned from the peace conference in Versailles with a plan for U.S. membership in a League of Nations, a global organization that would attempt to resolve international disputes without resort to war. The League of Nations plan was part of the peace treaty, which Wilson submitted to the Senate for its approval. Opposition mounted quickly, leading Wilson to spend every ounce of his energy on a speaking tour around the country. He refused to consider compromises and refused to pay heed to concerns about his health. He suffered a stroke in October, disabling him and effectively ending any chance that the treaty, with its commitment to the league, would pass. As he lay in bed in Washington, the Senate rejected the treaty on November 19, 1919.

All areas of society, it seemed, were unraveling. Race riots broke out in Chicago and the Ku Klux Klan revived its hate campaign in the South. Reformers, appalled by the use of alcohol, especially in immigrant communities, were on the verge of passing a constitutional amendment banning its production, purchase, and consumption. Even the nation's favorite pasttime, baseball, was not immune to shock and scandal. In 1919, members of the Chicago White Sox threw the World Series at the behest of gamblers.

Harding's call to "normalcy" resonated with a nation that saw only continued trouble abroad and frightening changes at home. Although Harding traveled during the campaign, he is best known for conducting mini campaign rallies on the porch of his home in Ohio.

The Democratic candidate was another Ohioan, James Cox, the state's governor. He never stood a chance, for he seemed not to represent normalcy but a continuation of the frightening new world Woodrow Wilson had delivered to American shores. In one of the more lopsided elections in U.S. history, Harding took about 61 percent of the popular vote and 404 votes in the Electoral College. Cox and his running mate, a young New Yorker named Franklin Roosevelt, received 36 percent, and 127, respectively.

---

1. Eliot Asinof, *America's Loss of Innocence* (New York: Donald I. Fine, 1990), 139.

Warren Harding was singularly ill suited for the presidency. While he assembled some superb cabinet members, such as Secretary of Commerce Herbert Hoover and Secretary of State Charles Evan Hughes, he was a flawed judge of character. If he knew what he hoped to achieve, if he had a sense of where he might take the nation over the next four or even eight years, he surely did not share those thoughts with his listeners on Inauguration Day.

Harding's speech of more than three thousand words contained little eloquence and few specifics. Perhaps the most moving part of the speech was his personal address to "the maimed and wounded soldiers" who were gathered in the audience. Thousands of soldiers returned from France scarred for life, some of them suffering from mental ailments brought on by the shells and bombs of industrial-strength war. Harding expressed the hope that his government would produce no "maimed successors."

The new President, the first to receive votes from women, recognized the "nationwide induction of womanhood into our political life," saying that the nation "may count upon her intuitions, her refinements, her intelligence, and her influence to exalt the social order."

Though bland and littered with platitudes, Harding's speech implicitly recognized the arguments that were taking place not only in the United States but all over the world. The Bolshevik triumph in Russia concentrated the capitalist mind on a defense of the status quo. "There never can be equality of rewards or possessions so long as the human plan contains varied talents and differing degrees of industry and thrift, but ours ought to be a country free from the great blotches of distressed poverty," he said, adding that the nation "ought to find a way to guard against the perils and penalties of unemployment."

Those perils seemed less pressing as the nation settled into normalcy. In short order, the trauma of 1919 and early 1920 gave way to the Roaring Twenties, a decade of prosperity, mass consumption, and easy money on Wall Street. Warren Harding, however, did not live long enough to enjoy all this. At the age of fifty-seven, he died of a heart attack in 1923, his administration awash in scandal. He was the first president to die in office of natural causes since Zachary Taylor in 1850.

Bibliographic Note
Francis Russell, *The Shadow of Blooming Grove: Warren G. Harding in His Times*
(New York: McGraw Hill, 1968).

When one surveys the world about him after the great storm, noting the marks of destruction and yet rejoicing in the ruggedness of the things which withstood it, if he is an American he breathes the clarified atmosphere with a strange mingling of regret and new hope. We have seen a world passion spend its fury, but we contemplate our Republic unshaken, and hold our civilization secure. Liberty—liberty within the law—and civilization are inseparable, and though both were threatened we find them now secure; and there comes to Americans the profound assurance that our representative government is the highest expression and surest guaranty of both.

Standing in this presence, mindful of the solemnity of this occasion, feeling the emotions which no one may know until he senses the great weight of responsibility for himself, I must utter my belief in the divine inspiration of the Founding Fathers. Surely there must have been God's intent in the making of this New World Republic. Ours is an organic law which had but one ambiguity, and we saw that effaced in a baptism of sacrifice and blood, with union maintained, the nation supreme, and its concord inspiring. We have seen the world rivet its hopeful gaze on the great truths on which the founders wrought. We have seen civil, human, and religious liberty verified and glorified. In the beginning the Old World scoffed at our experiment; today our foundations of political and social belief stand unshaken, a precious inheritance to ourselves, an inspiring example of freedom and civilization to all mankind. Let us express renewed and strengthened devotion, in grateful reverence for the immortal beginning, and utter our confidence in the supreme fulfillment.

The recorded progress of our Republic, materially and spiritually, in itself proves the wisdom of the inherited policy of noninvolvement in Old World affairs. Confident of our ability to work out our own destiny, and jealously guarding our right to do so, we seek no part in directing the destinies of the Old World. We do not mean to be entangled. We will accept no responsibility except as our own conscience and judgment, in each instance, may determine.

Our eyes never will be blind to a developing menace, our ears never deaf to the call of civilization. We recognize the new order in the world, with the closer contacts which progress has wrought. We sense the call of the human heart for fellowship, fraternity, and cooperation. We crave friendship and harbor no hate. But America, our America, the America builded on the foundation laid by the inspired fathers, can be a party to no permanent military alliance. It can enter into no political commitments, nor assume any economic obligations which will subject our decisions to any other than our own authority.

I am sure our own people will not misunderstand, nor will the world misconstrue. We have no thought to impede the paths to closer relationship. We wish to promote understanding. We want to do our part in making offensive warfare so hateful that governments and peoples who resort to it must prove the righteousness of their cause or stand as outlaws before the bar of civilization.

We are ready to associate ourselves with the nations of the world, great and small, for conference, for counsel; to seek the expressed views of world opinion; to recommend a way to approximate disarmament and relieve the crushing burdens of military and naval establishments. We elect to participate in suggesting plans for mediation, conciliation, and arbitration, and would gladly join in that expressed conscience of progress, which seeks to clarify and write the laws of international relationship, and establish a world court for the disposition of such justiciable questions as nations are agreed to submit thereto. In expressing aspirations, in seeking practical plans, in translating humanity's new concept of righteousness and justice and its hatred of war into recommended action we are ready most heartily to unite, but every commitment must be made in the exercise of our national sovereignty. Since freedom impelled, and independence inspired, and nationality exalted, a world supergovernment is contrary to everything we cherish and can have no sanction by our Republic. This is not selfishness, it is sanctity. It is not aloofness, it is security. It is not suspicion of others, it is patriotic adherence to the things which made us what we are.

Today, better than ever before, we know the aspirations of humankind, and share them. We have come to a new realization of our place in the world and a new appraisal of our nation by the world. The unselfishness of these United States is a thing proven; our devotion to peace for ourselves and for the world is well established; our concern for preserved civilization has had its impassioned and heroic expression. There was no American failure to resist the attempted reversion of civilization; there will be no failure today or tomorrow.

The success of our popular government rests wholly upon the correct interpretation of the deliberate, intelligent, dependable popular will of America. In a deliberate questioning of a suggested change of national policy, where internationality was to supersede nationality, we turned to a referendum, to the American people. There was ample discussion, and there is a public mandate in manifest understanding.

America is ready to encourage, eager to initiate, anxious to participate in any seemly program likely to lessen the probability of war, and promote that brotherhood of mankind which must be God's highest conception of human relationship. Because we cherish ideals of justice and peace, because we appraise international comity and helpful relationship no less highly than any people of the world, we aspire to a high place in the moral leadership of civilization, and we hold a maintained America, the proven Republic, the unshaken temple of representative democracy, to be not only an inspiration and example, but the highest agency of strengthening good will and promoting accord on both continents.

Mankind needs a worldwide benediction of understanding. It is needed among individuals, among peoples, among governments, and it will inaugurate an era of good feeling to make the birth of a new order. In such understanding men will strive confidently for the promotion of their better relationships and nations will promote the comities so essential to peace.

We must understand that ties of trade bind nations in closest intimacy, and none may receive except as he gives. We have not strengthened ours in accordance with our resources or our genius, notably on our own continent, where a galaxy of republics reflects the glory of New World democracy, but in the new order of finance and trade we mean to promote enlarged activities and seek expanded confidence.

Perhaps we can make no more helpful contribution by example than prove a republic's capacity to emerge from the wreckage of war. While the world's embittered travail did not leave us devastated lands nor desolated cities, left no gaping wounds, no breast with hate, it did involve us in the delirium of expenditure, in expanded currency and credits, in unbalanced industry, in unspeakable waste, and disturbed relationships. While it uncovered our portion of hateful selfishness at home, it also revealed the heart of America as sound and fearless, and beating in confidence unfailing.

Amid it all we have riveted the gaze of all civilization to the unselfishness and the righteousness of representative democracy, where our freedom never has made offensive warfare, never has sought territorial aggrandizement through force, never has turned to the arbitrament of arms until reason has been exhausted. When the governments of the earth shall have established a freedom like our own and shall have sanctioned the pursuit of peace as we have practiced it, I believe the last sorrow and the final sacrifice of international warfare will have been written.

Let me speak to the maimed and wounded soldiers who are present today, and through them convey to their comrades the gratitude of the Republic for their sacrifices in its defense. A generous country will never forget the services you rendered, and you may hope for a policy under government that will relieve any maimed successors from taking your places on another such occasion as this.

Our supreme task is the resumption of our onward, normal way. Reconstruction, readjustment, restoration all these must follow. I would like to hasten them. If it will lighten the spirit and add to the resolution with which we take up the task, let me repeat for our nation, we shall give no people just cause to make war upon us; we hold no national prejudices; we entertain no spirit of revenge; we do not hate; we do not covet; we dream of no conquest, nor boast of armed prowess.

If, despite this attitude, war is again forced upon us, I earnestly hope a way may be found which will unify our individual and collective strength and consecrate all America, materially and spiritually, body and soul, to national defense. I can vision the ideal republic, where every man and woman is called under the flag for assignment to duty for whatever service, military or civic, the individual is best fitted; where we may call to universal service every plant, agency, or facility, all in the sublime sacrifice for country, and not one penny of war profit shall inure to the benefit of private individual, corporation, or combination, but all above the normal shall flow into the defense chest of the nation. There is something inherently wrong, something out of accord with the ideals of representative democracy, when one portion of our citizenship turns its activities to private gain amid defensive war while another is fighting, sacrificing, or dying for national preservation.

Out of such universal service will come a new unity of spirit and purpose, a new confidence and consecration, which would make our defense impregnable, our triumph assured. Then we should have little or no disorganization of our economic, industrial, and commercial systems at home, no staggering war debts, no swollen fortunes to flout the sacrifices of our soldiers, no excuse for sedition, no pitiable slackerism, no outrage of treason. Envy and jealousy would have no soil for their menacing development, and revolution would be without the passion which engenders it.

A regret for the mistakes of yesterday must not, however, blind us to the tasks of today. War never left such an aftermath. There has been staggering loss of life and measureless wastage of materials. Nations are still groping for return to stable ways. Discouraging indebtedness confronts us like all the war-torn nations, and these obligations must be provided for. No civilization can survive repudiation.

We can reduce the abnormal expenditures, and we will. We can strike at war taxation, and we must. We must face the grim necessity, with full knowledge that the task is to be solved, and we must proceed with a full realization that no statute enacted by man can repeal the inexorable laws of nature. Our most dangerous tendency is to expect too much of government, and at the same time do for it too little. We contemplate the immediate task of putting our public household in order. We need a rigid and yet sane economy, combined with fiscal justice, and it must be attended by individual prudence and thrift, which are so essential to this trying hour and reassuring for the future.

The business world reflects the disturbance of war's reaction. Herein flows the lifeblood of material existence. The economic mechanism is intricate and its parts interdependent, and has suffered the shocks and jars incident to abnormal demands, credit inflations, and price upheavals. The normal balances have been impaired, the channels of distribution have been clogged, the relations of labor and management have been strained. We must seek the readjustment with care and courage. Our people must give and take. Prices must reflect the receding fever of war activities. Perhaps we never shall know the old levels of wages again, because war invariably readjusts compensations, and the necessaries of life will show their inseparable relationship, but we must strive for normalcy to reach stability. All the penalties will not be light, nor evenly distributed. There is no way of making them so. There is no instant step from disorder to order. We must face a condition of grim reality, charge off our losses, and start afresh. It is the oldest lesson of civilization. I would like government to do all it can to mitigate; then, in understanding, in mutuality of interest, in concern for the common good, our tasks will be solved. No altered system will work a miracle. Any wild experiment will only add to the confusion. Our best assurance lies in efficient administration of our proven system.

The forward course of the business cycle is unmistakable. Peoples are turning from destruction to production. Industry has sensed the changed order and our own people are turning to resume their normal, onward way. The call is for productive America to go on. I know that Congress and the administration will favor every wise government policy to aid the resumption and encourage continued progress.

I speak for administrative efficiency, for lightened tax burdens, for sound commercial practices, for adequate credit facilities, for sympathetic concern for all agricultural problems, for the omission of unnecessary interference of government with business, for an end to government's experiment in business, and for more efficient business in government administration. With all of this must attend a mindfulness of the human side of all activities, so that social, industrial, and economic justice will be squared with the purposes of a righteous people.

With the nationwide induction of womanhood into our political life, we may count upon her intuitions, her refinements, her intelligence, and her influence to exalt the social order. We count upon her exercise of the full privileges and the performance of the duties of citizenship to speed the attainment of the highest state.

I wish for an America no less alert in guarding against dangers from within than it is watchful against enemies from without. Our fundamental law recognizes no class, no group, no section; there must be none in legislation or administration. The supreme inspiration is the common weal. Humanity hungers for international peace, and we crave it with all mankind. My most reverent prayer for America is for industrial peace, with its rewards, widely and generally distributed, amid the inspirations of equal opportunity. No one justly may deny the equality of opportunity which made us what we are. We have mistaken unpreparedness to embrace it to be a challenge of the reality, and due concern for making all citizens fit for participation will give added strength of citizenship and magnify our achievement.

If revolution insists upon overturning established order, let other peoples make the tragic experiment. There is no place for it in America. When world war threatened civilization we pledged our resources and our lives to its preservation, and when revolution threatens we unfurl the flag of law and order and renew our consecration. Ours is a constitutional freedom where the popular will is the law supreme and minorities are sacredly protected. Our revisions, reformations, and evolutions reflect a deliberate judgment and an orderly progress, and we mean to cure our ills, but never destroy or permit destruction by force.

I had rather submit our industrial controversies to the conference table in advance than to a settlement table after conflict and suffering. The earth is thirsting for the cup of good will, understanding is its fountain source. I would like to acclaim an era of good feeling amid dependable prosperity and all the blessings which attend.

It has been proved again and again that we cannot, while throwing our markets open to the world, maintain American standards of living and opportunity, and hold our industrial eminence in such unequal competition. There is a luring fallacy in the theory of banished barriers of trade, but preserved American standards require our higher production costs to be reflected in our tariffs on imports. Today, as never before, when peoples are seeking trade restoration and expansion, we must adjust our tariffs to the new order. We seek participation in the world's exchanges, because therein lies our way to widened influence and the triumphs of peace. We know full well we cannot sell where we do not buy, and we cannot sell successfully where we do not carry. Opportunity is calling not alone for the restoration, but for a new era in production, transportation, and trade. We shall answer it best by meeting the demand of a surpassing home market, by promoting self-reliance in production, and by bidding enterprise, genius, and efficiency to carry our cargoes in American bottoms to the marts of the world.

We would not have an America living within and for herself alone, but we would have her self-reliant, independent, and ever nobler, stronger, and richer. Believing in our higher standards, reared through constitutional liberty and maintained opportunity, we invite the world to the same heights. But pride in things wrought is no reflex of a completed task. Common welfare is the goal of our national endeavor. Wealth is not inimical to welfare; it ought to be its friendliest agency. There never can be equality of rewards or possessions so long as the human plan contains varied talents and differing degrees of industry and thrift, but ours ought to be a country free from the great blotches of distressed poverty. We ought to find a way to guard against the perils and penalties of unemployment. We want an America of homes, illumined with hope and happiness, where mothers, freed from the necessity for long hours of toil beyond their own doors, may preside as befits the hearthstone of American citizenship. We want the cradle of American childhood rocked under conditions so wholesome and so hopeful that no blight may touch it in its development, and we want to provide that no selfish interest, no material necessity, no lack of opportunity shall prevent the gaining of that education so essential to best citizenship.

There is no shortcut to the making of these ideals into glad realities. The world has witnessed again and again the futility and the mischief of ill-considered remedies for social and economic disorders. But we are mindful today as never before of the friction of modern industrialism, and we must learn its causes and reduce its evil consequences by sober and tested methods. Where genius has made for great possibilities, justice and happiness must be reflected in a greater common welfare.

Service is the supreme commitment of life. I would rejoice to acclaim the era of the Golden Rule and crown it with the autocracy of service. I pledge an administration wherein all the agencies of government are called to serve, and ever promote an understanding of government purely as an expression of the popular will.

One cannot stand in this presence and be unmindful of the tremendous responsibility. The world upheaval has added heavily to our tasks. But with the realization comes the surge of high resolve, and there is reassurance in belief in the God-given destiny of our Republic. If I felt that there is to be sole responsibility in the executive for the America of tomorrow I should shrink from the burden. But here are a hundred millions, with common concern and shared responsibility, answerable to God and country. The Republic summons them to their duty, and I invite cooperation.

I accept my part with single-mindedness of purpose and humility of spirit, and implore the favor and guidance of God in His Heaven. With these I am unafraid, and confidently face the future.

I have taken the solemn oath of office on that passage of Holy Writ wherein it is asked: "What doth the Lord require of thee but to do justly, and to love mercy, and to walk humbly with thy God?" This I plight to God and country.

Calvin Coolidge

· INAUGURAL ADDRESS ·

Wednesday, March 4, 1925

$\mathscr{B}$y the time the two main political parties and the revived Progressive Party met in the summer of 1924 to choose their candidates for President, Warren Harding was a forgotten figure. The corruption associated with his tenure, most famously the Teapot Dome scandal involving private deals to drill oil on federal lands, did not cast a shadow over his vice president and successor, Calvin Coolidge. He moved quickly to rid the administration of tainted officials.

Remarkably, then, the Republican Party showed no ill effects from Harding's tenure when it convened in Cleveland to nominate Coolidge for a term in his own right. Prosperity and good times made it seem as though Teapot Dome had never taken place.

The early 1920s saw the beginnings of a mass consumer society that replaced the old Victorian social order with its rigid hierarchy and emphasis on self-sacrifice, modesty, and restraint. Mass media, in the form of radio and widely circulated newspapers and magazines, broke down sectional barriers and helped create a national popular culture. Leisure time, a new concept for most people, created a market for spectator sports and mass entertainment. Movie stars and famous athletes became the idols of the crowd, from Mary Pickford to Babe Ruth, Rudolph Valentino to Jack Dempsey.

While the country seemed satisfied with its direction, discontent and dissent had not disappeared. An anti-immigrant backlash led to the passage, in 1924, of new immigration restrictions that effectively closed the door to newcomers. Writers despaired of a nation they viewed as hopelessly anti-intellectual and déclassé, a nation that had banned alcohol in a puritanical frenzy in 1920. Ernest Hemingway, Gertrude Stein, John Dos Passos, and F. Scott Fitzgerald crossed the Atlantic in search of inspiration. A generation of African American artists such as Langston Hughes, Zora Hurston, and Duke Ellington found their voices despite intense racial discrimination.

The prosperity of the 1920s masked a culture war that finally revealed itself at the Democratic National Convention of 1924 in New York City. It lasted from June 24 to July 9 and became a stage on which tensions between North and South, dry (in favor of Prohibition) and wet (against Prohibition), native born and immigrant, rural and urban, were played out in bitter, raucous soliloquies. A proposal to condemn the KKK, supported by the convention's urban, predominately Catholic delegates, was bitterly opposed by the party's powerful Southern bloc. New York governor Alfred E. Smith was the candidate of choice for the party's northern and urban delegates, while southern and western delegates supported William McAdoo. After 102 inconclusive ballots, John Davis, a former congressman, received the party's nomination as a compromise candidate.

President Coolidge, the beneficiary of the Democrats' utter disarray, was in no position to gloat about the fiasco in New York. On July 7, two days before the exhausted Democrats settled on Davis as their nominee, the President's sixteen-year-old son, Calvin Coolidge Jr., died of blood poisoning. He had developed a blister on his foot while playing tennis on the grounds of the White House, and when it became infected, there were no medicines to cure him. Years later, Coolidge, known as a reserved and even cold man, wrote of his heartbreak and the guilt he felt as he watched his son die. "We do not know what might have happened to him under other circumstances, but if I had not been President he would not have raised the blister on his toe which resulted in the blood poisoning," he wrote. "When he went the power and glory of the Presidency went with him. I do not know why such a price was exacted for occupying the White House."[1]

Coolidge barely campaigned in the fall. But like Harding, the man who brought him to the national stage, Coolidge won in a walk in November, taking 54 percent of the popular vote and 382 electoral votes to Davis's 28 percent and 136, respectively. Robert La Follette, running on the Progressive line, won 16 percent of the popular vote and 13 electoral votes.

On Inauguration Day 1925, however, Calvin Coolidge let loose with a meandering, four thousand word inaugural address that surely must have

---

1. Jules Abels, *In the Time of Silent Cal* (New York: G. P. Putnam, 1969), 39.

shocked those who thought they knew Silent Cal well. There was little that escaped his urge to articulate. He reiterated his past opposition to membership in the League of Nations but spoke favorably of efforts to arbitrate international disputes. He praised the effects of the Washington Naval Treaty, which limited the size of the world's most powerful navies, including America's. He seemed to suggest his support for international efforts to outlaw war: "Much may be hoped for from the earnest studies of those who advocate the outlawing of aggressive war." But, in a touch of eloquence, he noted that peace cannot simply be brought about by fiat. "Peace will come when there is realization that only under a reign of law, based on righteousness and supported by the religious conviction of the brotherhood of man, can there be any hope of a complete and satisfying life. Parchment will fail, the sword will fail, it is only the spiritual nature of man that can be triumphant."

Woodrow Wilson had told Americans that they were "provincials no longer." But Coolidge saw some dangers in the nation's new cosmopolitan outlook. America, he said, should not "become implicated in the political controversies of the Old World." The same applied to other parts of the New World. It was only after "a great deal of hesitation," Coolidge said, that the United States agreed to "help to maintain order, protect life and property, and establish responsible government in some of the small countries of the Western Hemisphere."

On domestic issues, Coolidge pledged to lower taxes, reaffirmed his party's support for protective tariffs, and countered collectivist complaints against private property with the assertion that the "rights and duties" of property owners have "divine sanction." The stability of society, he said, depended upon "production and conservation."

During the next four years, Calvin Coolidge did little to change the nation he led, and the nation he led showed little interest in changing. Times were just too good.

Bibliographic Note
Nathan Miller, *New World Coming: The 1920s and the Making of Modern America*
(New York: Scribner, 2003).

My countrymen:

No one can contemplate current conditions without finding much that is satisfying and still more that is encouraging. Our own country is leading the world in the general readjustment to the results of the great conflict. Many of its burdens will bear heavily upon us for years, and the secondary and indirect effects we must expect to experience for some time. But we are beginning to comprehend more definitely what course should be pursued, what remedies ought to be applied, what actions should be taken for our deliverance, and are clearly manifesting a determined will faithfully and conscientiously to adopt these methods of relief. Already we have sufficiently rearranged our domestic affairs so that confidence has returned, business has revived, and we appear to be entering an era of prosperity which is gradually reaching into every part of the nation. Realizing that we cannot live unto ourselves alone, we have contributed of our resources and our counsel to the relief of the suffering and the settlement of the disputes among the European nations. Because of what America is and what America has done, a firmer courage, a higher hope, inspires the heart of all humanity.

These results have not occurred by mere chance. They have been secured by a constant and enlightened effort marked by many sacrifices and extending over many generations. We cannot continue these brilliant successes in the future, unless we continue to learn from the past. It is necessary to keep the former experiences of our country both at home and abroad continually before us, if we are to have any science of government. If we wish to erect new structures, we must have a definite knowledge of the old foundations. We must realize that human nature is about the most constant thing in the universe and that the essentials of human relationship do not change. We must frequently take our bearings from these fixed stars of our political firmament if we expect to hold a true course. If we examine carefully what we have done, we can determine the more accurately what we can do.

We stand at the opening of the one hundred and fiftieth year since our national consciousness first asserted itself by unmistakable action with an array of force. The old sentiment of detached and dependent colonies disappeared in the new sentiment of a united and independent nation. Men began to discard the narrow confines of a local charter for the broader opportunities of a national constitution. Under the eternal urge of freedom we became an independent nation. A little less than fifty years later that freedom and independence were reasserted in the face of all the world, and guarded, supported, and secured by the Monroe Doctrine. The narrow fringe of states along the Atlantic seaboard advanced its frontiers across the hills and plains of an intervening continent until it passed down the golden slope to the Pacific. We made freedom a birthright. We extended our domain over distant islands in order to safeguard our own interests and accepted the consequent obligation to bestow justice and liberty upon less favored peoples. In the defense of our own ideals and in the general cause of liberty we entered the Great War. When victory had been fully secured, we withdrew to our own shores unrecompensed save in the consciousness of duty done.

Throughout all these experiences we have enlarged our freedom, we have strengthened our independence. We have been, and propose to be, more and more American. We believe that we can best serve our own country and most successfully discharge our obligations to humanity by continuing to be openly and candidly, intensely and scrupulously, American. If we have any heritage, it has been that. If we have any destiny, we have found it in that direction.

But if we wish to continue to be distinctively American, we must continue to make that term comprehensive enough to embrace the legitimate desires of a civilized and enlightened people determined in all their relations to pursue a conscientious and religious life. We cannot permit ourselves to be narrowed and dwarfed by slogans and phrases. It is not the adjective, but the substantive, which is of real importance. It is not the name of the action, but the result of the action, which is the chief concern. It will be well not to be too much disturbed by the thought of either isolation or entanglement of pacifists and militarists. The physical configuration of the earth has separated us from all of the Old World, but the common brotherhood of man, the highest law of all our being, has united us by inseparable bonds with all humanity. Our country represents nothing but peaceful intentions toward all the earth, but it ought not to fail to maintain such a military force as comports with the dignity and security of a great people. It ought to be a balanced force, intensely modern, capable of defense by sea and land, beneath the surface and in the air. But it should be so conducted that all the world may see in it, not a menace, but an instrument of security and peace.

This nation believes thoroughly in an honorable peace under which the rights of its citizens are to be everywhere protected. It has never found that the necessary enjoyment of such a peace could be maintained only by a great and threatening array of arms. In common with other nations, it is now more determined than ever to promote peace through friendliness and good will, through mutual understandings and mutual forbearance. We have never practiced the policy of competitive armaments. We have recently committed ourselves by covenants with the other great nations to a limitation of our sea power. As one result of this, our navy ranks larger, in comparison, than it ever did before. Removing the burden of expense and jealousy, which must always accrue from a keen rivalry, is one of the most effective methods of diminishing that unreasonable hysteria and misunderstanding which are the most potent means of fomenting war. This policy represents a new departure in the world. It is a thought, an ideal, which has led to an entirely new line of action. It will not be easy to maintain. Some never moved from their old positions, some are constantly slipping back to the old ways of thought and the old action of seizing a musket and relying on force. America has taken the lead in this new direction, and that lead America must continue to hold. If we expect others to rely on our fairness and justice we must show that we rely on their fairness and justice.

If we are to judge by past experience, there is much to be hoped for in international relations from frequent conferences and consultations. We have before us the beneficial results of the Washington conference and the various consultations recently held upon European affairs, some of which were in response to our suggestions and in some of which we were active participants. Even the failures cannot but be accounted useful and an immeasurable advance over threatened or actual warfare. I am strongly in favor of continuation of this policy, whenever conditions are such that there is even a promise that practical and favorable results might be secured.

In conformity with the principle that a display of reason rather than a threat of force should be the determining factor in the intercourse among nations, we have long advocated the peaceful settlement of disputes by methods of arbitration and have negotiated many treaties to secure that result. The same considerations should lead to our adherence to the Permanent Court of International Justice. Where great principles are involved, where great movements are under way which promise much for the welfare of humanity by reason of the very fact that many other nations have given such movements their actual support, we ought not to withhold our own sanction because of any small and inessential difference, but only upon the ground of the most important and compelling fundamental reasons. We cannot barter away our independence or our sovereignty, but we ought to engage in no refinements of logic, no sophistries, and no subterfuges, to argue away the undoubted duty of this country by reason of the might of its numbers, the power of its resources, and its position of leadership in the world, actively and comprehensively to signify its approval and to bear its full share of the responsibility of a candid and disinterested attempt at the establishment of a tribunal for the administration of evenhanded justice between nation and nation. The weight of our enormous influence must be cast upon the side of a reign not of force but of law and trial, not by battle but by reason.

We have never any wish to interfere in the political conditions of any other countries. Especially are we determined not to become implicated in the political controversies of the Old World. With a great deal of hesitation, we have responded to appeals for help to maintain order, protect life and property, and establish responsible government in some of the small countries of the Western Hemisphere. Our private citizens have advanced large sums of money to assist in the necessary financing and relief of the Old World. We have not failed, nor shall we fail to respond, whenever necessary to mitigate human suffering and assist in the rehabilitation of distressed nations. These, too, are requirements which must be met by reason of our vast powers and the place we hold in the world.

Some of the best thought of mankind has long been seeking for a formula for permanent peace. Undoubtedly the clarification of the principles of international law would be helpful, and the efforts of scholars to prepare such a work for adoption by the various nations should have our sympathy and support. Much may be hoped for from the earnest studies of those who advocate the outlawing of aggressive war. But all these plans and preparations, these treaties and covenants, will not of themselves be adequate. One of the greatest dangers to peace lies in the economic pressure to which people find themselves subjected. One of the most practical things to be done in the world is to seek arrangements under which such pressure may be removed, so that opportunity may be renewed and hope may be revived. There must be some assurance that effort and endeavor will be followed by success and prosperity. In the making and financing of such adjustments there is not only an opportunity, but a real duty, for

America to respond with her counsel and her resources. Conditions must be provided under which people can make a living and work out of their difficulties. But there is another element, more important than all, without which there cannot be the slightest hope of a permanent peace. That element lies in the heart of humanity. Unless the desire for peace be cherished there, unless this fundamental and only natural source of brotherly love be cultivated to its highest degree, all artificial efforts will be in vain. Peace will come when there is realization that only under a reign of law, based on righteousness and supported by the religious conviction of the brotherhood of man, can there be any hope of a complete and satisfying life. Parchment will fail, the sword will fail, it is only the spiritual nature of man that can be triumphant.

It seems altogether probable that we can contribute most to these important objects by maintaining our position of political detachment and independence. We are not identified with any Old World interests. This position should be made more and more clear in our relations with all foreign countries. We are at peace with all of them. Our program is never to oppress, but always to assist. But while we do justice to others, we must require that justice be done to us. With us a treaty of peace means peace, and a treaty of amity means amity. We have made great contributions to the settlement of contentious differences in both Europe and Asia. But there is a very definite point beyond which we cannot go. We can only help those who help themselves. Mindful of these limitations, the one great duty that stands out requires us to use our enormous powers to trim the balance of the world.

While we can look with a great deal of pleasure upon what we have done abroad, we must remember that our continued success in that direction depends upon what we do at home. Since its very outset, it has been found necessary to conduct our government by means of political parties. That system would not have survived from generation to generation if it had not been fundamentally sound and provided the best instrumentalities for the most complete expression of the popular will. It is not necessary to claim that it has always worked perfectly. It is enough to know that nothing better has been devised. No one would deny that there should be full and free expression and an opportunity for independence of action within the party. There is no salvation in a narrow and bigoted partisanship. But if there is to be responsible party government, the party label must be something more than a mere device for securing office. Unless those who are elected under the same party designation are willing to assume sufficient responsibility and exhibit sufficient loyalty and coherence, so that they can cooperate with each other in the support of the broad general principles of the party platform, the election is merely a mockery, no decision is made at the polls, and there is no representation of the popular will. Common honesty and good faith with the people who support a party at the polls require that party, when it enters office, to assume the control of that portion of the government to which it has been elected. Any other course is bad faith and a violation of the party pledges.

When the country has bestowed its confidence upon a party by making it a majority in the Congress, it has a right to expect such unity of action as will make the party majority an effective instrument of government. This administration has come into power with a very clear and definite mandate from the people. The expression of the popular will in favor of maintaining our constitutional guarantees was overwhelming and decisive. There was a manifestation of such faith in the integrity of the courts that we can consider that issue rejected for some time to come. Likewise, the policy of public ownership of railroads and certain electric utilities met with unmistakable defeat. The people declared that they wanted their rights to have not a political but a judicial determination, and their independence and freedom continued and supported by having the ownership and control of their property, not in the government, but in their own hands. As they always do when they have a fair chance, the people demonstrated that they are sound and are determined to have a sound government.

When we turn from what was rejected to inquire what was accepted, the policy that stands out with the greatest clearness is that of economy in public expenditure with reduction and reform of taxation. The principle involved in this effort is that of conservation. The resources of this country are almost beyond computation. No mind can comprehend them. But the cost of our combined governments is likewise almost beyond definition. Not only those who are now making their tax returns, but those who meet the enhanced cost of existence in their monthly bills, know by hard experience what this great burden is and what it does. No matter what others may want, these people want a drastic economy. They are opposed to waste. They know that extravagance lengthens the hours and diminishes the rewards of their labor. I favor the policy of economy, not because I wish to save money, but because I wish to save people. The men and women of this country who toil are the ones who bear the cost of the government. Every dollar that we carelessly waste means that their life will be so much the more meager. Every dollar that we prudently save means that their life will be so much the more abundant. Economy is idealism in its most practical form.

If extravagance were not reflected in taxation, and through taxation both directly and indirectly injuriously affecting the people, it would not be of so much consequence. The wisest and soundest method of solving our tax problem is through economy. Fortunately, of all the great nations this country is best in a position to adopt that simple remedy. We do not any longer need wartime revenues. The collection of any taxes which are not absolutely required, which do not beyond reasonable doubt contribute to the public welfare, is only a species of legalized larceny. Under this Republic the rewards of industry belong to those who earn them. The only constitutional tax is the tax which ministers to public necessity. The property of the country belongs to the people of the country. Their title is absolute. They do not support any privileged class; they do not need to maintain great military forces; they ought not to be burdened with a great array of public employees. They are not required to make any contribution to government expenditures except that which they voluntarily assess upon themselves through the action of their own representatives. Whenever taxes become burdensome a remedy can be applied by the people; but if they do not act for themselves, no one can be very successful in acting for them.

The time is arriving when we can have further tax reduction, when, unless we wish to hamper the people in their right to earn a living, we must have tax reform. The method of raising revenue ought not to impede the transaction of business; it ought to encourage it. I am opposed to extremely high rates, because they produce little or no revenue, because they are bad for the country, and, finally, because they are wrong. We cannot finance the country, we cannot improve social conditions, through any system of injustice, even if we attempt to inflict it upon the rich. Those who suffer the most harm will be the poor. This country believes in prosperity. It is absurd to suppose that it is envious of those who are already prosperous. The wise and correct course to follow in taxation and all other economic legislation is not to destroy those who have already secured success but to create conditions under which everyone will have a better chance to be successful. The verdict of the country has been given on this question. That verdict stands. We shall do well to heed it.

These questions involve moral issues. We need not concern ourselves much about the rights of property if we will faithfully observe the rights of persons. Under our institutions their rights are supreme. It is not property but the right to hold property, both great and small, which our Constitution guarantees. All owners of property are charged with a service. These rights and duties have been revealed, through the conscience of society, to have a divine sanction. The very stability of our society rests upon production and conservation. For individuals or for governments to waste and squander their resources is to deny these rights and disregard these obligations. The result of economic dissipation to a nation is always moral decay.

These policies of better international understandings, greater economy, and lower taxes have contributed largely to peaceful and prosperous industrial relations. Under the helpful influences of restrictive immigration and a protective tariff, employment is plentiful, the rate of pay is high, and wage earners are in a state of contentment seldom before seen. Our transportation systems have been gradually recovering and have been able to meet all the requirements of the service. Agriculture has been very slow in reviving, but the price of cereals at last indicates that the day of its deliverance is at hand.

We are not without our problems, but our most important problem is not to secure new advantages but to maintain those which we already possess. Our system of government made up of three separate and independent departments, our divided sovereignty composed of nation and state, the matchless wisdom that is enshrined in our Constitution, all these need constant effort and tireless vigilance for their protection and support.

In a republic the first rule for the guidance of the citizen is obedience to law. Under a despotism the law may be imposed upon the subject. He has no voice in its making, no influence in its administration, it does not represent him. Under a free government the citizen makes his own laws, chooses his own administrators, which do represent him. Those who want their rights respected under the Constitution and the law ought to set the example themselves of observing the Constitution and the law. While there may be those of high intelligence who violate the law at times, the barbarian and the defective always violate it. Those who disregard the rules of society are not exhibiting a superior intelligence, are not promoting freedom and independence, are not following the path of civilization, but are displaying the traits of ignorance, of servitude, of savagery, and treading the way that leads back to the jungle.

The essence of a republic is representative government. Our Congress represents the people and the states. In all legislative affairs it is the natural collaborator with the President. In spite of all the criticism which often falls to its lot, I do not hesitate to say that there is no more independent and effective legislative body in the world. It is, and should be, jealous of its prerogative. I welcome its cooperation, and expect to share with it not only the responsibility, but the credit, for our common effort to secure beneficial legislation.

These are some of the principles which America represents. We have not by any means put them fully into practice, but we have strongly signified our belief in them. The encouraging feature of our country is not that it has reached its destination, but that it has overwhelmingly expressed its determination to proceed in the right direction. It is true that we could, with profit, be less sectional and more national in our thought. It would be well if we could replace much that is only a false and ignorant prejudice with a true and enlightened pride of race. But the last election showed that appeals to class and nationality had little effect. We were all found loyal to a common citizenship. The fundamental precept of liberty is toleration. We cannot permit any inquisition either within or without the law or apply any religious test to the holding of office. The mind of America must be forever free.

It is in such contemplations, my fellow countrymen, which are not exhaustive but only representative, that I find ample warrant for satisfaction and encouragement. We should not let the much that is to do obscure the much which has been done. The past and present show faith and hope and courage fully justified. Here stands our country, an example of tranquillity at home, a patron of tranquillity abroad. Here stands its government, aware of its might but obedient to its conscience. Here it will continue to stand, seeking peace and prosperity, solicitous for the welfare of the wage earner, promoting enterprise, developing waterways and natural resources, attentive to the intuitive counsel of womanhood, encouraging education, desiring the advancement of religion, supporting the cause of justice and honor among the nations. America seeks no earthly empire built on blood and force. No ambition, no temptation, lures her to thought of foreign dominions. The legions which she sends forth are armed, not with the sword, but with the cross. The higher state to which she seeks the allegiance of all mankind is not of human, but of divine origin. She cherishes no purpose save to merit the favor of Almighty God.

Herbert Hoover

· INAUGURAL ADDRESS ·

Monday, March 4, 1929

*F*ew new Presidents have entered office with higher expectations than Herbert Hoover. He was well educated (a graduate not of an elite East Coast school, however, but of Stanford University in California), a humanitarian, and an engineer. Like many fellow Progressives, he believed in government regulation of private enterprise, in the peaceful arbitration of international problems, in the application of science and reason in public affairs, and in the efficient organization of government. "We have long since abandoned . . . laissez-faire," Hoover wrote. "We have confirmed its abandonment in terms of legislation, of social and economic justice."[1]

Before becoming President, he had never held elective office. Instead, he built a reputation as a superb administrator, a man who organized emergency-relief supplies during and after World War I in a tour de force of diplomacy, efficiency, and simple human decency. After the war, Warren Harding named Hoover to head the Commerce Department, which he promptly reorganized to great effect. When Calvin Coolidge chose not to run for a second full term, Hoover was very much the heir apparent. His nomination at the Republican National Convention in Kansas City, Missouri, was virtually uncontested.

Hoover's opponent was Alfred E. Smith, the first Roman Catholic to win a major party's presidential nomination. Smith was all that Hoover was not. A self-educated, professional politician, and a son of Manhattan's teeming Lower East Side, Smith was a wet at a time when the heartland was dry, he was a Catholic of immigrant stock at a time when the Ellis Island generation had not gained acceptance in mainstream politics, and he was seen as an agent of change at a time when all seemed right in the

---

1. Robert Slayton, *Empire Statesman* (New York: Free Press, 2001), 272.

country. What's more, he was a product of the Tammany Hall machine, known more for its corruption than its support for Smith's progressive policies as governor of New York.

It is hard to imagine any Democrat beating Herbert Hoover in 1928. Hoover was the beneficiary of what was called "Coolidge prosperity," and Hoover was careful to make sure that the public associated him with that prosperity and its presumed author even though he privately regarded Coolidge as too conservative. Workers had disposable income to buy the automobiles that were rolling off Henry Ford's assembly lines and the consumer goods that were being advertised on the new mass medium of radio. Young people transgressed the last remnants of Victorian culture with their style of dress, their independence, and their embrace of mass culture. Most of all, the nation was at peace.

While Herbert Hoover conducted a superb campaign in 1928 and proved to be a dream candidate for the Republican Party, the election is remembered not as a personal triumph for the victor but for the ugliness and bigotry engendered by Smith's religion. When Smith spoke in Oklahoma, a revived KKK burned crosses in the distance. Despite the seemingly homogenizing effects of mass culture, the nation was deeply divided by section, by religion, by place of origin, and especially by race. Democrats and Republicans alike played the race card, with some Democrats criticizing Hoover for desegregating the Commerce Department while Republicans portrayed Smith as a supporter of racial intermarriage.

Herbert Hoover's campaign did not seek to exacerbate those tensions, but Smith's mere presence in the race—as a Catholic, a wet, a northerner, and a city dweller—did exactly that. His support for changes in the nation's Prohibition laws, his religion, his thick New York accent (broadcast via radio to parts of the nation that had never heard such a speech pattern), and his ties to Tammany Hall made him seem dangerous and exotic.

On Election Day, more than 67 percent of registered voters turned out, reversing a downward trend in participation. When those votes were counted, Hoover collected twenty-one million popular votes and 444 votes in the Electoral College. Smith won fifteen million and 88, respectively.

The election did not turn on Smith's Catholicism—he lost, for example, his home state of New York, a Catholic bastion. Smith lost because the nation saw nothing but good times ahead and saw in Hoover a man who would do nothing to spoil the party.

Herbert Hoover's inaugural address was heard by more people than all of his predecessors' inaugurals combined—and then some. Spectators gathered in the rain that afternoon heard the speech on loudspeakers, while millions more listened on radio. What they heard was a long policy paper delivered in an efficient if rather dull monotone. The loudest cheers came when the President reiterated his dry bona fides to an audience that was wet in body though not in spirit. The Eighteenth Amendment—which prohibited the sale and consumption of alcohol—took up several opening minutes in the half-hour address. The new President declared that the "most malign" danger facing the Republic was the "disregard and disobedience of law."

The law he had in mind was the ban on alcohol. Hoover committed his presidency to the enforcement of the law, but he recognized that the law made criminals out of ordinary citizens. "There would be little traffic in illegal liquor if only criminals patronized it," he said. "We must awake to the fact that this patronage from large numbers of law-abiding citizens is supplying the rewards and stimulating crime." Aware that his speech was being heard in millions of home, Hoover spoke directly to his large audience, saying that the judicial system depended "upon the moral support which you, as citizens, extend."

Though stripped of decorative language, Hoover's speech contained a grace note without precedent in inaugural oratory: he acknowledged the outgoing President, Coolidge, by name and thanked him for his service. Other Presidents acknowledged their predecessors—Adams in 1797, Madison in 1809, Monroe in 1817, and Van Buren in 1837—but never by name. Hoover, however, gave Coolidge a final moment in the sun in the very opening of his speech, saying that the nation was "deeply indebted to Calvin Coolidge" for his "wise guidance."

Though he insisted that Inauguration Day was "not the time and place for extended discussion," Hoover in fact delivered a long speech filled with optimistic ideas about government and business working in tandem to achieve progressive goals. Neither he nor the nation had any reason to doubt his ability to deliver.

Bibliographic Note
Martin L. Fausold, ed., *The Hoover Presidency: A Reappraisal*
(Albany, N.Y.: State University of New York Press, 1974).

My countrymen:

This occasion is not alone the administration of the most sacred oath which can be assumed by an American citizen. It is a dedication and consecration under God to the highest office in service of our people. I assume this trust in the humility of knowledge that only through the guidance of Almighty Providence can I hope to discharge its ever-increasing burdens.

It is in keeping with tradition throughout our history that I should express simply and directly the opinions which I hold concerning some of the matters of present importance.

Our Progress

If we survey the situation of our nation both at home and abroad, we find many satisfactions; we find some causes for concern. We have emerged from the losses of the Great War and the reconstruction following it with increased virility and strength. From this strength we have contributed to the recovery and progress of the world. What America has done has given renewed hope and courage to all who have faith in government by the people. In the large view, we have reached a higher degree of comfort and security than ever existed before in the history of the world. Through liberation from widespread poverty we have reached a higher degree of individual freedom than ever before. The devotion to and concern for our institutions are deep and sincere. We are steadily building a new race— a new civilization great in its own attainments. The influence and high purposes of our nation are respected among the peoples of the world. We aspire to distinction in the world, but to a distinction based upon confidence in our sense of justice as well as our accomplishments within our own borders and in our own lives. For wise guidance in this great period of recovery the nation is deeply indebted to Calvin Coolidge.

But all this majestic advance should not obscure the constant dangers from which self-government must be safeguarded. The strong man must at all times be alert to the attack of insidious disease.

THE FAILURE OF OUR SYSTEM OF CRIMINAL JUSTICE

The most malign of all these dangers today is disregard and disobedience of law. Crime is increasing. Confidence in rigid and speedy justice is decreasing. I am not prepared to believe that this indicates any decay in the moral fiber of the American people. I am not prepared to believe that it indicates an impotence of the federal government to enforce its laws.

It is only in part due to the additional burdens imposed upon our judicial system by the Eighteenth Amendment. The problem is much wider than that. Many influences had increasingly complicated and weakened our law enforcement organization long before the adoption of the Eighteenth Amendment.

To reestablish the vigor and effectiveness of law enforcement we must critically consider the entire federal machinery of justice, the redistribution of its functions, the simplification of its procedure, the provision of additional special tribunals, the better selection of juries, and the more effective organization of our agencies of investigation and prosecution that justice may be sure and that it may be swift. While the authority of the federal government extends to but part of our vast system of national, state, and local justice, yet the standards which the federal government establishes have the most profound influence upon the whole structure.

We are fortunate in the ability and integrity of our federal judges and attorneys. But the system which these officers are called upon to administer is in many respects ill adapted to present-day conditions. Its intricate and involved rules of procedure have become the refuge of both big and little criminals. There is a belief abroad that by invoking technicalities, subterfuge, and delay, the ends of justice may be thwarted by those who can pay the cost.

Reform, reorganization and strengthening of our whole judicial and enforcement system, both in civil and criminal sides, have been advocated for years by statesmen, judges, and bar associations. First steps toward that end should not longer be delayed. Rigid and expeditious justice is the first safeguard of freedom, the basis of all ordered liberty, the vital force of progress. It must not come to be in our Republic that it can be defeated by the indifference of the citizen, by exploitation of the delays and entanglements of the law, or by combinations of criminals. Justice must not fail because the agencies of enforcement are either delinquent or inefficiently organized. To consider these evils, to find their remedy, is the most sore necessity of our times.

Enforcement of The Eighteenth Amendment

Of the undoubted abuses which have grown up under the Eighteenth Amendment, part are due to the causes I have just mentioned; but part are due to the failure of some states to accept their share of responsibility for concurrent enforcement and to the failure of many state and local officials to accept the obligation under their oath of office zealously to enforce the laws. With the failures from these many causes has come a dangerous expansion in the criminal elements who have found enlarged opportunities in dealing in illegal liquor.

But a large responsibility rests directly upon our citizens. There would be little traffic in illegal liquor if only criminals patronized it. We must awake to the fact that this patronage from large numbers of law-abiding citizens is supplying the rewards and stimulating crime.

I have been selected by you to execute and enforce the laws of the country. I propose to do so to the extent of my own abilities, but the measure of success that the government shall attain will depend upon the moral support which you, as citizens, extend. The duty of citizens to support the laws of the land is coequal with the duty of their government to enforce the laws which exist. No greater national service can be given by men and women of good will—who, I know, are not unmindful of the responsibilities of citizenship—than that they should, by their example, assist in stamping out crime and outlawry by refusing participation in and condemning all transactions with illegal liquor. Our whole system of self-government will crumble either if officials elect what laws they will enforce or citizens elect what laws they will support. The worst evil of disregard for some law is that it destroys respect for all law. For our citizens to patronize the violation of a particular law on the ground that they are opposed to it is destructive of the very basis of all that protection of life, of homes and property which they rightly claim under other laws. If citizens do not like a law, their duty as honest men and women is to discourage its violation; their right is openly to work for its repeal.

To those of criminal mind there can be no appeal but vigorous enforcement of the law. Fortunately they are but a small percentage of our people. Their activities must be stopped.

A National Investigation

I propose to appoint a national commission for a searching investigation of the whole structure of our federal system of jurisprudence, to include the method of enforcement of the Eighteenth Amendment and the causes of abuse under it. Its purpose will be to make such recommendations for reorganization of the administration of federal laws and court procedure as may be found desirable. In the meantime it is essential that a large part of the enforcement activities be transferred from the Treasury Department to the Department of Justice as a beginning of more effective organization.

THE RELATION OF GOVERNMENT TO BUSINESS

The election has again confirmed the determination of the American people that regulation of private enterprise and not government ownership or operation is the course rightly to be pursued in our relation to business. In recent years we have established a differentiation in the whole method of business regulation between the industries which produce and distribute commodities on the one hand and public utilities on the other. In the former, our laws insist upon effective competition; in the latter, because we substantially confer a monopoly by limiting competition, we must regulate their services and rates. The rigid enforcement of the laws applicable to both groups is the very base of equal opportunity and freedom from domination for all our people, and it is just as essential for the stability and prosperity of business itself as for the protection of the public at large. Such regulation should be extended by the federal government within the limitations of the Constitution and only when the individual states are without power to protect their citizens through their own authority. On the other hand, we should be fearless when the authority rests only in the federal government.

COOPERATION BY THE GOVERNMENT

The larger purpose of our economic thought should be to establish more firmly stability and security of business and employment and thereby remove poverty still further from our borders. Our people have in recent years developed a newfound capacity for cooperation among themselves to effect high purposes in public welfare. It is an advance toward the highest conception of self-government. Self-government does not and should not imply the use of political agencies alone. Progress is born of cooperation in the community—not from governmental restraints. The government should assist and encourage these movements of collective self-help by itself cooperating with them. Business has by cooperation made great progress in the advancement of service, in stability, in regularity of employment, and in the correction of its own abuses. Such progress, however, can continue only so long as business manifests its respect for law.

There is an equally important field of cooperation by the federal government with the multitude of agencies, state, municipal, and private, in the systematic development of those processes which directly affect public health, recreation, education, and the home. We have need further to perfect the means by which government can be adapted to human service.

EDUCATION

Although education is primarily a responsibility of the states and local communities, and rightly so, yet the nation as a whole is vitally concerned in its development everywhere to the highest standards and to complete universality. Self-government can succeed only through an instructed electorate. Our objective is not simply to overcome illiteracy. The nation has marched far beyond that. The more complex the problems of the nation become, the greater is the need for more and more advanced instruction. Moreover, as our numbers increase and as our life expands with science and invention, we must discover more and more leaders for every walk of life. We cannot hope to succeed in directing this increasingly complex civilization unless we can draw all the talent of leadership from the whole people. One civilization after another has been wrecked upon the attempt to secure sufficient leadership from a single group or class. If we would prevent the growth of class distinctions and would constantly refresh our leadership with the ideals of our people, we must draw constantly from the general mass. The full opportunity for every boy and girl to rise through the selective processes of education can alone secure to us this leadership.

PUBLIC HEALTH

In public health the discoveries of science have opened a new era. Many sections of our country and many groups of our citizens suffer from diseases the eradication of which are mere matters of administration and moderate expenditure. Public health service should be as fully organized and as universally incorporated into our governmental system as is public education. The returns are a thousandfold in economic benefits, and infinitely more in reduction of suffering and promotion of human happiness.

WORLD PEACE

The United States fully accepts the profound truth that our own progress, prosperity, and peace are interlocked with the progress, prosperity, and peace of all humanity. The whole world is at peace. The dangers to a continuation of this peace today are largely the fear and suspicion which still haunt the world. No suspicion or fear can be rightly directed toward our country.

Those who have a true understanding of America know that we have no desire for territorial expansion, for economic or other domination of other peoples. Such purposes are repugnant to our ideals of human freedom. Our form of government is ill adapted to the responsibilities which inevitably follow permanent limitation of the independence of other peoples. Superficial observers seem to find no destiny for our abounding increase in population, in wealth, and in power except that of imperialism. They fail to see that the American people are engrossed in the building for themselves of a new economic system, a new social system, a new political system all of which are characterized by aspirations of freedom of opportunity and thereby are the negation of imperialism. They fail to realize that because of our abounding prosperity our youth are pressing more and more into our institutions of learning; that our people are seeking a larger vision through art, literature, science, and travel; that they are moving toward stronger moral and spiritual life—that from these things our sympathies are broadening beyond the bounds of our nation and race toward their true expression in a real brotherhood of man. They fail to see that the idealism of America will lead it to no narrow or selfish channel, but inspire it to do its full share as a nation toward the advancement of civilization. It will do that not by mere declaration but by taking a practical part in supporting all useful international undertakings. We not only desire peace with the world, but to see peace maintained throughout the world. We wish to advance the reign of justice and reason toward the extinction of force.

The recent treaty for the renunciation of war as an instrument of national policy sets an advanced standard in our conception of the relations of nations. Its acceptance should pave the way to greater limitation of armament, the offer of which we sincerely extend to the world. But its full realization also implies a greater and greater perfection in the instrumentalities for pacific settlement of controversies between nations. In the creation and use of these instrumentalities we should support every sound method of conciliation, arbitration, and judicial settlement. American statesmen were among the first to propose and they have constantly urged upon the world, the establishment of a tribunal for the settlement of controversies of a justiciable character. The Permanent Court of International Justice in its major purpose is thus peculiarly identified with American ideals and with American statesmanship. No more potent instrumentality for this purpose has ever been conceived and no other is practicable of establishment. The reservations placed upon our adherence should not be misinterpreted. The United States seeks by these reservations no special privilege or advantage but only to clarify our relation to advisory opinions and other matters which are subsidiary to the major purpose of the court. The way should, and I believe will, be found by which we may take our proper place in a movement so fundamental to the progress of peace.

Our people have determined that we should make no political engagements such as membership in the League of Nations, which may commit us in advance as a nation to become involved in the settlements of controversies between other countries. They adhere to the belief that the independence of America from such obligations increases its ability and availability for service in all fields of human progress.

I have lately returned from a journey among our sister republics of the Western Hemisphere. I have received unbounded hospitality and courtesy as their expression of friendliness to our country. We are held by particular bonds of sympathy and common interest with them. They are each of them building a racial character and a culture which is an impressive contribution to human progress. We wish only for the maintenance of their independence, the growth of their stability, and their prosperity. While we have had wars in the Western Hemisphere, yet on the whole the record is in encouraging contrast with that of other parts of the world. Fortunately the New World is largely free from the inheritances of fear and distrust which have so troubled the Old World. We should keep it so.

It is impossible, my countrymen, to speak of peace without profound emotion. In thousands of homes in America, in millions of homes around the world, there are vacant chairs. It would be a shameful confession of our unworthiness if it should develop that we have abandoned the hope for which all these men died. Surely civilization is old enough, surely mankind is mature enough so that we ought in our own lifetime to find a way to permanent peace. Abroad, to west and east, are nations whose sons mingled their blood with the blood of our sons on the battlefields. Most of these nations have contributed to our race, to our culture, our knowledge, and our progress. From one of them we derive our very language and from many of them much of the genius of our institutions. Their desire for peace is as deep and sincere as our own.

Peace can be contributed to by respect for our ability in defense. Peace can be promoted by the limitation of arms and by the creation of the instrumentalities for peaceful settlement of controversies. But it will become a reality only through self-restraint and active effort in friendliness and helpfulness. I covet for this administration a record of having further contributed to advance the cause of peace.

PARTY RESPONSIBILITIES

In our form of democracy the expression of the popular will can be effected only through the instrumentality of political parties. We maintain party government not to promote intolerant partisanship but because opportunity must be given for expression of the popular will, and organization provided for the execution of its mandates and for accountability of government to the people. It follows that the government both in the executive and the legislative branches must carry out in good faith the platforms upon which the party was entrusted with power. But the government is that of the whole people; the party is the instrument through which policies are determined and men chosen to bring them into being. The animosities of elections should have no place in our government, for government must concern itself alone with the common weal.

SPECIAL SESSION OF THE CONGRESS

Action upon some of the proposals upon which the Republican Party was returned to power, particularly further agricultural relief and limited changes in the tariff, cannot in justice to our farmers, our labor, and our manufacturers be postponed. I shall therefore request a special session of Congress for the consideration of these two questions. I shall deal with each of them upon the assembly of the Congress.

OTHER MANDATES FROM THE ELECTION

It appears to me that the more important further mandates from the recent election were the maintenance of the integrity of the Constitution; the vigorous enforcement of the laws; the continuance of economy in public expenditure; the continued regulation of business to prevent domination in the community; the denial of ownership or operation of business by the government in competition with its citizens; the avoidance of policies which would involve us in the controversies of foreign nations; the more effective reorganization of the departments of the federal government; the expansion of public works; and the promotion of welfare activities affecting education and the home.

These were the more tangible determinations of the election, but beyond them was the confidence and belief of the people that we would not neglect the support of the embedded ideals and aspirations of America. These ideals and aspirations are the touchstones upon which the day-to-day administration and legislative acts of government must be tested. More than this, the government must, so far as lies within its proper powers, give leadership to the realization of these ideals and to the fruition of these aspirations. No one can adequately reduce these things of the spirit to phrases or to a catalogue of definitions. We do know what the attainments of these ideals should be: the preservation of self-government and its full foundations in local government; the perfection of justice whether in economic or in social fields; the maintenance of ordered liberty; the denial of domination by any group or class; the building up and preservation of equality of opportunity; the stimulation of initiative and individuality; absolute integrity in public affairs; the choice of officials for fitness to office; the direction of economic progress toward prosperity for the further lessening of poverty; the freedom of public opinion; the sustaining of education and of the advancement of knowledge; the growth of religious spirit and the tolerance of all faiths; the strengthening of the home; the advancement of peace.

There is no short road to the realization of these aspirations. Ours is a progressive people, but with a determination that progress must be based upon the foundation of experience. Ill-considered remedies for our faults bring only penalties after them. But if we hold the faith of the men in our mighty past who created these ideals, we shall leave them heightened and strengthened for our children.

CONCLUSION

This is not the time and place for extended discussion. The questions before our country are problems of progress to higher standards; they are not the problems of degeneration. They demand thought and they serve to quicken the conscience and enlist our sense of responsibility for their settlement. And that responsibility rests upon you, my countrymen, as much as upon those of us who have been selected for office.

Ours is a land rich in resources; stimulating in its glorious beauty; filled with millions of happy homes; blessed with comfort and opportunity. In no nation are the institutions of progress more advanced. In no nation are the fruits of accomplishment more secure. In no nation is the government more worthy of respect. No country is more loved by its people. I have an abiding faith in their capacity, integrity, and high purpose. I have no fears for the future of our country. It is bright with hope.

In the presence of my countrymen, mindful of the solemnity of this occasion, knowing what the task means and the responsibility which it involves, I beg your tolerance, your aid, and your cooperation. I ask the help of Almighty God in this service to my country to which you have called me.

Franklin D. Roosevelt

Saturday, March 4, 1933

As Franklin D. Roosevelt delivered his inaugural address as the nation's thirty-third President, the nation to which he spoke had good reason to fear for the future. More than twelve million people, nearly 40 percent of the country's workforce, were unemployed. Banks were failing by the thousands, taking with them the life savings of people who thought the prosperity of the 1920s would never end. Thriving downtowns had become pathetic ghost towns as eighty-five thousand businesses, large and small, failed. Farmers were thrown off lands they had tilled for years. Millionaires were suddenly penniless; the poorest of the poor lived in ramshackle settlements in parks and other public spaces, places they called Hoovervilles.

The stock market crash of 1929 set off the worst economic catastrophe in U.S. history. National income fell from eighty-one billion dollars to forty-one billion dollars between 1929 and 1932. The country's industrial output fell by nearly half during those same years. Workers in some industries saw their paychecks shrink by more than half—but at least they had jobs. The millions who were out of work had no social safety net to help them.

The United States was not alone in its misery. Economic hard times plagued much of the developed world, destabilizing governments and encouraging those who offered alternatives to industrial capitalism and liberal democracy. In the United States, shocked and demoralized citizens turned against the incumbent President, Herbert Hoover. As the economic crisis worsened, Hoover seemed, perhaps unfairly, to be in over his head, unable to find creative solutions to the burgeoning disaster.

From New York came the voice of a man who exuded confidence and optimism. Franklin Delano Roosevelt was a wealthy aristocrat from New

York's Hudson Valley who was elected the state's governor in 1928, succeeding Al Smith. Polio had left him paralyzed from the waist down, but Roosevelt's withered legs mattered less than his huge smile and buoyant demeanor. In 1932, he became a voice for change, not simply in the White House, but in government's very relationship with its citizens.

FDR broke precedent by accepting his nomination in person at the Democratic National Convention in Chicago. In that speech, he pledged a "new deal" for his fellow citizens. He was vague on specifics, but they weren't necessary. Hoover's standing was so low that Roosevelt didn't need concrete proposals to persuade Americans that a change was necessary. All he needed to be was somebody other than Herbert Hoover. The incumbent President, desperate to retain office, compared Roosevelt and his supporters to "the fumes of the witch's caldron which boiled in Russia."[1]

Roosevelt won in a landslide, with twenty-three million votes to Hoover's sixteen million, and 472 votes in the Electoral College, compared with Hoover's 59. But before the president-elect took office, he was nearly assassinated in Miami when a gunman opened fire on his motorcade. The mayor of Chicago, Anton Cermak, who was in Miami at the time, was killed.

Four weeks later, about 150,000 spectators gathered on a dark, rainy Saturday morning to witness an inauguration under circumstances unlike any since the Civil War. Not coincidentally, Roosevelt's inaugural was a virtual war speech. He explicitly compared the nation's economic emergency to a military crisis and used the language of war—he talked about "lines of attack" and his role as leader of "this great army of our people"—to marshal support for his new administration. He warned his audience, which included millions listening on radio, that he might require special powers to cope with the calamity.

The speech is best known, however, not for its martial themes nor for its scapegoating of the nation's banking industry, but for a single phrase that seemed to comfort a beleaguered nation. In the very first paragraph of his speech, Roosevelt said, "So, first of all, let me assert my firm belief that the only thing we have to fear is fear itself." The "fear itself" line has become one of the best-remembered phrases from presidential inaugurals. Interestingly enough, however, it did not prompt applause from FDR's

---

1. Arthur Schlesinger, *The Crisis of the Old Order: 1919–1933, the Age of Roosevelt* (New York: Houghton Mifflin, 2003), 437.

audience, perhaps because the sentence actually continued without a pause to invite the audience's approval. After the famous phrase, Roosevelt further described the fear he wished to banish: "nameless, unreasoning, unjustified terror which paralyzes needed efforts to convert retreat into advance."

Roosevelt's inaugural had little of the new President's exuberance and brio. There was no sense that happy days were here again, as his audacious campaign song insisted. Rather, a stern Roosevelt demanded that the nation face "our common difficulties" frankly. "Only a foolish optimist can deny the dark realities of the moment," he said.

He did not offer many specifics about how he might turn back the night of despair and bring back the sunlight of prosperity. But he was more than happy to point fingers at those he blamed for the nation's plight: the "money changers."

The financiers of Wall Street were an easy target for the new President, for wild financial speculation had been one of the main reasons for the stock market crash and the subsequent Depression. Roosevelt was a shrewd politician who also happened to have a landed gentleman's disdain for the grasping speculators of Wall Street. So his denunciation of the "money changers" was not out of character, nor was it politically unpopular. Still, after cautioning the nation about "fear itself," Roosevelt's use of the loaded term "money changers" seems surprisingly divisive and demagogic.

If he was short on specifics, Roosevelt made his top priority clear. "Our greatest primary task is to put people to work," he said. Over the next hundred days, Roosevelt sought to achieve that goal with a blizzard of proposals and initiatives that changed Washington's relationship to the economy and to its citizens.

One other issue, once so pressing, now almost a sidelight, was resolved on March 23, when Roosevelt signed a bill that permitted the production and sale of low-alcohol beer and wine. It was the beginning of the end of Prohibition. Later in the year, the Eighteenth Amendment, which barred the sale of alcoholic beverages, became the first constitutional amendment to be repealed, with passage of the Twenty-first Amendment.

Beer, wine, and spirits were legal again. If this brought some joy to consumers, it was tempered by the knowledge that drink often was no match for despair.

I AM CERTAIN THAT MY FELLOW AMERICANS EXPECT that on my induction into the presidency I will address them with a candor and a decision which the present situation of our nation impels. This is preeminently the time to speak the truth, the whole truth, frankly and boldly. Nor need we shrink from honestly facing conditions in our country today. This great nation will endure as it has endured, will revive and will prosper. So, first of all, let me assert my firm belief that the only thing we have to fear is fear itself—nameless, unreasoning, unjustified terror which paralyzes needed efforts to convert retreat into advance. In every dark hour of our national life a leadership of frankness and vigor has met with that understanding and support of the people themselves which is essential to victory. I am convinced that you will again give that support to leadership in these critical days.

In such a spirit on my part and on yours we face our common difficulties. They concern, thank God, only material things. Values have shrunken to fantastic levels; taxes have risen; our ability to pay has fallen; government of all kinds is faced by serious curtailment of income; the means of exchange are frozen in the currents of trade; the withered leaves of industrial enterprise lie on every side; farmers find no markets for their produce; the savings of many years in thousands of families are gone.

More important, a host of unemployed citizens face the grim problem of existence, and an equally great number toil with little return. Only a foolish optimist can deny the dark realities of the moment.

Yet our distress comes from no failure of substance. We are stricken by no plague of locusts. Compared with the perils which our forefathers conquered because they believed and were not afraid, we have still much to be thankful for. Nature still offers her bounty and human efforts have multiplied it. Plenty is at our doorstep, but a generous use of it languishes in the very sight of the supply. Primarily this is because the rulers of the exchange of mankind's goods have failed, through their own stubbornness and their own incompetence, have admitted their failure, and abdicated. Practices of the unscrupulous money changers stand indicted in the court of public opinion, rejected by the hearts and minds of men.

True they have tried, but their efforts have been cast in the pattern of an outworn tradition. Faced by failure of credit they have proposed only the lending of more money. Stripped of the lure of profit by which to induce our people to follow their false leadership, they have resorted to exhortations, pleading tearfully for restored confidence. They know only the rules of a generation of self-seekers. They have no vision, and when there is no vision the people perish.

The money changers have fled from their high seats in the temple of our civilization. We may now restore that temple to the ancient truths. The measure of the restoration lies in the extent to which we apply social values more noble than mere monetary profit.

Happiness lies not in the mere possession of money; it lies in the joy of achievement, in the thrill of creative effort. The joy and moral stimulation of work no longer must be forgotten in the mad chase of evanescent profits. These dark days will be worth all they cost us if they teach us that our true destiny is not to be ministered unto but to minister to ourselves and to our fellow men.

Recognition of the falsity of material wealth as the standard of success goes hand in hand with the abandonment of the false belief that public office and high political position are to be valued only by the standards of pride of place and personal profit; and there must be an end to a conduct in banking and in business which too often has given to a sacred trust the likeness of callous and selfish wrongdoing. Small wonder that confidence languishes, for it thrives only on honesty, on honor, on the sacredness of obligations, on faithful protection, on unselfish performance; without them it cannot live. Restoration calls, however, not for changes in ethics alone. This nation asks for action, and action now.

Our greatest primary task is to put people to work. This is no unsolvable problem if we face it wisely and courageously. It can be accomplished in part by direct recruiting by the government itself, treating the task as we would treat the emergency of a war, but at the same time, through this employment, accomplishing greatly needed projects to stimulate and reorganize the use of our natural resources.

Hand in hand with this we must frankly recognize the overbalance of population in our industrial centers and, by engaging on a national scale in a redistribution, endeavor to provide a better use of the land for those best fitted for the land. The task can be helped by definite efforts to raise the values of agricultural products and with this the power to purchase the output of our cities. It can be helped by preventing realistically the tragedy of the growing loss through foreclosure of our small homes and our farms. It can be helped by insistence that the federal, state, and local governments act forthwith on the demand that their cost be drastically reduced. It can be helped by the unifying of relief activities which today are often scattered, uneconomical, and unequal. It can be helped by national planning for and supervision of all forms of transportation and of communications and other utilities which have a definitely public character. There are many ways in which it can be helped, but it can never be helped merely by talking about it. We must act and act quickly.

Finally, in our progress toward a resumption of work we require two safeguards against a return of the evils of the old order: there must be a strict supervision of all banking and credits and investments; there must be an end to speculation with other people's money, and there must be provision for an adequate but sound currency.

There are the lines of attack. I shall presently urge upon a new Congress in special session detailed measures for their fulfillment, and I shall seek the immediate assistance of the several states.

Through this program of action we address ourselves to putting our own national house in order and making income balance outgo. Our international trade relations, though vastly important, are in point of time and necessity secondary to the establishment of a sound national economy. I favor as a practical policy the putting of first things first. I shall spare no effort to restore world trade by international economic readjustment, but the emergency at home cannot wait on that accomplishment.

The basic thought that guides these specific means of national recovery is not narrowly nationalistic. It is the insistence, as a first consideration, upon the interdependence of the various elements in all parts of the United States—a recognition of the old and permanently important manifestation of the American spirit of the pioneer. It is the way to recovery. It is the immediate way. It is the strongest assurance that the recovery will endure.

In the field of world policy I would dedicate this nation to the policy of the good neighbor—the neighbor who resolutely respects himself and, because he does so, respects the rights of others—the neighbor who respects his obligations and respects the sanctity of his agreements in and with a world of neighbors.

If I read the temper of our people correctly, we now realize as we have never realized before our interdependence on each other; that we cannot merely take but we must give as well; that if we are to go forward, we must move as a trained and loyal army willing to sacrifice for the good of a common discipline, because without such discipline no progress is made, no leadership becomes effective. We are, I know, ready and willing to submit our lives and property to such discipline, because it makes possible a leadership which aims at a larger good. This I propose to offer, pledging that the larger purposes will bind upon us all as a sacred obligation with a unity of duty hitherto evoked only in time of armed strife.

With this pledge taken, I assume unhesitatingly the leadership of this great army of our people dedicated to a disciplined attack upon our common problems.

Action in this image and to this end is feasible under the form of government which we have inherited from our ancestors. Our Constitution is so simple and practical that it is possible always to meet extraordinary needs by changes in emphasis and arrangement without loss of essential form. That is why our constitutional system has proved itself the most superbly enduring political mechanism the modern world has produced. It has met every stress of vast expansion of territory, of foreign wars, of bitter internal strife, of world relations.

It is to be hoped that the normal balance of executive and legislative authority may be wholly adequate to meet the unprecedented task before us. But it may be that an unprecedented demand and need for undelayed action may call for temporary departure from that normal balance of public procedure.

I am prepared under my constitutional duty to recommend the measures that a stricken nation in the midst of a stricken world may require. These measures, or such other measures as the Congress may build out of its experience and wisdom, I shall seek, within my constitutional authority, to bring to speedy adoption.

But in the event that the Congress shall fail to take one of these two courses, and in the event that the national emergency is still critical, I shall not evade the clear course of duty that will then confront me. I shall ask the Congress for the one remaining instrument to meet the crisis—broad executive power to wage a war against the emergency, as great as the power that would be given to me if we were in fact invaded by a foreign foe.

For the trust reposed in me I will return the courage and the devotion that befit the time. I can do no less.

We face the arduous days that lie before us in the warm courage of the national unity; with the clear consciousness of seeking old and precious moral values; with the clean satisfaction that comes from the stern performance of duty by old and young alike. We aim at the assurance of a rounded and permanent national life.

We do not distrust the future of essential democracy. The people of the United States have not failed. In their need they have registered a mandate that they want direct, vigorous action. They have asked for discipline and direction under leadership. They have made me the present instrument of their wishes. In the spirit of the gift I take it.

In this dedication of a nation we humbly ask the blessing of God. May He protect each and every one of us. May He guide me in the days to come.

Franklin D. Roosevelt

· SECOND INAUGURAL ADDRESS ·

Wednesday, January 20, 1937

Washington was gray and wet again when Franklin D. Roosevelt took the oath of office for the second time, but this time around, it was cold too. The ceremony and speech took place on January 20, some six weeks earlier than the traditional date of March 4. The Twentieth Amendment to the Constitution, ratified in 1933, set the new date to shorten the lethargic interval between Election Day and Inauguration Day. The change was well timed: given the urgent pace of Roosevelt's first term, lethargy no longer was in fashion in Washington.

As he faced a shivering audience sheltered under umbrellas, Franklin Roosevelt was as dominant a political figure as any of his storied predecessors. He had easily defeated Kansas governor Alfred M. Landon in the 1936 election, winning 27.7 million popular votes to Landon's 16.6 million, and taking 523 votes in the Electoral College to Landon's 8 (the Republican challenger won only Maine and Vermont).

With his charisma and his ability to bring disparate people together, Roosevelt built a coalition that would dominate American politics for half a century. Under Roosevelt, the Democrats brought together white southern segregationists as well as African Americans who hoped Roosevelt would break the chains of Jim Crow, middle- and working-class Catholics and Jews who believed this Protestant aristocrat understood their problems, and farmers as well as unionized factory workers. The New Deal coalition would dominate U.S politics until the 1960s.

Roosevelt's skillful use of radio and his outgoing personality had a great deal to do with his extraordinary popularity. His "fireside chats"— informal speeches delivered to millions of living rooms via radio— connected voters to their President as never before. His listeners heard not a blue-blooded New Yorker with a patrician accent but the friendly voice of a neighbor who seemed to understand their problems.

But Roosevelt had more than personality to offer in 1936. New programs and agencies had put the unemployed to work on public works projects, constructed a safety net for the unemployed and the elderly, brought electricity to rural communities, guaranteed workers the right to join unions, and sought to build a new partnership between the private and public sectors. Initiatives such as the Social Security Act and the National Labor Relations Act, both passed in 1935, challenged the cult of individualism that was at the heart of America's self-image.

As Roosevelt prepared for his new term, the Depression was in its seventh year and remained a fact of life in the bleak American landscape. But the nation clearly believed it was moving in the right direction. The unemployment rate in 1936 was 16.9 percent, a figure that would be seen as catastrophic at almost any other time in U.S. history, but which, in 1936, was a profound improvement over the 25 percent jobless rate of 1933. The nation's gross national product grew by an astonishing 14.1 percent in 1936 (the GNP had declined by almost as much, 13.4 percent, in 1932).

But even as Roosevelt cruised to reelection, the improvised, experimental structure of reform and relief seemed in danger of collapse. As early as 1935, the Supreme Court ruled, in *Schecter Poultry Corp. v. United States*, that the National Industrial Recovery Act, which established the National Recovery Administration, was unconstitutional because it gave the government power over intrastate—as opposed to interstate—commerce. The Court delivered another blow to reform and relief in 1936 when it struck down the Agricultural Adjustment Act.

The decisions infuriated Roosevelt and his New Dealers, who saw the Court as a bastion of political opposition that no campaign, no election, could defeat. As Roosevelt prepared to deliver his second inaugural address, he and his advisers were working on a plan that they hoped would circumvent the Court: If the President were given the power to appoint a new Supreme Court justice for every one over the age of seventy (six of the nine justices fit that description), he could overcome the judicial barrier to further New Deal reforms. Roosevelt would call it a reorganization of the judicial branch. His critics, and they would include members of his own party, called it a court-packing scheme. It was destined to fail spectacularly early in the new term.

Roosevelt said nothing specific about the Court plan in his inaugural address. He focused instead on speaking to the millions of Americans who might be inclined to believe that the nation's calamity was drawing to a close. The nation, he said, had come a long way in four years. But was that enough? "Have we found our happy valley?" the President asked. His answer provided the speech with its most remembered lines.

"I see millions of families trying to live on incomes so meager that the pall of family disaster hangs over them day by day . . . I see millions denied education, recreation, and the opportunity to better their lot and the lot of their children," he said as his litany neared its climactic phrase. "I see one-third of a nation ill-housed, ill-clad, ill-nourished."

No great nation, Roosevelt said, would be satisfied with such a picture. Roosevelt said he drew it so that it might be erased.

Less remembered, but just as powerful, were Roosevelt's opening comments about the changes he believed the nation had achieved during his first term. Americans were no longer willing to "leave the problems of our common welfare to be solved by the winds of chance and the hurricanes of disaster." At issue was not just a return to prosperity, but morality itself. A moral nation put in place "practical controls over blind economic forces and blindly selfish men." Americans had always known that "heedless self-interest was bad morals; we know now that it is bad economics." The nation was "fashioning an instrument of unimagined power for the establishment of a morally better world."

Just as Woodrow Wilson articulated a moral basis for America's expanded role in the world, Franklin Roosevelt, in this speech, established a moral critique for government's expanded role in the economy. His second inaugural address marked a high point for the reforms and creativity associated with the New Deal, for within months of this speech, the economy collapsed again, leading to a recession and a spike in unemployment. Republicans, left for dead in 1936, revived themselves and made impressive gains in the congressional elections in 1938, further frustrating Roosevelt and his team. And not long afterward, the New Deal gave way to an even greater emergency.

Bibliographic Note
William E. Leuchtenburg, *Franklin Roosevelt and the New Deal, 1932–1940*
(New York: Harper and Row, 1963).

WHEN FOUR YEARS AGO WE MET TO INAUGURATE A PRESIDENT, the Republic, single-minded in anxiety, stood in spirit here. We dedicated ourselves to the fulfillment of a vision—to speed the time when there would be for all the people that security and peace essential to the pursuit of happiness. We of the Republic pledged ourselves to drive from the temple of our ancient faith those who had profaned it; to end by action, tireless and unafraid, the stagnation and despair of that day. We did those first things first.

Our covenant with ourselves did not stop there. Instinctively we recognized a deeper need—the need to find through government the instrument of our united purpose to solve for the individual the ever-rising problems of a complex civilization. Repeated attempts at their solution without the aid of government had left us baffled and bewildered. For, without that aid, we had been unable to create those moral controls over the services of science which are necessary to make science a useful servant instead of a ruthless master of mankind. To do this we knew that we must find practical controls over blind economic forces and blindly selfish men.

We of the Republic sensed the truth that democratic government has innate capacity to protect its people against disasters once considered inevitable, to solve problems once considered unsolvable. We would not admit that we could not find a way to master economic epidemics just as, after centuries of fatalistic suffering, we had found a way to master epidemics of disease. We refused to leave the problems of our common welfare to be solved by the winds of chance and the hurricanes of disaster.

In this we Americans were discovering no wholly new truth; we were writing a new chapter in our book of self-government.

This year marks the one hundred and fiftieth anniversary of the Constitutional Convention which made us a nation. At that Convention our forefathers found the way out of the chaos which followed the Revolutionary War; they created a strong government with powers of united action sufficient then and now to solve problems utterly beyond individual or local solution. A century and a half ago they established the federal government in order to promote the general welfare and secure the blessings of liberty to the American people.

Today we invoke those same powers of government to achieve the same objectives.

Four years of new experience have not belied our historic instinct. They hold out the clear hope that government within communities, government within the separate states, and government of the United States can do the things the times require, without yielding its democracy. Our tasks in the last four years did not force democracy to take a holiday.

Nearly all of us recognize that as intricacies of human relationships increase, so power to govern them also must increase—power to stop evil; power to do good. The essential democracy of our nation and the safety of our people depend not upon the absence of power, but upon lodging it with those whom the people can change or continue at stated intervals through an honest and free system of elections. The Constitution of 1787 did not make our democracy impotent.

In fact, in these last four years, we have made the exercise of all power more democratic; for we have begun to bring private autocratic powers into their proper subordination to the public's government. The legend that they were invincible—above and beyond the processes of a democracy— has been shattered. They have been challenged and beaten.

Our progress out of the Depression is obvious. But that is not all that you and I mean by the new order of things. Our pledge was not merely to do a patchwork job with secondhand materials. By using the new materials of social justice we have undertaken to erect on the old foundations a more enduring structure for the better use of future generations.

In that purpose we have been helped by achievements of mind and spirit. Old truths have been relearned; untruths have been unlearned. We have always known that heedless self-interest was bad morals; we know now that it is bad economics. Out of the collapse of a prosperity whose builders boasted their practicality has come the conviction that in the long run economic morality pays. We are beginning to wipe out the line that divides the practical from the ideal; and in so doing we are fashioning an instrument of unimagined power for the establishment of a morally better world.

This new understanding undermines the old admiration of worldly success as such. We are beginning to abandon our tolerance of the abuse of power by those who betray for profit the elementary decencies of life.

In this process evil things formerly accepted will not be so easily condoned. Hardheadedness will not so easily excuse hard-heartedness. We are moving toward an era of good feeling. But we realize that there can be no era of good feeling save among men of good will.

For these reasons I am justified in believing that the greatest change we have witnessed has been the change in the moral climate of America.

Among men of good will, science and democracy together offer an ever-richer life and ever-larger satisfaction to the individual. With this change in our moral climate and our rediscovered ability to improve our economic order, we have set our feet upon the road of enduring progress.

Shall we pause now and turn our back upon the road that lies ahead? Shall we call this the promised land? Or, shall we continue on our way? For "each age is a dream that is dying, or one that is coming to birth."

Many voices are heard as we face a great decision. Comfort says, "Tarry awhile." Opportunism says, "This is a good spot." Timidity asks, "How difficult is the road ahead?"

True, we have come far from the days of stagnation and despair. Vitality has been preserved. Courage and confidence have been restored. Mental and moral horizons have been extended.

But our present gains were won under the pressure of more than ordinary circumstances. Advance became imperative under the goad of fear and suffering. The times were on the side of progress.

To hold to progress today, however, is more difficult. Dulled conscience, irresponsibility, and ruthless self-interest already reappear. Such symptoms of prosperity may become portents of disaster! Prosperity already tests the persistence of our progressive purpose.

Let us ask again: have we reached the goal of our vision of that fourth day of March 1933? Have we found our happy valley?

I see a great nation, upon a great continent, blessed with a great wealth of natural resources. Its 130 million people are at peace among themselves; they are making their country a good neighbor among the nations. I see a United States which can demonstrate that, under democratic methods of government, national wealth can be translated into a spreading volume of human comforts hitherto unknown, and the lowest standard of living can be raised far above the level of mere subsistence.

But here is the challenge to our democracy: in this nation I see tens of millions of its citizens—a substantial part of its whole population—who at this very moment are denied the greater part of what the very lowest standards of today call the necessities of life.

I see millions of families trying to live on incomes so meager that the pall of family disaster hangs over them day by day.

I see millions whose daily lives in city and on farm continue under conditions labeled indecent by a so-called polite society half a century ago.

I see millions denied education, recreation, and the opportunity to better their lot and the lot of their children.

I see millions lacking the means to buy the products of farm and factory and by their poverty denying work and productiveness to many other millions.

I see one-third of a nation ill-housed, ill-clad, ill-nourished.

It is not in despair that I paint you that picture. I paint it for you in hope—because the nation, seeing and understanding the injustice in it, proposes to paint it out. We are determined to make every American citizen the subject of his country's interest and concern; and we will never regard any faithful law-abiding group within our borders as superfluous. The test of our progress is not whether we add more to the abundance of those who have much; it is whether we provide enough for those who have too little.

If I know aught of the spirit and purpose of our nation, we will not listen to comfort, opportunism, and timidity. We will carry on.

Overwhelmingly, we of the Republic are men and women of good will; men and women who have more than warm hearts of dedication; men and women who have cool heads and willing hands of practical purpose as well. They will insist that every agency of popular government use effective instruments to carry out their will.

Government is competent when all who compose it work as trustees for the whole people. It can make constant progress when it keeps abreast of all the facts. It can obtain justified support and legitimate criticism when the people receive true information of all that government does.

If I know aught of the will of our people, they will demand that these conditions of effective government shall be created and maintained. They will demand a nation uncorrupted by cancers of injustice and, therefore, strong among the nations in its example of the will to peace.

Today we reconsecrate our country to long-cherished ideals in a suddenly changed civilization. In every land there are always at work forces that drive men apart and forces that draw men together. In our personal ambitions we are individualists. But in our seeking for economic and political progress as a nation, we all go up, or else we all go down, as one people.

To maintain a democracy of effort requires a vast amount of patience in dealing with differing methods, a vast amount of humility. But out of the confusion of many voices rises an understanding of dominant public need. Then political leadership can voice common ideals, and aid in their realization.

In taking again the oath of office as President of the United States, I assume the solemn obligation of leading the American people forward along the road over which they have chosen to advance.

While this duty rests upon me I shall do my utmost to speak their purpose and to do their will, seeking Divine guidance to help us each and every one to give light to them that sit in darkness and to guide our feet into the way of peace.

Franklin D. Roosevelt

· THIRD INAUGURAL ADDRESS ·

Monday, January 20, 1941

*U*ntil Franklin D. Roosevelt braced himself against a podium on January 20, 1941, a sunny, cold morning in Washington, no President had ever given a third inaugural address, and none ever will again, unless the constitutional limit of two terms is repealed. But in 1940, Roosevelt did what none of his predecessors dared—he ignored the tradition, first articulated by Thomas Jefferson—that no President ought to serve longer than George Washington.

Eight years after his first inauguration, Roosevelt took the oath of office in the midst of another crisis, this one overseas. Germany, Italy, and Japan were on the march. Poland, Holland, Belgium, Denmark, Norway, and, most spectacularly, France, had fallen to Adolf Hitler's Germany. Italy had invaded Greece and North Africa. Japanese troops were fighting in China and Indochina. And the Soviet Union had invaded Finland and swallowed up a portion of Poland and the Baltic states of Lithuania, Estonia, and Latvia.

The world was at war, again. And as was the case in the great conflict of 1914–18, the United States attempted to remain on the sidelines, although its sympathies were clear.

In the summer of 1940, as the forces of totalitarianism toppled nation after nation, America's two major political parties convened to choose its presidential candidates. Ironically, the Democrats seemed to be in disarray because of their uncertainty over the third-term issue. They weren't even sure if Roosevelt wanted to run again—he did, but he didn't want to ask for it. The party was happy to give FDR the nomination, but was less than thrilled by his choice of a new vice president to replace John Garner: Secretary of Agriculture Henry Wallace, a onetime Republican who had little in common with the big-city bosses who dominated the party.

The Republicans, coming off a catastrophic defeat in 1936, were positively enthused after their convention nominated Wendell Wilkie, a businessman and a former Democrat. Wilkie was a galvanizing speaker who brought his campaign to the industrial cities of the Northeast and Midwest, bastions of support for Roosevelt and places where he was not always given a polite response. His courage and tenacity won him admiration even from Democrats.

During the 1940 presidential campaign, both Roosevelt and Wilkie expressed support for Great Britain as the island nation defended itself against Hitler's air bombardment. While both candidates pledged to stay out of the war, both also promised to help the British—short of sending troops.

In the middle of the campaign, Roosevelt signed the Selective Training and Service Act of 1940, authorizing the nation's first peacetime draft. Crafted as a measure to defend American sovereignty and not as the first step toward intervention overseas, the bill nevertheless barely made it out of the House of Representatives, passing by a single vote. The American public and its representatives remained wary of Roosevelt's intentions. Although isolationist sentiment was widespread in 1940, there was more general agreement on the nation's need to build up its defenses. Congress approved nearly two billion dollars in new spending to improve the nation's armed services, particularly the Navy and the Army Air Corps. Nearly seventeen million males between the ages of twenty-one and thirty-five registered for the draft. More controversially, Roosevelt approved the transfer of fifty destroyers to Great Britain in September, when a German invasion seemed imminent. Britain "paid" for the ships by offering the United States long-term leases of naval and air facilities in Newfoundland, Bermuda, the Bahamas and several other locations.

The shift to a quasi-war economy helped the nation finally shake the economic doldrums that had lingered since the nation fell into recession in 1937. Unemployment, which rose to 19 percent in 1938, fell to about 14 percent in 1940, and would shrink to just below 10 percent (9.9) in 1941—the first time since 1930 that the unemployment rate could be measured in single digits. America in 1940 had not yet returned to widespread prosperity, but the worst seemed past.

Roosevelt was the beneficiary of the improved economy and the deteriorating situation overseas. Both situations seemed to argue against change, especially when the challenger, Wilkie, was running for political office for the first time. Still, while Roosevelt piled up an impressive victory in the

Electoral College, taking 449 votes to Wilkie's 82, the popular vote showed some slippage in Roosevelt's popularity. FDR won his third term with twenty-seven million votes, while Wilkie received twenty-two million.

During the campaign, Roosevelt assured the nation that he would not lead America into war. But war surely was coming. Just after Christmas 1940, Roosevelt used his fireside chat to tell Americans that their country ought to become "the great arsenal of democracy," a bold attempt to rally support for arming Britain and other nations fighting fascism and militarism.

Democracy and its defense were the themes of his inaugural address. Unlike his first two addresses, Roosevelt had nothing to say about the nation's economy or about any domestic issue at all. There was no hint of any policy initiatives, nor were there any bitter denunciations of those he had blamed—the money changers, among others—for the Great Depression.

Equally absent was the word *war*. Roosevelt avoided any specific reference to the cataclysm abroad. When Roosevelt said, "Democracy is not dying," his listeners knew where and how democracy was under attack.

At the very beginning of the speech and near its conclusion, Roosevelt cited the example of George Washington, quoting from the first President's first inaugural—Roosevelt, who had one-upped Washington by running for a third term, became the first President to quote another inaugural address in his own inaugural. Just as the nation faced a crisis in 1789, Roosevelt said, it faced another one in 1941. But he argued that democracies were capable of facing crises, despite the seeming triumph of dictators in Europe and Asia. As proof of democracy's resilience, he offered an example from recent history. "Eight years ago, when the life of this Republic seemed frozen by a fatalistic terror, we proved that this is not true," he said. "We were in the midst of shock—but we acted. We acted quickly, boldly, decisively."

Inaction, he pointedly argued, solved nothing in 1932 and would solve nothing in 1941. Roosevelt was preparing Americans to defend their values and democracy itself. But even this speech did little to prepare the country for the shock of December 7, 1941, when the Japanese attacked Pearl Harbor, plunging America into a war it sought to avoid.

ON EACH NATIONAL DAY OF INAUGURATION SINCE 1789, the people have renewed their sense of dedication to the United States.

In Washington's day the task of the people was to create and weld together a nation.

In Lincoln's day the task of the people was to preserve that nation from
disruption from within.

In this day the task of the people is to save that nation and its institutions from disruption from without.

To us there has come a time, in the midst of swift happenings, to pause
for a moment and take stock—to recall what our place in history has been,
and to rediscover what we are and what we may be. If we do not, we risk
the real peril of inaction.

Lives of nations are determined not by the count of years, but by the
lifetime of the human spirit. The life of a man is three-score years and ten:
a little more, a little less. The life of a nation is the fullness of the measure
of its will to live.

There are men who doubt this. There are men who believe that democracy, as a form of government and a frame of life, is limited or measured
by a kind of mystical and artificial fate, that for some unexplained reason,
tyranny and slavery have become the surging wave of the future—and
that freedom is an ebbing tide.

But we Americans know that this is not true.

Eight years ago, when the life of this Republic seemed frozen by a fatalistic terror, we proved that this is not true. We were in the midst of shock—
but we acted. We acted quickly, boldly, decisively.

These later years have been living years—fruitful years for the people
of this democracy. For they have brought to us greater security and, I hope,
a better understanding that life's ideals are to be measured in other than
material things.

Most vital to our present and our future is this experience of a democracy which successfully survived crisis at home; put away many evil things;
built new structures on enduring lines; and, through it all, maintained the
fact of its democracy.

For action has been taken within the three-way framework of the Constitution of the United States. The coordinate branches of the government

continue freely to function. The Bill of Rights remains inviolate. The freedom of elections is wholly maintained. Prophets of the downfall of American democracy have seen their dire predictions come to naught.

No, democracy is not dying.

We know it because we have seen it revive—and grow.

We know it cannot die—because it is built on the unhampered initiative of individual men and women joined together in a common enterprise—an enterprise undertaken and carried through by the free expression of a free majority.

We know it because democracy alone, of all forms of government, enlists the full force of men's enlightened will.

We know it because democracy alone has constructed an unlimited civilization capable of infinite progress in the improvement of human life.

We know it because, if we look below the surface, we sense it still spreading on every continent—for it is the most humane, the most advanced, and in the end the most unconquerable of all forms of human society.

A nation, like a person, has a body—a body that must be fed and clothed and housed, invigorated and rested, in a manner that measures up to the objectives of our time.

A nation, like a person, has a mind—a mind that must be kept informed and alert, that must know itself, that understands the hopes and the needs of its neighbors—all the other nations that live within the narrowing circle of the world.

And a nation, like a person, has something deeper, something more permanent, something larger than the sum of all its parts. It is that something which matters most to its future—which calls forth the most sacred guarding of its present.

It is a thing for which we find it difficult—even impossible—to hit upon a single, simple word.

And yet we all understand what it is—the spirit—the faith of America. It is the product of centuries. It was born in the multitudes of those who came from many lands—some of high degree, but mostly plain people, who sought here, early and late, to find freedom more freely.

The democratic aspiration is no mere recent phase in human history. It is human history. It permeated the ancient life of early peoples. It blazed anew in the middle ages. It was written in Magna Charta.

In the Americas its impact has been irresistible. America has been the New World in all tongues, to all peoples, not because this continent was a newfound land, but because all those who came here believed they could create upon this continent a new life—a life that should be new in freedom.

Its vitality was written into our own Mayflower Compact, into the Declaration of Independence, into the Constitution of the United States, into the Gettysburg Address.

Those who first came here to carry out the longings of their spirit, and the millions who followed, and the stock that sprang from them—all have moved forward constantly and consistently toward an ideal which in itself has gained stature and clarity with each generation.

The hopes of the Republic cannot forever tolerate either undeserved poverty or self-serving wealth.

We know that we still have far to go; that we must more greatly build the security and the opportunity and the knowledge of every citizen, in the measure justified by the resources and the capacity of the land.

But it is not enough to achieve these purposes alone. It is not enough to clothe and feed the body of this nation, and instruct and inform its mind. For there is also the spirit. And of the three, the greatest is the spirit.

Without the body and the mind, as all men know, the nation could not live.

But if the spirit of America were killed, even though the nation's body and mind, constricted in an alien world, lived on, the America we know would have perished.

That spirit—that faith—speaks to us in our daily lives in ways often unnoticed, because they seem so obvious. It speaks to us here in the capital of the nation. It speaks to us through the processes of governing in the sovereignties of forty-eight states. It speaks to us in our counties, in our cities, in our towns, and in our villages. It speaks to us from the other nations of the hemisphere, and from those across the seas—the enslaved, as well as the free. Sometimes we fail to hear or heed these voices of freedom because to us the privilege of our freedom is such an old, old story.

The destiny of America was proclaimed in words of prophecy spoken by our first President in his first inaugural in 1789—words almost directed, it would seem, to this year of 1941: "The preservation of the sacred fire of liberty and the destiny of the republican model of government are justly considered . . . deeply, . . . finally, staked on the experiment intrusted to the hands of the American people."

If we lose that sacred fire—if we let it be smothered with doubt and fear—then we shall reject the destiny which Washington strove so valiantly and so triumphantly to establish. The preservation of the spirit and faith of the nation does, and will, furnish the highest justification for every sacrifice that we may make in the cause of national defense.

In the face of great perils never before encountered, our strong purpose is to protect and to perpetuate the integrity of democracy.

For this we muster the spirit of America, and the faith of America.

We do not retreat. We are not content to stand still. As Americans, we go forward, in the service of our country, by the will of God.

Franklin D. Roosevelt

Saturday, January 20, 1945

*W*eeks after Franklin D. Roosevelt won a fourth term as President, the press inquired about his plans for yet another inauguration. Would there be, as tradition had it, a grand parade in front of the Capitol after the swearing-in ceremonies?

No, the President said, there would be no parade. He explained why, with a simple, poignant question. "Who is there to parade?"[1] he asked. The young men and women of the armed forces were preparing for climactic battles in Europe and in the Pacific. They were not available for ceremonial parades.

Roosevelt's speech on January 20, 1945, was the first explicit wartime inaugural address since Abraham Lincoln's second inaugural in 1865. In both cases, however, victory was in sight, allowing both men to speak not of ongoing sacrifices, but of hopes for peace and reconciliation.

Franklin Roosevelt was a sick and exhausted man on January 20, 1945. Gaunt, with great circles under his eyes, he appeared to be but a shadow of the exuberant, optimistic politician who revived the nation's spirits in the early 1930s. His condition was hardly a secret to his aides and associates. Democratic Party leaders feared that FDR would not survive a fourth term, leading them to block the renomination of vice president Henry Wallace, whom they didn't trust. Harry Truman, a senator from Missouri, replaced Wallace on the ticket.

Roosevelt faced an energetic challenge from Republican Thomas E. Dewey, who occupied the post Roosevelt himself once held—governor of New York. Overshadowing the campaign, despite Dewey's best efforts, was the war. Dewey argued against the centralization of government that

---

1. Samuel Rosenman, *Working with Roosevelt* (New York: Harper & Row, 1952), 516.

had taken place during FDR's tenure, while other Republicans contended that Roosevelt was beholden to corrupt urban bosses. Dewey's attacks late in the campaign energized Roosevelt, who had been content to let Truman carry the burden of travel and speech making. During a campaign swing through Boston, he noted that Dewey had charged that Communists were infiltrating the New Deal and that Roosevelt's long tenure as President posed "the threat of monarchy in the United States."

"Now, really," Roosevelt said, his timing and cadence no worse for the wear on his body, "which is it—communism or monarchy?"[2]

Roosevelt's greatest asset in 1944 was the Allied cause. American, British, and Canadian forces successfully invaded Nazi-occupied France on June 6, leading to the liberation of Paris in August. From the east, the Red Army was pushing forward into Romania. In the Pacific theater, American troops successfully invaded the Philippine Islands in August, leading to the Allied victory at the Battle of Leyte Gulf in October.

In the wartime election, Roosevelt won 432 electoral votes and 25.6 million popular votes, while Dewey won 99 and 22 million, respectively.

During the weeks between his reelection and his inauguration, Roosevelt devoted his declining energy to postwar planning. He and his aides would join British Prime Minister Winston Churchill and Soviet leader Joseph Stalin for a summit meeting in Yalta on the Black Sea in early 1945, where they would attempt to agree on a postwar world order, including plans to establish the successor to the failed League of Nations, the United Nations.

But before that climactic meeting, with troops still in the field and in danger, Roosevelt decided on a low-key, simple inauguration ceremony. It took place on the South Porch of the White House, rather than at the Capitol, and was witnessed by only a few thousand people rather than the tens of thousands that traditionally gathered at the Capitol and along Pennsylvania Avenue. As Roosevelt promised, there were no parades or elaborate ceremonies. After a prayer and the playing of "The Star-Spangled Banner," Truman took the oath as vice president and Roosevelt stood, without his heavy leg braces and despite his frail health, to take the oath as President for the fourth time and deliver his fourth inaugural.

It was, he said at the very beginning, "my wish that the form of this inauguration be simple and its words brief." Brief it was—he took less than

2. *Ibid.* 503.

five minutes to deliver the shortest inaugural address since George Washington's second in 1793. Like many other Roosevelt speeches, this one was the work of Sam Rosenman, Robert Sherwood, and poet Archibald MacLeish. But Roosevelt's personal touch is evident as well in his reference to a childhood teacher who taught him that "life will not always run smoothly."

The speech is a postwar document delivered in a time of war, intended not to rouse martial spirits but to invite thoughts of peace and to reflect on the meaning of sacrifice. The United States, he said, had "learned lessons—at a fearful cost—and we shall profit from them.

"We have learned that we cannot live alone," he continued, a clear reference to the isolationist sentiment that overtook the nation after the last world war. The world had shrunk; sovereign nations no longer could take shelter behind the outdated protection of oceans or mountains. The nation's "well-being," he said, "is dependent on the well-being of other nations far away." Ironically, he quoted the American philosopher Ralph Waldo Emerson, who celebrated self-reliance and individualism. "We have learned the simple truth, as Emerson said, that 'The only way to have a friend is to be one,'" Roosevelt said. "We can gain no lasting peace if we approach it with suspicion and mistrust or with fear."

To help achieve that lasting peace, Roosevelt flew to Yalta two days after his inaugural. He returned, according to speechwriter Sam Rosenman, "certain" that his "prayer" for an enduring peace had been answered.

He died of a cerebral hemorrhage on April 13, 1945, three months into his term, four weeks before the Germans surrendered, and four months before atomic bombs fell on Hiroshima and Nagasaki.

MR. CHIEF JUSTICE, MR. VICE PRESIDENT, MY FRIENDS, you will understand and, I believe, agree with my wish that the form of this inauguration be simple and its words brief.

We Americans of today, together with our allies, are passing through a period of supreme test. It is a test of our courage—of our resolve—of our wisdom—of our essential democracy.

If we meet that test—successfully and honorably—we shall perform a service of historic importance which men and women and children will honor throughout all time.

As I stand here today, having taken the solemn oath of office in the presence of my fellow countrymen—in the presence of our God—I know that it is America's purpose that we shall not fail.

In the days and in the years that are to come we shall work for a just and honorable peace, a durable peace, as today we work and fight for total victory in war.

We can and we will achieve such a peace.

We shall strive for perfection. We shall not achieve it immediately— but we still shall strive. We may make mistakes—but they must never be mistakes which result from faintness of heart or abandonment of moral principle.

I remember that my old schoolmaster, Dr. Peabody, said, in days that seemed to us then to be secure and untroubled: "Things in life will not always run smoothly. Sometimes we will be rising toward the heights—then all will seem to reverse itself and start downward. The great fact to remember is that the trend of civilization itself is forever upward; that a line drawn through the middle of the peaks and the valleys of the centuries always has an upward trend."

Our Constitution of 1787 was not a perfect instrument; it is not perfect yet. But it provided a firm base upon which all manner of men, of all races and colors and creeds, could build our solid structure of democracy.

And so today, in this year of war, 1945, we have learned lessons—at a fearful cost—and we shall profit by them.

We have learned that we cannot live alone, at peace; that our own well-being is dependent on the well-being of other nations far away. We have learned that we must live as men, not as ostriches, nor as dogs in the manger.

We have learned to be citizens of the world, members of the human community.

We have learned the simple truth, as Emerson said, that "The only way to have a friend is to be one." We can gain no lasting peace if we approach it with suspicion and mistrust or with fear.

We can gain it only if we proceed with the understanding, the confidence, and the courage which flow from conviction.

The Almighty God has blessed our land in many ways. He has given our people stout hearts and strong arms with which to strike mighty blows for freedom and truth. He has given to our country a faith which has become the hope of all peoples in an anguished world.

So we pray to Him now for the vision to see our way clearly—to see the way that leads to a better life for ourselves and for all our fellow men—to the achievement of His will to peace on earth.

Harry S. Truman

Thursday, January 20, 1949

*O*n a sun-splashed January afternoon in the nation's capital, Harry S. Truman did something few Americans believed possible only four or five months earlier: he took the oath of office as President for a second time.

Even Truman's fellow Democrats had seemed resigned to losing the White House for the first time since 1928. He seemed too small for the office that his larger-than-life predecessor filled for more than a decade. He was not a particularly good speaker, he lacked the late President's gravitas and charisma, and he seemed too much the small-town politician. Making matters worse, strikes wracked the nation, and returning veterans had a hard time finding housing thanks to a national shortage. Truman became a lightning rod for the nation's postwar discontents.

Some Democrats desperately sought an alternative to Truman. Taking a page out of nineteenth-century politics, they considered a military hero as a possible candidate—Dwight Eisenhower. Nobody knew much about Eisenhower's politics or his party affiliation, but for Democrats like Senator Claude Pepper of Florida and Mayors Frank Hague of Jersey City and William O'Dwyer of New York, union activists like Walter Reuther of the United Automobile Workers, the liberal advocacy group Americans for Democratic Action, and even FDR's son, James—it didn't seem to matter. Truman was regarded as a certain loser, and for these prominent party members, better to be rid of him than suffer defeat in November.

Eisenhower made it clear on the eve of the convention that he would not be a candidate for President, leaving the Democrats with no real alternative to Truman. The convention was a disaster. Hubert H. Humphrey, the young mayor of Minneapolis, delivered a stirring speech on behalf of a strong civil rights plank in the party platform, leading some southern Democrats to bolt the party and nominate their own candidate for President, Strom Thurmond of South Carolina. Meanwhile, some liberals went off on their own, too, rallying behind the candidacy of former vice president Henry A. Wallace, the nominee of the Progressive Party. So Truman was left with a split on his right and on his left.

Republicans, with victory seemingly at hand, already had rallied behind Governor Thomas Dewey of New York, who ran a credible race against Roosevelt in 1944. Ironically, the Republicans passed a platform that indicated how ingrained the New Deal had become in U.S. politics: it supported federal spending on housing (a critical issue in 1948 as returned veterans and their families sought places to live), an expansion of Social Security benefits, and new civil rights legislation.

The public's discontent with Truman seemed at odds with America's position in the world in 1948. It was a global colossus. The nation's industrial output was the envy of the world, and its unemployment rate was below 4 percent. Overseas, U.S. dollars were rebuilding parts of the Old World under a plan outlined by Secretary of State George C. Marshall, while U.S. troops occupied the defeated Axis powers and established bases in friendly nations, projecting American power abroad as never before.

There was, however, a challenge to U.S. domination, and it came from a former ally—the Soviet Union. Fear of Communist expansion and Soviet intentions had led, during Truman's term, to the beginning of what came to be known as the Cold War between Washington and Moscow. Truman was the architect of an aggressively anti-Communist foreign policy, setting the tone for Cold War diplomacy for the next quarter century.

Truman had some issues to exploit in the summer of 1948, even if his own party was split and demoralized. He traveled the country by train, delivering speeches from the back of railroad cars and taunting Republicans for leading a "do-nothing Congress." Still, Dewey held a commanding lead in the polls, and many major media outlets carried stories speculating about how President Dewey would govern.

Truman never gave up, and on Election Day, in one of the biggest upsets in U.S. political history, he won convincingly, taking 24 million popular votes and 303 electoral votes to Dewey's 21.9 million and 189, respectively. Strom Thurmond recorded the strongest third-party performance since Theodore Roosevelt's 1912 campaign, taking about 1.1 million votes and 39 electoral votes. Truman's Democrats also regained both houses of Congress.

And so, on a cloudless afternoon in January, it was Harry Truman, not Thomas Dewey, who took the oath of office and who delivered the inaugural address. Truman's speech was an unambiguous Cold War document establishing the parameters of the ideological struggle between West and East. Delivered crisply—and broadcast over the latest media innovation, television—the speech divided the world into democratic and Communist, good and evil, and established four policy goals to achieve "peace and freedom" in the world. The speech became known as the "point four" speech because of Truman's enunciation of those four goals. The first three were not especially surprising: Truman pledged to support the United Nations, to continue programs like the Marshall Plan to help rebuild war-torn nations, and to strengthen alliances with non-Communist nations. Point four, however, caught the attention of his audience, the press, and politicians around the globe. The United States, he said, "must embark on a bold new program for making the benefits of our scientific advances and industrial progress available for the improvement and growth of undeveloped areas.

"More than half the people of the world are living in conditions approaching misery," he said, painting a picture of a world very different from that of postwar America. "They are victims of disease. Their economic life is primitive and stagnant. Their poverty is a handicap and a threat both to them and to more prosperous areas." The United States, he said, "should make available" to such nations "our store of technical knowledge." He added, "The old imperialism—exploitation for foreign profit—has no place in our plans. What we envisage is a program of development based on concepts of democratic fair dealing."

As a document, Truman's inaugural set the rhetorical standard for future cold war inaugurals. It saw the United States engaged in a titanic struggle for its freedom, and for the freedom of its allies. But Truman also asserted that such a war could not be won by arms alone. There were hearts and minds to consider as well.

Bibliographic Note

Harry Truman, *Memoirs*, vol. 2 (New York: Doubleday, 1956).

MR. VICE PRESIDENT, MR. CHIEF JUSTICE, AND FELLOW CITIZENS, I accept with humility the honor which the American people have conferred upon me. I accept it with a deep resolve to do all that I can for the welfare of this nation and for the peace of the world.

In performing the duties of my office, I need the help and prayers of every one of you. I ask for your encouragement and your support. The tasks we face are difficult, and we can accomplish them only if we work together.

Each period of our national history has had its special challenges. Those that confront us now are as momentous as any in the past. Today marks the beginning not only of a new administration, but of a period that will be eventful, perhaps decisive, for us and for the world.

It may be our lot to experience, and in large measure to bring about, a major turning point in the long history of the human race. The first half of this century has been marked by unprecedented and brutal attacks on the rights of man, and by the two most frightful wars in history. The supreme need of our time is for men to learn to live together in peace and harmony.

The peoples of the earth face the future with grave uncertainty, composed almost equally of great hopes and great fears. In this time of doubt, they look to the United States as never before for good will, strength, and wise leadership.

It is fitting, therefore, that we take this occasion to proclaim to the world the essential principles of the faith by which we live, and to declare our aims to all peoples.

The American people stand firm in the faith which has inspired this nation from the beginning. We believe that all men have a right to equal justice under law and equal opportunity to share in the common good. We believe that all men have the right to freedom of thought and expression. We believe that all men are created equal because they are created in the image of God.

From this faith we will not be moved.

The American people desire, and are determined to work for, a world in which all nations and all peoples are free to govern themselves as they see fit, and to achieve a decent and satisfying life. Above all else, our people desire, and are determined to work for, peace on earth—a just and lasting peace—based on genuine agreement freely arrived at by equals.

In the pursuit of these aims, the United States and other like-minded nations find themselves directly opposed by a regime with contrary aims and a totally different concept of life.

That regime adheres to a false philosophy which purports to offer freedom, security, and greater opportunity to mankind. Misled by this philosophy, many peoples have sacrificed their liberties only to learn to their sorrow that deceit and mockery, poverty and tyranny, are their reward.

That false philosophy is communism.

Communism is based on the belief that man is so weak and inadequate that he is unable to govern himself, and therefore requires the rule of strong masters.

Democracy is based on the conviction that man has the moral and intellectual capacity, as well as the inalienable right, to govern himself with reason and justice.

Communism subjects the individual to arrest without lawful cause, punishment without trial, and forced labor as the chattel of the state. It decrees what information he shall receive, what art he shall produce, what leaders he shall follow, and what thoughts he shall think.

Democracy maintains that government is established for the benefit of the individual, and is charged with the responsibility of protecting the rights of the individual and his freedom in the exercise of his abilities.

Communism maintains that social wrongs can be corrected only by violence.

Democracy has proved that social justice can be achieved through peaceful change.

Communism holds that the world is so deeply divided into opposing classes that war is inevitable.

Democracy holds that free nations can settle differences justly and maintain lasting peace.

These differences between communism and democracy do not concern the United States alone. People everywhere are coming to realize that what is involved is material well-being, human dignity, and the right to believe in and worship God.

I state these differences, not to draw issues of belief as such, but because the actions resulting from the Communist philosophy are a threat to the efforts of free nations to bring about world recovery and lasting peace.

Since the end of hostilities, the United States has invested its substance and its energy in a great constructive effort to restore peace, stability, and freedom to the world.

We have sought no territory and we have imposed our will on none. We have asked for no privileges we would not extend to others.

We have constantly and vigorously supported the United Nations and related agencies as a means of applying democratic principles to international relations. We have consistently advocated and relied upon peaceful settlement of disputes among nations.

We have made every effort to secure agreement on effective international control of our most powerful weapon, and we have worked steadily for the limitation and control of all armaments.

We have encouraged, by precept and example, the expansion of world trade on a sound and fair basis.

Almost a year ago, in company with sixteen free nations of Europe, we launched the greatest cooperative economic program in history. The purpose of that unprecedented effort is to invigorate and strengthen democracy in Europe, so that the free people of that continent can resume their rightful place in the forefront of civilization and can contribute once more to the security and welfare of the world.

Our efforts have brought new hope to all mankind. We have beaten back despair and defeatism. We have saved a number of countries from losing their liberty. Hundreds of millions of people all over the world now agree with us, that we need not have war—that we can have peace.

The initiative is ours.

We are moving on with other nations to build an even stronger structure of international order and justice. We shall have as our partners countries which, no longer solely concerned with the problem of national survival, are now working to improve the standards of living of all their people. We are ready to undertake new projects to strengthen the free world.

In the coming years, our program for peace and freedom will emphasize four major courses of action.

First, we will continue to give unfaltering support to the United Nations and related agencies, and we will continue to search for ways to strengthen their authority and increase their effectiveness. We believe that the United Nations will be strengthened by the new nations which are being formed in lands now advancing toward self-government under democratic principles.

Second, we will continue our programs for world economic recovery.

This means, first of all, that we must keep our full weight behind the European recovery program. We are confident of the success of this major venture in world recovery. We believe that our partners in this effort will achieve the status of self-supporting nations once again.

In addition, we must carry out our plans for reducing the barriers to world trade and increasing its volume. Economic recovery and peace itself depend on increased world trade.

Third, we will strengthen freedom-loving nations against the dangers of aggression.

We are now working out with a number of countries a joint agreement designed to strengthen the security of the North Atlantic area. Such an agreement would take the form of a collective defense arrangement within the terms of the United Nations Charter.

We have already established such a defense pact for the Western Hemisphere by the Treaty of Rio de Janeiro.

The primary purpose of these agreements is to provide unmistakable proof of the joint determination of the free countries to resist armed attack from any quarter. Each country participating in these arrangements must contribute all it can to the common defense.

If we can make it sufficiently clear, in advance, that any armed attack affecting our national security would be met with overwhelming force, the armed attack might never occur.

I hope soon to send to the Senate a treaty respecting the North Atlantic security plan.

In addition, we will provide military advice and equipment to free nations which will cooperate with us in the maintenance of peace and security.

Fourth, we must embark on a bold new program for making the benefits of our scientific advances and industrial progress available for the improvement and growth of underdeveloped areas.

More than half the people of the world are living in conditions approaching misery. Their food is inadequate. They are victims of disease. Their economic life is primitive and stagnant. Their poverty is a handicap and a threat both to them and to more prosperous areas.

For the first time in history, humanity possesses the knowledge and the skill to relieve the suffering of these people.

The United States is preeminent among nations in the development of industrial and scientific techniques. The material resources which we can afford to use for the assistance of other peoples are limited. But our imponderable resources in technical knowledge are constantly growing and are inexhaustible.

I believe that we should make available to peace-loving peoples the benefits of our store of technical knowledge in order to help them realize their aspirations for a better life. And, in cooperation with other nations, we should foster capital investment in areas needing development.

Our aim should be to help the free peoples of the world, through their own efforts, to produce more food, more clothing, more materials for housing, and more mechanical power to lighten their burdens.

We invite other countries to pool their technological resources in this undertaking. Their contributions will be warmly welcomed. This should be a cooperative enterprise in which all nations work together through the United Nations and its specialized agencies wherever practicable. It must be a worldwide effort for the achievement of peace, plenty, and freedom.

With the cooperation of business, private capital, agriculture, and labor in this country, this program can greatly increase the industrial activity in other nations and can raise substantially their standards of living.

Such new economic developments must be devised and controlled to benefit the peoples of the areas in which they are established. Guarantees to the investor must be balanced by guarantees in the interest of the people whose resources and whose labor go into these developments.

The old imperialism—exploitation for foreign profit—has no place in our plans. What we envisage is a program of development based on the concepts of democratic fair dealing.

All countries, including our own, will greatly benefit from a constructive program for the better use of the world's human and natural resources. Experience shows that our commerce with other countries expands as they progress industrially and economically.

Greater production is the key to prosperity and peace. And the key to greater production is a wider and more vigorous application of modern scientific and technical knowledge.

Only by helping the least fortunate of its members to help themselves can the human family achieve the decent, satisfying life that is the right of all people.

Democracy alone can supply the vitalizing force to stir the peoples of the world into triumphant action, not only against their human oppressors, but also against their ancient enemies—hunger, misery, and despair.

On the basis of these four major courses of action we hope to help create the conditions that will lead eventually to personal freedom and happiness for all mankind.

If we are to be successful in carrying out these policies, it is clear that we must have continued prosperity in this country and we must keep ourselves strong.

Slowly but surely we are weaving a world fabric of international security and growing prosperity.

We are aided by all who wish to live in freedom from fear—even by those who live today in fear under their own governments.

We are aided by all who want relief from the lies of propaganda—who desire truth and sincerity.

We are aided by all who desire self-government and a voice in deciding their own affairs.

We are aided by all who long for economic security—for the security and abundance that men in free societies can enjoy.

We are aided by all who desire freedom of speech, freedom of religion, and freedom to live their own lives for useful ends.

Our allies are the millions who hunger and thirst after righteousness.

In due time, as our stability becomes manifest, as more and more nations come to know the benefits of democracy and to participate in growing abundance, I believe that those countries which now oppose us will abandon their delusions and join with the free nations of the world in a just settlement of international differences.

Events have brought our American democracy to new influence and new responsibilities. They will test our courage, our devotion to duty, and our concept of liberty.

But I say to all men, what we have achieved in liberty, we will surpass in greater liberty.

Steadfast in our faith in the Almighty, we will advance toward a world where man's freedom is secure.

To that end we will devote our strength, our resources, and our firmness of resolve. With God's help, the future of mankind will be assured in a world of justice, harmony, and peace.

Dwight D. Eisenhower

Tuesday, January 20, 1953

$\mathcal{T}$he United States was at war again when Dwight D. Eisenhower, retired five-star general and one of World War II's most famous soldiers,
took the oath of office as the nation's thirty-fourth President in 1953.
American troops were fighting in Korea even as official Washington gathered for ceremonies marking the first transfer of power from one party to
another in twenty years.

On that chilly afternoon in the nation's capital, the conflict in Korea
had been raging for nearly three years and had cost the lives of more than
thirty thousand American soldiers. Its fading popularity and uncertain
prospects helped contribute to Harry Truman's decision not to run for a
second full term in 1952.

Korea was part of an overall increase in tensions between the Communist world and the West, led by the United States. During Truman's second
term, the Soviet Union successfully exploded an atomic bomb, breaking
the U.S. monopoly on nuclear weapons, and China fell to a Communist
insurgency. In 1949, ten Western European nations joined with the United
States and Canada to form the North Atlantic Treaty Organization. Each
nation promised to consider an attack on one as an attack on all. In response, the Soviets engineered a collective security arrangement with the
Communist nations of Eastern Europe, called the Warsaw Pact.

The first shots of the Cold War, however, were fired not in Europe but in
Asia. In 1950, the Communist government of North Korea invaded pro-West
South Korea, setting in motion the first test of the new United Nations organization. The U.N. declared the invasion illegal and requested member nations to come to South Korea's aid. Under the command of legendary U.S.
general Douglas MacArthur, a U.S.-led coalition pushed the North Koreans
back over the Cold War boundary that partitioned the peninsula. Commu-

nist China then intervened on North Korea's side, pushing back the U.S. coalition. When MacArthur repeatedly proposed an attack on China itself, a
move that would have widened the war, Harry Truman fired him. MacArthur returned home a hero, adding to Truman's unpopularity.

The threat of Communist aggression abroad and infiltration at home
dominated foreign affairs and domestic politics in the early 1950s. A State
Department official, Alger Hiss, was accused of spying for the Soviets, and
a federal worker named Julius Rosenberg and his wife, Ethel, were indicted
for passing atomic secrets to Russia. Republicans in Congress, especially
Senator Joseph McCarthy of Wisconsin, accused the Truman administration of being soft on communism.

Anxiety over events abroad contrasted with the nation's post-World
War II domestic bliss. Unemployment was below 5 percent, the nation's
gross national product would grow from $205 billion in 1940 to $500 billion in 1960, and a building boom outside the nation's cities allowed millions to buy their own homes, sometimes with government-backed loans.
Consumer goods flooded the nation's stores, the new medium of television began to change the nation's entertainment habits, and thousands of
veterans used government benefits to obtain a college education and a
professional career.

Not everybody, however, had a place at the table. Blacks remained subject to legal segregation and routine injustice in the South, leading hundreds of thousands to leave the region and move to what they hoped would
be a better life, with better jobs, in the industrial cities of the Northeast
and Midwest. Truman desegregated the nation's military, but broader civil
rights remained heavily contested.

Truman's withdrawal from the 1952 campaign created a political vac-
uum, setting the stage for Eisenhower. He was a reluctant candidate—he
thought, at the age of sixty-two, that he was too old to be President—but
an appealing one, with a wide smile, genial disposition, and impeccable
military record. He won the Republican Party's nomination in Chicago on
the first ballot.

The Democrats chose Governor Adlai Stevenson of Illinois as their
nominee. Stevenson, whose grandfather had been vice president during
Grover Cleveland's first administration, was a cerebral, witty man who
vowed to "talk sense" to the American people. But late in the campaign,
Stevenson's sensible appeals were overshadowed by Eisenhower's promise
to go to Korea to seek a negotiated settlement to the unpopular war.

Eisenhower's promise, combined with the nation's desire for change, led to a Republican sweep of the White House and Capitol Hill. Eisenhower won thirty-four million votes, about 55 percent of the total, and 442 electoral votes to Stevenson's twenty-seven million and 89, respectively.

The transition from Harry Truman to Dwight Eisenhower was not smooth. Truman resented Eisenhower's criticism of his foreign policy, his refusal to condemn Joseph McCarthy's attacks on former secretary of state George Marshall (who had been Eisenhower's boss during World War II), and his pledge to go to Korea, which Truman regarded as political grandstanding. Eisenhower, according to biographer Stephen Ambrose, regarded Truman as overly partisan, crude, and undignified. During their drive together from the White House to the inaugural ceremonies on Capitol Hill, the two men barely spoke.

After taking the oath of office, Eisenhower did something no other new President had ever done. Rather than immediately begin his inaugural address, he delivered a prayer he had written himself earlier in the morning. The speech itself was replete with references to communism's embrace of atheism. The nation's enemies, he said, "know no god but force, no devotion but its use. They tutor men in treason. They feed upon the hunger of others. Whatever defies them, they torture, especially the truth."

Eisenhower's speech depicted the United States as the defender of universal freedoms abroad, although it had nothing to say about more complex issues at home. As the 1950s progressed, African Americans would demand the same rights and liberties to which the United States was committed abroad, but Eisenhower's speech did not address those demands.

It did, however, foresee a global and interdependent struggle between free nations and the Communist world. "Freedom is pitted against slavery," he said, "lightness against the dark."

The anti-Communist tenor of Eisenhower's speech reflected a bipartisan consensus. But conservative Republicans who hoped to hear a ringing denunciation of the New Deal and Truman's Fair Deal or a repudiation of government's new and expanded role in society were disappointed. The new President did not pledge to roll back the programs put in place since the Great Depression.

He did, however, help bring about an end to the war that helped elect him. After Eisenhower visited Korea, as promised, in December 1952, peace negotiations took on a new urgency, leading to an armistice in July 1953.

Bibliographic Note

Stephen Ambrose, *Eisenhower: The President* (New York: Simon & Schuster, 1984).

MY FRIENDS, before I begin the expression of those thoughts that I deem appropriate to this moment, would you permit me the privilege of uttering a little private prayer of my own. And I ask that you bow your heads:

Almighty God, as we stand here at this moment my future associates in the executive branch of government join me in beseeching that Thou will make full and complete our dedication to the service of the people in this throng, and their fellow citizens everywhere.

Give us, we pray, the power to discern clearly right from wrong, and allow all our words and actions to be governed thereby, and by the laws of this land. Especially we pray that our concern shall be for all the people regardless of station, race, or calling.

May cooperation be permitted and be the mutual aim of those who, under the concepts of our Constitution, hold to differing political faiths; so that all may work for the good of our beloved country and Thy glory. Amen.

My fellow citizens:

The world and we have passed the midway point of a century of continuing challenge. We sense with all our faculties that forces of good and evil are massed and armed and opposed as rarely before in history.

This fact defines the meaning of this day. We are summoned by this honored and historic ceremony to witness more than the act of one citizen swearing his oath of service, in the presence of God. We are called as a people to give testimony in the sight of the world to our faith that the future shall belong to the free.

Since this century's beginning, a time of tempest has seemed to come upon the continents of the earth. Masses of Asia have awakened to strike off shackles of the past. Great nations of Europe have fought their bloodiest wars. Thrones have toppled and their vast empires have disappeared. New nations have been born.

For our own country, it has been a time of recurring trial. We have grown in power and in responsibility. We have passed through the anxieties of depression and of war to a summit unmatched in man's history. Seeking to secure peace in the world, we have had to fight through the forests of the Argonne, to the shores of Iwo Jima, and to the cold mountains of Korea.

In the swift rush of great events, we find ourselves groping to know the full sense and meaning of these times in which we live. In our quest of understanding, we beseech God's guidance. We summon all our knowledge of the past and we scan all signs of the future. We bring all our wit and all our will to meet the question:

How far have we come in man's long pilgrimage from darkness toward light? Are we nearing the light—a day of freedom and of peace for all mankind? Or are the shadows of another night closing in upon us?

Great as are the preoccupations absorbing us at home, concerned as we are with matters that deeply affect our livelihood today and our vision of the future, each of these domestic problems is dwarfed by, and often even created by, this question that involves all humankind.

This trial comes at a moment when man's power to achieve good or to inflict evil surpasses the brightest hopes and the sharpest fears of all ages. We can turn rivers in their courses, level mountains to the plains. Oceans and land and sky are avenues for our colossal commerce. Disease diminishes and life lengthens.

Yet the promise of this life is imperiled by the very genius that has made it possible. Nations amass wealth. Labor sweats to create—and turns out devices to level not only mountains but also cities. Science seems ready to confer upon us, as its final gift, the power to erase human life from this planet.

At such a time in history, we who are free must proclaim anew our faith. This faith is the abiding creed of our fathers. It is our faith in the deathless dignity of man, governed by eternal moral and natural laws.

This faith defines our full view of life. It establishes, beyond debate, those gifts of the Creator that are man's inalienable rights, and that make all men equal in His sight.

In the light of this equality, we know that the virtues most cherished by free people—love of truth, pride of work, devotion to country—all are treasures equally precious in the lives of the most humble and of the most exalted. The men who mine coal and fire furnaces and balance ledgers and turn lathes and pick cotton and heal the sick and plant corn—all serve as proudly, and as profitably, for America as the statesmen who draft treaties and the legislators who enact laws.

This faith rules our whole way of life. It decrees that we, the people, elect leaders not to rule but to serve. It asserts that we have the right to choice of our own work and to the reward of our own toil. It inspires the initiative that makes our productivity the wonder of the world. And it warns that any man who seeks to deny equality among all his brothers betrays the spirit of the free and invites the mockery of the tyrant.

It is because we, all of us, hold to these principles that the political changes accomplished this day do not imply turbulence, upheaval, or disorder. Rather this change expresses a purpose of strengthening our dedication and devotion to the precepts of our founding documents, a conscious renewal of faith in our country and in the watchfulness of a Divine Providence.

The enemies of this faith know no god but force, no devotion but its use. They tutor men in treason. They feed upon the hunger of others. Whatever defies them, they torture, especially the truth.

Here, then, is joined no argument between slightly differing philosophies. This conflict strikes directly at the faith of our fathers and the lives of our sons. No principle or treasure that we hold, from the spiritual knowledge of our free schools and churches to the creative magic of free labor and capital, nothing lies safely beyond the reach of this struggle.

Freedom is pitted against slavery; lightness against the dark.

The faith we hold belongs not to us alone but to the free of all the world. This common bond binds the grower of rice in Burma and the planter of wheat in Iowa, the shepherd in southern Italy and the mountaineer in the Andes. It confers a common dignity upon the French soldier who dies in Indochina, the British soldier killed in Malaya, the American life given in Korea.

We know, beyond this, that we are linked to all free peoples not merely by a noble idea but by a simple need. No free people can for long cling to any privilege or enjoy any safety in economic solitude. For all our own material might, even we need markets in the world for the surpluses of our farms and our factories. Equally, we need for these same farms and factories vital materials and products of distant lands. This basic law of interdependence, so manifest in the commerce of peace, applies with thousandfold intensity in the event of war.

So we are persuaded by necessity and by belief that the strength of all free peoples lies in unity; their danger, in discord.

To produce this unity, to meet the challenge of our time, destiny has laid upon our country the responsibility of the free world's leadership.

So it is proper that we assure our friends once again that, in the discharge of this responsibility, we Americans know and we observe the difference between world leadership and imperialism; between firmness and truculence; between a thoughtfully calculated goal and spasmodic reaction to the stimulus of emergencies.

We wish our friends the world over to know this above all: we face the threat—not with dread and confusion—but with confidence and conviction.

We feel this moral strength because we know that we are not helpless prisoners of history. We are free men. We shall remain free, never to be proven guilty of the one capital offense against freedom, a lack of stanch faith.

In pleading our just cause before the bar of history and in pressing our labor for world peace, we shall be guided by certain fixed principles.

These principles are:

1. Abhorring war as a chosen way to balk the purposes of those who threaten us, we hold it to be the first task of statesmanship to develop the strength that will deter the forces of aggression and promote the conditions of peace. For, as it must be the supreme purpose of all free men, so it must be the dedication of their leaders, to save humanity from preying upon itself.

   In the light of this principle, we stand ready to engage with any and all others in joint effort to remove the causes of mutual fear and distrust among nations, so as to make possible drastic reduction of armaments. The sole requisites for undertaking such effort are that—in their purpose—they be aimed logically and honestly toward secure peace for all; and that—in their result—they provide methods by which every participating nation will prove good faith in carrying out its pledge.

2. Realizing that common sense and common decency alike dictate the futility of appeasement, we shall never try to placate an aggressor by the false and wicked bargain of trading honor for security. Americans, indeed all free men, remember that in the final choice a soldier's pack is not so heavy a burden as a prisoner's chains.

3. Knowing that only a United States that is strong and immensely productive can help defend freedom in our world, we view our nation's strength and security as a trust upon which rests the hope of free men everywhere. It is the firm duty of each of our free citizens and of every free citizen everywhere to place the cause of his country before the comfort, the convenience of himself.

4. Honoring the identity and the special heritage of each nation in the world, we shall never use our strength to try to impress upon another people our own cherished political and economic institutions.

5. Assessing realistically the needs and capacities of proven friends of freedom, we shall strive to help them to achieve their own security and well-being. Likewise, we shall count upon them to assume, within the limits of their resources, their full and just burdens in the common defense of freedom.

6. Recognizing economic health as an indispensable basis of military strength and the free world's peace, we shall strive to foster everywhere, and to practice ourselves, policies that encourage productivity and profitable trade. For the impoverishment of any single people in the world means danger to the well-being of all other peoples.

7. Appreciating that economic need, military security, and political wisdom combine to suggest regional groupings of free peoples, we hope, within the framework of the United Nations, to help strengthen such special bonds the world over. The nature of these ties must vary with the different problems of different areas.

   In the Western Hemisphere, we enthusiastically join with all our neighbors in the work of perfecting a community of fraternal trust and common purpose.

   In Europe, we ask that enlightened and inspired leaders of the Western nations strive with renewed vigor to make the unity of their peoples a reality. Only as free Europe unitedly marshals its strength can it effectively safeguard, even with our help, its spiritual and cultural heritage.

8. Conceiving the defense of freedom, like freedom itself, to be one and indivisible, we hold all continents and peoples in equal regard and honor. We reject any insinuation that one race or another, one people or another, is in any sense inferior or expendable.

9. Respecting the United Nations as the living sign of all people's hope for peace, we shall strive to make it not merely an eloquent symbol but an effective force. And in our quest for an honorable peace, we shall neither compromise, nor tire, nor ever cease.

By these rules of conduct, we hope to be known to all peoples.

By their observance, an earth of peace may become not a vision but a fact.

This hope—this supreme aspiration—must rule the way we live.

We must be ready to dare all for our country. For history does not long entrust the care of freedom to the weak or the timid. We must acquire proficiency in defense and display stamina in purpose.

We must be willing, individually and as a nation, to accept whatever sacrifices may be required of us. A people that values its privileges above its principles soon loses both.

These basic precepts are not lofty abstractions, far removed from matters of daily living. They are laws of spiritual strength that generate and define our material strength. Patriotism means equipped forces and a prepared citizenry. Moral stamina means more energy and more productivity, on the farm and in the factory. Love of liberty means the guarding of every resource that makes freedom possible—from the sanctity of our families and the wealth of our soil to the genius of our scientists.

And so each citizen plays an indispensable role. The productivity of our heads, our hands, and our hearts is the source of all the strength we can command, for both the enrichment of our lives and the winning of the peace.

No person, no home, no community can be beyond the reach of this call. We are summoned to act in wisdom and in conscience, to work with industry, to teach with persuasion, to preach with conviction, to weigh our every deed with care and with compassion. For this truth must be clear before us: whatever America hopes to bring to pass in the world must first come to pass in the heart of America.

The peace we seek, then, is nothing less than the practice and fulfillment of our whole faith among ourselves and in our dealings with others. This signifies more than the stilling of guns, easing the sorrow of war. More than escape from death, it is a way of life. More than a haven for the weary, it is a hope for the brave.

This is the hope that beckons us onward in this century of trial. This is the work that awaits us all, to be done with bravery, with charity, and with prayer to Almighty God.

Dwight D. Eisenhower

Monday, January 21, 1957

Dwight D. Eisenhower gave serious thought to retiring from the presidency after a single term. He believed that after decades of service to his country, he had done his duty. But that was before he suffered a heart attack.

Eisenhower was taken ill while on vacation in September 1955, initiating the first presidential medical crisis since Woodrow Wilson's stroke in 1920. After doctors stabilized him, he spent several weeks recovering. The President himself began to speculate about who might succeed him if he chose not to run again. The more he thought about retirement, the less he liked it.

In February 1956, after doctors told him his recovery was complete, Eisenhower announced that he would seek reelection. He later told his chief of staff, Sherman Adams, "You know, if it hadn't been for that heart attack, I doubt if I would have been a candidate again." He told another aide that he had been "close" to declining a second term until his illness.[1]

As the campaign began, the nation was midway through a decade that would later be criticized as dull, gray, and slow moving, but which, in fact, was eventful, dynamic, and dangerous. Two historic migrations were under way within the borders of the United States. African Americans from the rural South were continuing to move to the nation's industrialized cities, particularly in the Midwest and Northeast, while middle- and working-class whites, many of them World War II veterans, were moving from cities to new suburban communities. Eisenhower proposed an initiative that would accelerate the changes associated with those migrations: the inter-

---

1. Chester J. Pach Jr. and Elmo Richardson, *The Presidency of Dwight Eisenhower*, (Lawrence, KS: University Press of Kansas, 1979), 90.

state highway system. Created in 1956, the interstate system would literally reconfigure the map, linking far-flung cities to each other and to suburban areas.

Americans were on the move in the 1950s. And there were more Americans to move. The nation was in the midst of an extraordinary demographic bulge that became known as the baby boom. Just under seventy-six million children were born in the postwar years of 1946 to 1964.

At home and abroad, the world that existed before the war was giving way to something new, something quite unlike the old ways and old certainties of the early twentieth century. African Americans actively challenged the old racial order that subordinated them throughout the country but particularly in the South. They had an ally in the Supreme Court, which ruled in 1954, in the case of *Brown v. Board of Education of Topeka* that racial segregation in public schools was unconstitutional. In Montgomery, Alabama, in December 1955, an African American woman named Rosa Parks refused to surrender her seat to a white and move to the back of the bus, as the law required. Her defiance led to a black boycott of the city's segregated bus system, which led in turn to further protests throughout the South. The boycott lasted more than a year, ending when the U.S. Supreme Court struck down segregation on Montgomery's buses.

Overseas, the old order and the old power arrangements came crashing down in the weeks before Election Day. As nationalist movements in Africa, Asia, and the Middle East began challenging the old European imperial powers, Egypt seized the Suez Canal, which was owned by a private Western company and controlled by France and Great Britain. Enraged, the British and French invaded Egypt to retake the canal, while the eight-year-old state of Israel joined the invasion to stamp out paramilitary bases near its border. Eisenhower adamantly opposed the actions of America's allies, and, faced with that opposition, the old colonial powers had little choice but to withdraw.

But the United States was not so powerful that it could bend the world to its will. As America sought to rein in its allies in the Middle East, anti-Communist insurgents in the Hungarian capital of Budapest threatened to overthrow the nation's pro-Soviet leaders. The Russians sent in tanks to crush the fledgling rebellion. The Eisenhower administration conceded it was powerless to oppose the Soviet move despite America's sympathy for the rebels.

The crisis in the Middle East was reaching its climax when Americans went to the polls in November to reelect Eisenhower in rematch with Ad-lai Stevenson. Eisenhower won 35.5 million popular votes and 457 electoral votes to Stevenson's 26 million and 73, respectively.

January 20, the day designated as the start of a new presidential term, fell on a Sunday in 1957, so Eisenhower took the oath of office in a private ceremony. The public ceremonies took place on Monday. Eisenhower began his speech with the words, "Mr. Chairman," an oddly prominent mention of the man who organized the inaugural ceremonies, Robert Fleming.

According to biographer Stephen Ambrose, Eisenhower consciously crafted his inaugural to draw the nation's attention to the developing world. The old colonial powers of Europe, devastated by two world wars, were shedding the last of their colonies in Africa and Asia, but many of the newly independent states were desperately poor. Eisenhower, according to Ambrose, wanted them to view the United States as a friend and an ally— in part, of course, so that they would not turn to the Soviet Union and communism, forces he describes as "divisive" but "shaken" by the recent uprising in Hungary.

After celebrating his own nation's postwar prosperity, Eisenhower noted that in "too much of the earth there is want, discord, danger. New forces and new nations stir and strive across the earth . . . From the deserts of North Africa to the islands of the South Pacific, one-third of all mankind has entered upon an historic struggle for a new freedom: freedom from grinding poverty."

Peace would only come where there was justice, he said, and in order to help bring about justice, Americans must be willing to pay what Eisenhower called "the full price" of peace.

"We must use our skills and knowledge and, at times, our substance, to help others rise from misery, however far the scene of suffering may be from our shores," he said. "For wherever in the world a people knows desperate want, there must appear at least the spark of hope, the hope of progress—or there will surely rise at last the flames of conflict."

Eisenhower ended his second inaugural as he began his first, with a prayer. He would continue to preside over an unparalleled prosperity at home and continued confrontations with communism abroad.

Bibliographic Note
Stephen Ambrose, *Eisenhower: The President*
(New York: Simon & Schuster, 1984).

Mr. Chairman, Mr. Vice President, Mr. Chief Justice, Mr. Speaker, members of my family and friends, my countrymen, and the friends of my country, wherever they may be, we meet again, as upon a like moment four years ago, and again you have witnessed my solemn oath of service to you.

I, too, am a witness, today testifying in your name to the principles and purposes to which we, as a people, are pledged.

Before all else, we seek, upon our common labor as a nation, the blessings of Almighty God. And the hopes in our hearts fashion the deepest prayers of our whole people.

May we pursue the right—without self-righteousness.

May we know unity—without conformity.

May we grow in strength—without pride in self.

May we, in our dealings with all peoples of the earth, ever speak truth and serve justice.

And so shall America—in the sight of all men of good will—prove true to the honorable purposes that bind and rule us as a people in all this time of trial through which we pass.

We live in a land of plenty, but rarely has this earth known such peril as today.

In our nation work and wealth abound. Our population grows. Commerce crowds our rivers and rails, our skies, harbors, and highways. Our soil is fertile, our agriculture productive. The air rings with the song of our industry—rolling mills and blast furnaces, dynamos, dams, and assembly lines—the chorus of America the bountiful.

This is our home—yet this is not the whole of our world. For our world is where our full destiny lies—with men, of all people, and all nations, who are or would be free. And for them—and so for us—this is no time of ease or of rest.

In too much of the earth there is want, discord, danger. New forces and new nations stir and strive across the earth, with power to bring, by their fate, great good or great evil to the free world's future. From the deserts of North Africa to the islands of the South Pacific one-third of all mankind has entered upon an historic struggle for a new freedom; freedom from grinding poverty. Across all continents, nearly a billion people seek, sometimes almost in desperation, for the skills and knowledge and assistance by which they may satisfy from their own resources, the material wants common to all mankind.

No nation, however old or great, escapes this tempest of change and turmoil. Some, impoverished by the recent world war, seek to restore their means of livelihood. In the heart of Europe, Germany still stands tragically divided. So is the whole continent divided. And so, too, is all the world.

The divisive force is international communism and the power that it controls.

The designs of that power, dark in purpose, are clear in practice. It strives to seal forever the fate of those it has enslaved. It strives to break the ties that unite the free. And it strives to capture—to exploit for its own greater power—all forces of change in the world, especially the needs of the hungry and the hopes of the oppressed.

Yet the world of international communism has itself been shaken by a fierce and mighty force: the readiness of men who love freedom to pledge their lives to that love. Through the night of their bondage, the unconquerable will of heroes has struck with the swift, sharp thrust of lightning. Budapest is no longer merely the name of a city; henceforth it is a new and shining symbol of man's yearning to be free.

Thus across all the globe there harshly blow the winds of change. And, we—though fortunate be our lot—know that we can never turn our backs to them.

We look upon this shaken earth, and we declare our firm and fixed purpose—the building of a peace with justice in a world where moral law prevails.

The building of such a peace is a bold and solemn purpose. To proclaim it is easy. To serve it will be hard. And to attain it, we must be aware of its full meaning—and ready to pay its full price.

We know clearly what we seek, and why.

We seek peace, knowing that peace is the climate of freedom. And now, as in no other age, we seek it because we have been warned, by the power of modern weapons, that peace may be the only climate possible for human life itself.

Yet this peace we seek cannot be born of fear alone: it must be rooted in the lives of nations. There must be justice, sensed and shared by all peoples, for, without justice the world can know only a tense and unstable truce. There must be law, steadily invoked and respected by all nations, for without law, the world promises only such meager justice as the pity of the strong upon the weak. But the law of which we speak, comprehending the values of freedom, affirms the equality of all nations, great and small.

Splendid as can be the blessings of such a peace, high will be its cost: in toil patiently sustained, in help honorably given, in sacrifice calmly borne.

We are called to meet the price of this peace.

To counter the threat of those who seek to rule by force, we must pay the costs of our own needed military strength, and help to build the security of others.

We must use our skills and knowledge and, at times, our substance, to help others rise from misery, however far the scene of suffering may be from our shores. For wherever in the world a people knows desperate want, there must appear at least the spark of hope, the hope of progress—or there will surely rise at last the flames of conflict.

We recognize and accept our own deep involvement in the destiny of men everywhere. We are accordingly pledged to honor, and to strive to fortify, the authority of the United Nations. For in that body rests the best hope of our age for the assertion of that law by which all nations may live in dignity.

And, beyond this general resolve, we are called to act a responsible role in the world's great concerns or conflicts—whether they touch upon the affairs of a vast region, the fate of an island in the Pacific, or the use of a canal in the Middle East. Only in respecting the hopes and cultures of others will we practice the equality of all nations. Only as we show willingness and wisdom in giving counsel—in receiving counsel—and in sharing burdens, will we wisely perform the work of peace.

For one truth must rule all we think and all we do. No people can live to itself alone. The unity of all who dwell in freedom is their only sure defense. The economic need of all nations—in mutual dependence—makes isolation an impossibility; not even America's prosperity could long survive if other nations did not also prosper. No nation can longer be a fortress, lone and strong and safe. And any people, seeking such shelter for themselves, can now build only their own prison.

Our pledge to these principles is constant, because we believe in their rightness.

We do not fear this world of change. America is no stranger to much of its spirit. Everywhere we see the seeds of the same growth that America itself has known. The American experiment has, for generations, fired the passion and the courage of millions elsewhere seeking freedom, equality, and opportunity. And the American story of material progress has helped excite the longing of all needy peoples for some satisfaction of their human wants. These hopes that we have helped to inspire, we can help to fulfill.

In this confidence, we speak plainly to all peoples.

We cherish our friendship with all nations that are or would be free. We respect, no less, their independence. And when, in time of want or peril, they ask our help, they may honorably receive it; for we no more seek to buy their sovereignty than we would sell our own. Sovereignty is never bartered among freemen.

We honor the aspirations of those nations which, now captive, long for freedom. We seek neither their military alliance nor any artificial imitation of our society. And they can know the warmth of the welcome that awaits them when, as must be, they join again the ranks of freedom.

We honor, no less in this divided world than in a less tormented time, the people of Russia. We do not dread, rather do we welcome, their progress in education and industry. We wish them success in their demands for more intellectual freedom, greater security before their own laws, fuller enjoyment of the rewards of their own toil. For as such things come to pass, the more certain will be the coming of that day when our peoples may freely meet in friendship.

So we voice our hope and our belief that we can help to heal this divided world. Thus may the nations cease to live in trembling before the menace of force. Thus may the weight of fear and the weight of arms be taken from the burdened shoulders of mankind.

This, nothing less, is the labor to which we are called and our strength dedicated.

And so the prayer of our people carries far beyond our own frontiers, to the wide world of our duty and our destiny.

May the light of freedom, coming to all darkened lands, flame brightly—until at last the darkness is no more.

May the turbulence of our age yield to a true time of peace, when men and nations shall share a life that honors the dignity of each, the brotherhood of all.

John F. Kennedy

· INAUGURAL ADDRESS ·

Friday, January 20, 1961

*E*ight inches of snow fell the night before John F. Kennedy was to be inaugurated as the nation's first Roman Catholic President, and the youngest person ever elected to the office. Troops worked through the bitterly cold night, clearing streets and intersections so the inaugural ceremonies could go on as scheduled. John Kennedy would hardly allow a mere snowstorm to overshadow a speech he hoped historians would remember as fondly as they remembered the inaugurals of Thomas Jefferson, Abraham Lincoln, and Franklin Roosevelt.

More than forty years later, the words Kennedy spoke on that cold afternoon in 1961 have been granted a place in the American canon of great speeches. Nearly every paragraph contains a phrase or a sentence that has been etched into the nation's memory. The speech defined the Cold War as a "twilight struggle" that required Americans to "pay any price, bear any burden, meet any hardship, support any friend, oppose any foe, in order to assure the survival and the success of liberty." Kennedy saw the inaugural ceremony itself as a milestone in history, a time when "the torch" of leadership was passed to "a new generation of Americans." And he implored Americans to "ask not what your country can do for you—ask what you can do for your country."

It was the speech of a confident young man ("I do not shrink from this responsibility—I welcome it.") calling on his fellow citizens and the world to "begin anew" by exploring "what problems unite us instead of belaboring those problems which divide us." It was a challenge to the nation's adversaries as well as an attempt to reach out to them. "Let both sides seek to invoke the wonders of science instead of its terrors," he said. "Together let us explore the stars, conquer the deserts, eradicate disease, tap the ocean depths, and encourage the arts and commerce."

Standing near Kennedy as he delivered his memorable lines was the man he defeated in the 1960 election, Richard Nixon. At the age of forty-eight, Nixon was slightly older than Kennedy but vastly more experienced by virtue of the eight years he spent as Dwight Eisenhower's vice president. Nixon had been the overwhelming choice of the Republican Party to succeed Eisenhower, who was barred from seeking a third term. Nixon's record as an anti-Communist crusader and his energetic and very public role as Eisenhower's No. 2 man made him a popular figure in the party.

The choice of Nixon and Kennedy as the major-party presidential nominees in 1960 represented a milestone in U.S. politics. Both men were World War II veterans—both, in fact, served in the navy and were stationed in the Pacific theater of operations—and both were the first major-party presidential candidates born in the twentieth century. They were the first presidential candidates to debate each other during the campaign, they were the first to make extensive use of television commercials, and they were the first to compete for votes in fifty states (Alaska and Hawaii had become states in 1959).

Foreign policy was the overriding concern of both candidates and of the campaign as a whole. Kennedy's main criticism was that under Eisenhower the Soviets had gained an advantage in the possession of intercontinental ballistic missiles—the rockets that could carry nuclear warheads to targets thousands of miles away. Domestic issues seemed to engage them less, even though the nation was on the verge of a social revolution in the form of the civil rights movement.

Kennedy and Nixon debated each other four times, but their first meeting—the first debate ever between two major-party presidential candidates—is remembered best. It took place on September 26, 1960, in Chicago, not long after Nixon was released from a hospital after injuring his knee on a campaign stop. In the merciless light of a television studio, he looked like a man who had been ill, as indeed he had been. Kennedy, by contrast, was tan, rested, and well prepared. Yet the exchanges themselves were not nearly so one-sided. While history remembers Nixon's haggard appearance, it forgets that his responses were as thoughtful and articulate as Kennedy's.

About 80 million people, in a nation of some 108 million, watched the first debate, and Kennedy's smooth performance may have been a decisive factor in his victory.

John Kennedy may have seen a stagnant country as he campaigned in 1960, but some of his fellow Americans had reason to disagree. Millions now lived in new homes—in 1960 about a quarter of the nation's housing stock was less than ten years old. Eighty-five percent of the nation's teens were enrolled in high school, compared with ten percent at the turn of the twentieth century. Millions took advantage of affluence and government aid to become the first in their families to attend college, and some of those students founded a group called Students for a Democratic Society in 1959. Popular culture, too, reflected a dynamism that candidate Kennedy failed to see in 1960: white and black recording artists were moving away from big band music toward something called rock and roll, while writers like Norman Mailer, athletes like Jackie Robinson, and ordinary citizens like Rosa Parks already had announced the end of an old order and the beginning of something new.

On Election Day, the nation chose Kennedy in one of the closest elections in U.S. history, with Kennedy taking 34.2 million votes to Nixon's 34.1 million. The Electoral College vote was not nearly as close, with Kennedy winning 303 to Nixon's 219. The result was a cultural breakthrough, for Kennedy, a Roman Catholic descended from nineteenth-century Irish immigrants, became the first non–Anglo-Saxon Protestant to win the nation's highest office.

During the campaign, Kennedy and his supporters had been confronted with remnants of the anti-Catholicism that helped doom Alfred Smith's presidential campaign in 1928. The candidate outlined his vision of church and state relations in a historic speech to a convention of Baptist ministers in Houston in September 1960. The speech helped defuse the issue, but even afterward, speechwriter Theodore Sorensen occasionally heard voters complain that under Kennedy, a Catholic Mass might be celebrated in the White House.

That happened only once, Sorensen later wrote. The date was November 23, 1963, one day after John Kennedy was murdered in Dallas.

Bibliographic Note

Robert Dallek, *An Unfinished Life: John F. Kennedy, 1917–1963*
(New York: Little Brown, 2003).

Theodore Sorensen, *Kennedy* (New York: Harper & Row, 1965).

VICE PRESIDENT JOHNSON, MR. SPEAKER, MR. CHIEF JUSTICE,
PRESIDENT EISENHOWER, VICE PRESIDENT NIXON,
PRESIDENT TRUMAN, REVEREND CLERGY, FELLOW CITIZENS:

We observe today not a victory of party, but a celebration of freedom—symbolizing an end, as well as a beginning—signifying renewal, as well as change. For I have sworn before you and Almighty God the same solemn oath our forebears prescribed nearly a century and three quarters ago.

The world is very different now. For man holds in his mortal hands the power to abolish all forms of human poverty and all forms of human life. And yet the same revolutionary beliefs for which our forebears fought are still at issue around the globe—the belief that the rights of man come not from the generosity of the state, but from the hand of God.

We dare not forget today that we are the heirs of that first revolution. Let the word go forth from this time and place, to friend and foe alike, that the torch has been passed to a new generation of Americans—born in this century, tempered by war, disciplined by a hard and bitter peace, proud of our ancient heritage—and unwilling to witness or permit the slow undoing of those human rights to which this nation has always been committed, and to which we are committed today at home and around the world.

Let every nation know, whether it wishes us well or ill, that we shall pay any price, bear any burden, meet any hardship, support any friend, oppose any foe, in order to assure the survival and the success of liberty.

This much we pledge—and more.

To those old allies whose cultural and spiritual origins we share, we pledge the loyalty of faithful friends. United, there is little we cannot do in a host of cooperative ventures. Divided, there is little we can do—for we dare not meet a powerful challenge at odds and split asunder.

To those new states whom we welcome to the ranks of the free, we pledge our word that one form of colonial control shall not have passed away merely to be replaced by a far more iron tyranny. We shall not always expect to find them supporting our view. But we shall always hope to find them strongly supporting their own freedom—and to remember that, in the past, those who foolishly sought power by riding the back of the tiger ended up inside.

To those peoples in the huts and villages across the globe struggling to break the bonds of mass misery, we pledge our best efforts to help them help themselves, for whatever period is required—not because the Communists may be doing it, not because we seek their votes, but because it is right. If a free society cannot help the many who are poor, it cannot save the few who are rich.

To our sister republics south of our border, we offer a special pledge—to convert our good words into good deeds—in a new alliance for progress—to assist free men and free governments in casting off the chains of poverty. But this peaceful revolution of hope cannot become the prey of hostile powers. Let all our neighbors know that we shall join with them to oppose aggression or subversion anywhere in the Americas. And let every other power know that this hemisphere intends to remain the master of its own house.

To that world assembly of sovereign states, the United Nations, our last best hope in an age where the instruments of war have far outpaced the instruments of peace, we renew our pledge of support—to prevent it from becoming merely a forum for invective—to strengthen its shield of the new and the weak—and to enlarge the area in which its writ may run.

Finally, to those nations who would make themselves our adversary, we offer not a pledge but a request: that both sides begin anew the quest for peace, before the dark powers of destruction unleashed by science engulf all humanity in planned or accidental self-destruction.

We dare not tempt them with weakness. For only when our arms are sufficient beyond doubt can we be certain beyond doubt that they will never be employed.

But neither can two great and powerful groups of nations take comfort from our present course—both sides overburdened by the cost of modern weapons, both rightly alarmed by the steady spread of the deadly atom, yet both racing to alter that uncertain balance of terror that stays the hand of mankind's final war.

So let us begin anew—remembering on both sides that civility is not a sign of weakness, and sincerity is always subject to proof. Let us never negotiate out of fear. But let us never fear to negotiate.

Let both sides explore what problems unite us instead of belaboring those problems which divide us.

Let both sides, for the first time, formulate serious and precise proposals for the inspection and control of arms—and bring the absolute power to destroy other nations under the absolute control of all nations.

Let both sides seek to invoke the wonders of science instead of its terrors. Together let us explore the stars, conquer the deserts, eradicate disease, tap the ocean depths, and encourage the arts and commerce.

Let both sides unite to heed in all corners of the earth the command of Isaiah—to "undo the heavy burdens . . . and to let the oppressed go free."

And if a beachhead of cooperation may push back the jungle of suspicion, let both sides join in creating a new endeavor, not a new balance of power, but a new world of law, where the strong are just and the weak secure and the peace preserved.

All this will not be finished in the first hundred days. Nor will it be finished in the first thousand days, nor in the life of this administration, nor even perhaps in our lifetime on this planet. But let us begin.

In your hands, my fellow citizens, more than in mine, will rest the final success or failure of our course. Since this country was founded, each generation of Americans has been summoned to give testimony to its national loyalty. The graves of young Americans who answered the call to service surround the globe.

Now the trumpet summons us again—not as a call to bear arms, though arms we need; not as a call to battle, though embattled we are—but a call to bear the burden of a long twilight struggle, year in and year out, "rejoicing in hope, patient in tribulation"—a struggle against the common enemies of man: tyranny, poverty, disease, and war itself.

Can we forge against these enemies a grand and global alliance, North and South, East and West, that can assure a more fruitful life for all mankind? Will you join in that historic effort?

In the long history of the world, only a few generations have been granted the role of defending freedom in its hour of maximum danger. I do not shrink from this responsibility—I welcome it. I do not believe that any of us would exchange places with any other people or any other generation. The energy, the faith, the devotion which we bring to this endeavor will light our country and all who serve it—and the glow from that fire can truly light the world.

And so, my fellow Americans: ask not what your country can do for you—ask what you can do for your country.

My fellow citizens of the world: ask not what America will do for you, but what together we can do for the freedom of man.

Finally, whether you are citizens of America or citizens of the world, ask of us the same high standards of strength and sacrifice which we ask of you. With a good conscience our only sure reward, with history the final judge of our deeds, let us go forth to lead the land we love, asking His blessing and His help, but knowing that here on earth God's work must truly be our own.

Lyndon B. Johnson

Wednesday, January 20, 1965

Lyndon B. Johnson became the fourth vice president in the twentieth century to succeed a fallen predecessor, when John Kennedy was assassinated on November 22, 1963, in Johnson's home state of Texas. Kennedy's murder stunned the nation, and, for many, it symbolized the beginning of a national nervous breakdown that would continue for the next ten years and wreck two presidencies.

Civil rights dominated Johnson's agenda as he filled out the remainder of Kennedy's term. Like many other politicians and millions of Americans, Johnson had watched as tens of thousands of civil rights marchers converged on Washington, D.C., in June, 1963, to hear Martin Luther King deliver a rousing oration about his dream of a nation where character, not skin color, mattered most.

The campaign for civil rights provoked a violent reaction in the South. In 1963, four children were killed when an African American church was bombed in Birmingham, Alabama, and civil rights activist Medgar Evers, a World War II veteran, was shot and killed in front of his family in Mississippi. On college campuses a new organization, the Student Nonviolent Coordinating Committee, advocated civil disobedience to protest continued racial discrimination. In Alabama, Governor George Wallace vowed to support segregation forever.

At Johnson's insistence, Congress passed a new civil rights act in 1964. It banned discrimination in public accommodations such as hotels and restaurants, gave the Justice Department new powers to enforce school desegregation, required that the federal government cease funding of local programs that discriminated by race, established new procedures to guard against discrimination in the workplace, and strengthened voting rights. It was the most sweeping civil rights legislation since Reconstruction.

Overseas, the new President inherited a burgeoning crisis in Southeast Asia. President Kennedy had dispatched thousands of military advisers to South Vietnam's pro-West government, which was fighting a Communist insurgency within its borders and infiltration by the Communist government of North Vietnam. The nation had been partitioned, like Korea and Germany, as part of the Cold War power arrangements, and Kennedy believed the United States ought to help defend South Vietnam just as it had come to the defense of South Korea in the early 1950s. Johnson recommitted the United States to the defense of South Vietnam.

Johnson's political skills and his legislative achievements made him a formidable candidate for election in his own right in 1964. Although the party was united behind him, the Democratic National Convention in Atlantic City saw a battle break out over the seating of pro–civil rights delegates from Mississippi. The dispute led to a walkout by some white delegates.

The Republican Party chose Senator Barry Goldwater, an unabashed conservative from Arizona, as its nominee, rejecting the early favorite, Governor Nelson Rockefeller of New York. Rockefeller, a moderate Republican, led Goldwater in the early primaries, but when his second wife gave birth to a child midway through the campaign, voters were reminded that Rockefeller had been married before—and no President had ever been divorced (although Andrew Jackson married a divorced woman). Goldwater won a critical primary in California and went on to win the nomination at the party's convention in San Francisco.

Goldwater presented himself as an authentic Republican who opposed what he saw as the party's collaboration with New Deal Democrats in expanding the government's role in the economy and in society itself. His message received the support of a former movie actor named Ronald Reagan, who delivered a speech on Goldwater's behalf that was broadcast throughout the country and helped make Reagan a political star. But it didn't change very many minds. Johnson's supporters successfully portrayed Goldwater as a right-wing radical who was liable to start a nuclear war.

Johnson defeated Goldwater in a historic landslide, taking forty-three million popular votes to Goldwater's twenty-seven million. Johnson also captured 486 votes in the Electoral College to Goldwater's 52, but the vote totals disguised the beginnings of a profound political realignment.

Goldwater captured six states in the South: Louisiana, Mississippi, Alabama, Georgia, South Carolina, and Arizona. From the end of the Civil War to 1964, the South had been a Democratic Party bastion. After 1964, however, the white vote in the South moved those states into the Republican column.

Given a term of his own and a powerful mandate, Lyndon Johnson was determined to implement one of the most ambitious domestic programs ever undertaken by an American President. He would declare a war on poverty, initiate a series of social programs he called the Great Society, and would continue to advocate for civil rights. But the war in Vietnam and violent dissent at home overshadowed those plans and overtook his presidency.

All of that, however, lay in the future when he took the oath of office on January 20, 1965. Johnson's inaugural contained the generational and forward-looking theme that he heard John Kennedy deliver in 1961. "For every generation, there is a destiny," he said. "For some, history decides. For this generation, the choice must be our own." Change was the overriding theme of his inaugural. Like Kennedy, Johnson portrayed himself as an agent of progress, and urged his fellow Americans to embrace change rather than embrace the status quo or pine for the past. "Is our world gone? We say 'Farewell.' Is a new world coming? We welcome it—and we will bend it to the hopes of man," he said.

Listeners who had heard Harry Truman, Dwight Eisenhower, and John Kennedy divide the world into free and Communist, good and evil, may have been surprised to hear no such rhetoric from Johnson, even though troops were in the field in Vietnam even as he spoke. Absent from Johnson's speech were the almost ritualistic condemnations of communism. Instead, his speech is explicitly and implicitly about domestic affairs. "In a land of great wealth, families must not live in hopeless poverty," he said. He promised that want and ignorance would be "conquered" by "this generation of Americans."

That promise, and many others, would be cast aside in the months to come as Johnson sent hundreds of thousands of troops to Southeast Asia. Dissent over the war and dissatisfaction with the pace of civil rights reform exploded in 1967 and 1968. Riots broke out in Watts, Detroit, Newark, and scores of other cities.

The dreams of Inauguration Day were shattered by 1968. And so was Lyndon Johnson's presidency.

Bibliographic Note

Lyndon Baines Johnson, *The Vantage Point*
(New York: Holt, Reinhart and Winston, 1971).

Theodore H. White, *The Making of the President, 1964*
(New York: Atheneum Publishers, 1965).

MY FELLOW COUNTRYMEN, on this occasion, the oath I have taken before you and before God is not mine alone, but ours together. We are one nation and one people. Our fate as a nation and our future as a people rest not upon one citizen, but upon all citizens.

This is the majesty and the meaning of this moment.

For every generation, there is a destiny. For some, history decides. For this generation, the choice must be our own.

Even now, a rocket moves toward Mars. It reminds us that the world will not be the same for our children, or even for ourselves in a short span of years. The next man to stand here will look out on a scene different from our own, because ours is a time of change—rapid and fantastic change bearing the secrets of nature, multiplying the nations, placing in uncertain hands new weapons for mastery and destruction, shaking old values, and uprooting old ways.

Our destiny in the midst of change will rest on the unchanged character of our people, and on their faith.

THE AMERICAN COVENANT

They came here—the exile and the stranger, brave but frightened—to find a place where a man could be his own man. They made a covenant with this land. Conceived in justice, written in liberty, bound in union, it was meant one day to inspire the hopes of all mankind; and it binds us still. If we keep its terms, we shall flourish.

JUSTICE AND CHANGE

First, justice was the promise that all who made the journey would share in the fruits of the land.

In a land of great wealth, families must not live in hopeless poverty. In a land rich in harvest, children just must not go hungry. In a land of healing miracles, neighbors must not suffer and die unattended. In a great land of learning and scholars, young people must be taught to read and write.

For the more than thirty years that I have served this nation, I have believed that this injustice to our people, this waste of our resources, was our real enemy. For thirty years or more, with the resources I have had, I have vigilantly fought against it. I have learned, and I know, that it will not surrender easily.

But change has given us new weapons. Before this generation of Americans is finished, this enemy will not only retreat—it will be conquered.

Justice requires us to remember that when any citizen denies his fellow, saying, "His color is not mine," or "His beliefs are strange and different," in that moment he betrays America, though his forebears created this nation.

LIBERTY AND CHANGE

Liberty was the second article of our covenant. It was self-government. It was our Bill of Rights. But it was more. America would be a place where each man could be proud to be himself: stretching his talents, rejoicing in his work, important in the life of his neighbors and his nation.

This has become more difficult in a world where change and growth seem to tower beyond the control and even the judgment of men. We must work to provide the knowledge and the surroundings which can enlarge the possibilities of every citizen.

The American covenant called on us to help show the way for the liberation of man. And that is today our goal. Thus, if as a nation there is much outside our control, as a people no stranger is outside our hope.

Change has brought new meaning to that old mission. We can never again stand aside, prideful in isolation. Terrific dangers and troubles that we once called "foreign" now constantly live among us. If American lives must end, and American treasure be spilled, in countries we barely know, that is the price that change has demanded of conviction and of our enduring covenant.

Think of our world as it looks from the rocket that is heading toward Mars. It is like a child's globe, hanging in space, the continents stuck to its side like colored maps. We are all fellow passengers on a dot of earth. And each of us, in the span of time, has really only a moment among our companions.

How incredible it is that in this fragile existence, we should hate and destroy one another. There are possibilities enough for all who will abandon mastery over others to pursue mastery over nature. There is world enough for all to seek their happiness in their own way.

Our nation's course is abundantly clear. We aspire to nothing that belongs to others. We seek no dominion over our fellow man, but man's dominion over tyranny and misery.

But more is required. Men want to be a part of a common enterprise—a cause greater than themselves. Each of us must find a way to advance the purpose of the nation, thus finding new purpose for ourselves. Without this, we shall become a nation of strangers.

UNION AND CHANGE

The third article was union. To those who were small and few against the wilderness, the success of liberty demanded the strength of union. Two centuries of change have made this true again.

No longer need capitalist and worker, farmer and clerk, city and countryside, struggle to divide our bounty. By working shoulder to shoulder, together we can increase the bounty of all. We have discovered that every child who learns, every man who finds work, every sick body that is made whole—like a candle added to an altar—brightens the hope of all the faithful.

So let us reject any among us who seek to reopen old wounds and to rekindle old hatreds. They stand in the way of a seeking nation.

Let us now join reason to faith and action to experience, to transform our unity of interest into a unity of purpose. For the hour and the day and the time are here to achieve progress without strife, to achieve change without hatred—not without difference of opinion, but without the deep and abiding divisions which scar the Union for generations.

THE AMERICAN BELIEF

Under this covenant of justice, liberty, and union we have become a nation—prosperous, great, and mighty. And we have kept our freedom. But we have no promise from God that our greatness will endure. We have been allowed by Him to seek greatness with the sweat of our hands and the strength of our spirit.

I do not believe that the Great Society is the ordered, changeless, and sterile battalion of the ants. It is the excitement of becoming—always becoming, trying, probing, falling, resting, and trying again—but always trying and always gaining.

In each generation, with toil and tears, we have had to earn our heritage again.

If we fail now, we shall have forgotten in abundance what we learned in hardship: that democracy rests on faith, that freedom asks more than it gives, and that the judgment of God is harshest on those who are most favored.

If we succeed, it will not be because of what we have, but it will be because of what we are; not because of what we own, but, rather because of what we believe.

For we are a nation of believers. Underneath the clamor of building and the rush of our day's pursuits, we are believers in justice and liberty and union, and in our own Union. We believe that every man must someday be free. And we believe in ourselves.

Our enemies have always made the same mistake. In my lifetime—in depression and in war—they have awaited our defeat. Each time, from the secret places of the American heart, came forth the faith they could not see or that they could not even imagine. It brought us victory. And it will again.

For this is what America is all about. It is the uncrossed desert and the unclimbed ridge. It is the star that is not reached and the harvest sleeping in the unplowed ground. Is our world gone? We say "Farewell." Is a new world coming? We welcome it—and we will bend it to the hopes of man.

To these trusted public servants and to my family and those close friends of mine who have followed me down a long, winding road, and to all the people of this Union and the world, I will repeat today what I said on that sorrowful day in November 1963: "I will lead and I will do the best I can."

But you must look within your own hearts to the old promises and to the old dream. They will lead you best of all.

For myself, I ask only, in the words of an ancient leader: "Give me now wisdom and knowledge, that I may go out and come in before this people: for who can judge this thy people, that is so great?"

Richard Nixon

Monday, January 20, 1969

*N*ot since the Civil War had the United States been so bitterly divided as it was when Richard Milhous Nixon became President in 1969. The divisions this time were not sectional but generational, as millions of baby boomers came of age in a time of war abroad and social conflict at home.

Nixon's inauguration followed several years of civil unrest and political assassinations. In 1965 and 1967, riots broke out throughout the country as African Americans demanded an end to racial discrimination and social injustice. In 1968, civil rights leader Dr. Martin Luther King Jr., winner of the Nobel Peace Prize in 1963, was murdered in Memphis. Two months later, in June, Senator Robert F. Kennedy of New York was shot and killed in Los Angeles after winning the California presidential primary.

The election year of 1968 had begun with bloodshed in Vietnam. A Communist offensive against South Vietnam stunned the American military and public alike. Throughout the country, even at the very gates of the U.S. embassy, Communist forces staged spectacular surprise attacks. The offensive was timed to coincide with the Vietnamese new year, Tet.

The Tet offensive did not achieve its goal of toppling the South Vietnamese government. But the war was being fought not only on battlefields but in pubic opinion polls in the United States. And Americans were appalled by the ferocity of the Tet offensive. Antiwar sentiment began to spread beyond the college campus, and when antiwar Senator Eugene McCarthy of Wisconsin nearly defeated incumbent President Lyndon Johnson in the 1968 New Hampshire primary, Johnson withdrew as a candidate for reelection.

The nation's anger and discontent had little to do with economics. With unemployment at just over 3 percent, the nation's economy was anything but unstable, although inflation was a worrisome 5.5 percent. The overriding issue was the Vietnam War. More than 25,000 U.S. troops had been killed in the conflict, most of them since Johnson began building up the American commitment after his election in 1964. More than 460,000 troops were fighting in Vietnam in 1968, up from just under 200,000 in 1966. Still, there was no end in sight.

After Robert Kennedy's murder, Democrats turned to vice president Hubert Humphrey as their candidate. But his nomination was overshadowed by horrifying conflicts between war protesters and police officers outside the convention hall in Chicago.

For the first time in several years, luck seemed to be flowing in Richard Nixon's direction. The former vice president easily defeated his Republican rivals, including the first-term governor of California Ronald Reagan, at the GOP's convention in Miami. During the general election, Nixon refused Humphrey's challenge to debate, in part because he calculated that he had nothing to gain, in part because he knew that his debates with Kennedy in 1960 might have cost him the presidency.

With Humphrey trailing in the polls in late October, President Johnson announced that he would suspend the U.S. bombing campaign in North Vietnam. Rumors spread of a possible peace deal, prompting Nixon to send word to the South Vietnamese leader, Nguyễn Văn Thiệu, that he should not be party to any peace talks until after the election.

The election was close, and historic. Nixon won with 31.7 million popular votes to Humphrey's 31.2 million. Third-party candidate George Wallace, the prosegregation governor of Alabama, won 9 million votes. Nixon did far better in the Electoral College, winning 301 electoral votes to Humphrey's 191 and Wallace's 46. The election of 1968 split apart the New Deal coalition and signaled the end of Democratic dominance in the South. By the same token, Nixon's campaign further alienated black voters from the Republican Party. In 1960, Nixon had received nearly a third of the black vote. In 1968, he received about 12 percent.[1]

According to Nixon speechwriter Raymond Price, the new President read the inaugural addresses of all his predecessors as he prepared for his

---

1. Tom Wicker, *One of Us: Richard Nixon and the American Dream,* (New York: Random House, 1991), 384.

own swearing-in ceremony. He concluded, not without reason, that "only the short ones are remembered."[2] He concluded that Lincoln's second inaugural was among the best, although he was fond of Woodrow Wilson's two, Franklin Roosevelt's first, and Theodore Roosevelt's one and only. Like Lincoln, Nixon would be speaking to a bitterly divided nation, and his role would be to speak words of healing.

Nixon, according to Price, was intimately involved in drafting his inaugural address, but Price's hand is evident in the short, terse paragraphs—the work of a newspaper journalist, Price's former profession. It is a short speech, just over two thousand words, with several attempts at soaring Kennedy-like rhetoric. "No man can be fully free while his neighbor is not," he said. "To go forward at all is to go forward together. This means black and white together, as one nation, not two. The laws have caught up with our conscience . . . Let us take as our goal: where peace is unknown, make it welcome; where peace is fragile, make it strong; where peace is temporary, make it permanent." The echo of Kennedy is plainer still in the first four words of another sentence: "Let all nations know that during this administration our lines of communication will be open." And again in his call to national greatness: "I ask you to join in a high adventure—one as rich as humanity itself, and as exciting as the times we live in."

The centerpiece of the speech was Nixon's attempt to bring the nation together in a time of tumult. He urged Americans to "lower" their voices, for they "cannot learn from one another" until they stopped "shouting at one another."

"For its part, government will listen," he said. "We will strive to listen in new ways—to the voices of quiet anguish, the voices that speak without words, the voices of the heart—to the injured voices, the anxious voices, the voices that have despaired of being heard."

Nixon closed by quoting the words of Archibald MacLeish, the Pulitzer Prize–winning poet who also happened to have been one of Franklin Roosevelt's speechwriters. MacLeish imagined a world in which all people would see themselves as "riders on the earth together." Nixon seized upon this theme as a way to bind up the nation's divisions, and the world's.

Those divisions, however, were destined to get worse in the years to come.

---

2. Raymond Price, *With Nixon* (New York: Viking, 1977), 42.

SENATOR DIRKSEN, MR. CHIEF JUSTICE, MR. VICE PRESIDENT,
PRESIDENT JOHNSON, VICE PRESIDENT HUMPHREY, MY FELLOW
AMERICANS—AND MY FELLOW CITIZENS OF THE WORLD COMMUNITY:

I ask you to share with me today the majesty of this moment. In the orderly transfer of power, we celebrate the unity that keeps us free.

Each moment in history is a fleeting time, precious and unique. But some stand out as moments of beginning, in which courses are set that shape decades or centuries.

This can be such a moment.

Forces now are converging that make possible, for the first time, the hope that many of man's deepest aspirations can at last be realized. The spiraling pace of change allows us to contemplate, within our own lifetime, advances that once would have taken centuries.

In throwing wide the horizons of space, we have discovered new horizons on earth.

For the first time, because the people of the world want peace, and the leaders of the world are afraid of war, the times are on the side of peace.

Eight years from now America will celebrate its two hundredth anniversary as a nation. Within the lifetime of most people now living, mankind will celebrate that great new year which comes only once in a thousand years—the beginning of the third millennium.

What kind of nation we will be, what kind of world we will live in, whether we shape the future in the image of our hopes, is ours to determine by our actions and our choices.

The greatest honor history can bestow is the title of peacemaker. This honor now beckons America—the chance to help lead the world at last out of the valley of turmoil, and onto that high ground of peace that man has dreamed of since the dawn of civilization.

If we succeed, generations to come will say of us now living that we mastered our moment, that we helped make the world safe for mankind.

This is our summons to greatness.

I believe the American people are ready to answer this call.

The second third of this century has been a time of proud achievement. We have made enormous strides in science and industry and agriculture.

We have shared our wealth more broadly than ever. We have learned at last to manage a modern economy to assure its continued growth.

We have given freedom new reach, and we have begun to make its promise real for black as well as for white.

We see the hope of tomorrow in the youth of today. I know America's youth. I believe in them. We can be proud that they are better educated, more committed, more passionately driven by conscience than any generation in our history.

No people has ever been so close to the achievement of a just and abundant society, or so possessed of the will to achieve it. Because our strengths are so great, we can afford to appraise our weaknesses with candor and to approach them with hope.

Standing in this same place a third of a century ago, Franklin Delano Roosevelt addressed a nation ravaged by depression and gripped in fear. He could say in surveying the nation's troubles: "They concern, thank God, only material things."

Our crisis today is the reverse.

We have found ourselves rich in goods, but ragged in spirit; reaching with magnificent precision for the moon, but falling into raucous discord on earth.

We are caught in war, wanting peace. We are torn by division, wanting unity. We see around us empty lives, wanting fulfillment. We see tasks that need doing, waiting for hands to do them.

To a crisis of the spirit, we need an answer of the spirit.

To find that answer, we need only look within ourselves.

When we listen to "the better angels of our nature," we find that they celebrate the simple things, the basic things—such as goodness, decency, love, kindness.

Greatness comes in simple trappings.

The simple things are the ones most needed today if we are to surmount what divides us, and cement what unites us.

To lower our voices would be a simple thing.

In these difficult years, America has suffered from a fever of words; from inflated rhetoric that promises more than it can deliver; from angry rhetoric that fans discontents into hatreds; from bombastic rhetoric that postures instead of persuading.

We cannot learn from one another until we stop shouting at one another—until we speak quietly enough so that our words can be heard as well as our voices.

For its part, government will listen. We will strive to listen in new ways—to the voices of quiet anguish, the voices that speak without words, the voices of the heart—to the injured voices, the anxious voices, the voices that have despaired of being heard.

Those who have been left out, we will try to bring in.

Those left behind, we will help to catch up.

For all of our people, we will set as our goal the decent order that makes progress possible and our lives secure.

As we reach toward our hopes, our task is to build on what has gone before—not turning away from the old, but turning toward the new.

In this past third of a century, government has passed more laws, spent more money, initiated more programs, than in all our previous history.

In pursuing our goals of full employment, better housing, excellence in education; in rebuilding our cities and improving our rural areas; in protecting our environment and enhancing the quality of life—in all these and more, we will and must press urgently forward.

We shall plan now for the day when our wealth can be transferred from the destruction of war abroad to the urgent needs of our people at home.

The American Dream does not come to those who fall asleep.

But we are approaching the limits of what government alone can do.

Our greatest need now is to reach beyond government, and to enlist the legions of the concerned and the committed.

What has to be done, has to be done by government and people together or it will not be done at all. The lesson of past agony is that without the people we can do nothing; with the people we can do everything.

To match the magnitude of our tasks, we need the energies of our people—enlisted not only in grand enterprises, but more importantly in those small, splendid efforts that make headlines in the neighborhood newspaper instead of the national journal.

With these, we can build a great cathedral of the spirit—each of us raising it one stone at a time, as he reaches out to his neighbor, helping, caring, doing.

I do not offer a life of uninspiring ease. I do not call for a life of grim sacrifice. I ask you to join in a high adventure—one as rich as humanity itself, and as exciting as the times we live in.

The essence of freedom is that each of us shares in the shaping of his own destiny.

Until he has been part of a cause larger than himself, no man is truly whole.

The way to fulfillment is in the use of our talents; we achieve nobility in the spirit that inspires that use.

As we measure what can be done, we shall promise only what we know we can produce, but as we chart our goals we shall be lifted by our dreams.

No man can be fully free while his neighbor is not. To go forward at all is to go forward together.

This means black and white together, as one nation, not two. The laws have caught up with our conscience. What remains is to give life to what is in the law: to ensure at last that as all are born equal in dignity before God, all are born equal in dignity before man.

As we learn to go forward together at home, let us also seek to go forward together with all mankind.

Let us take as our goal: where peace is unknown, make it welcome; where peace is fragile, make it strong; where peace is temporary, make it permanent.

After a period of confrontation, we are entering an era of negotiation.

Let all nations know that during this administration our lines of communication will be open.

We seek an open world—open to ideas, open to the exchange of goods and people—a world in which no people, great or small, will live in angry isolation.

We cannot expect to make everyone our friend, but we can try to make no one our enemy.

Those who would be our adversaries, we invite to a peaceful competition—not in conquering territory or extending dominion, but in enriching the life of man.

As we explore the reaches of space, let us go to the new worlds together—not as new worlds to be conquered, but as a new adventure to be shared.

With those who are willing to join, let us cooperate to reduce the burden of arms, to strengthen the structure of peace, to lift up the poor and the hungry.

But to all those who would be tempted by weakness, let us leave no doubt that we will be as strong as we need to be for as long as we need to be.

Over the past twenty years, since I first came to this capital as a freshman Congressman, I have visited most of the nations of the world.

I have come to know the leaders of the world, and the great forces, the hatreds, the fears that divide the world.

I know that peace does not come through wishing for it—that there is no substitute for days and even years of patient and prolonged diplomacy.

I also know the people of the world.

I have seen the hunger of a homeless child, the pain of a man wounded in battle, the grief of a mother who has lost her son. I know these have no ideology, no race.

I know America. I know the heart of America is good.

I speak from my own heart, and the heart of my country, the deep concern we have for those who suffer, and those who sorrow.

I have taken an oath today in the presence of God and my countrymen to uphold and defend the Constitution of the United States. To that oath I now add this sacred commitment: I shall consecrate my office, my energies, and all the wisdom I can summon, to the cause of peace among nations.

Let this message be heard by strong and weak alike:

The peace we seek to win is not victory over any other people, but the peace that comes "with healing in its wings"; with compassion for those who have suffered; with understanding for those who have opposed us; with the opportunity for all the peoples of this earth to choose their own destiny.

Only a few short weeks ago, we shared the glory of man's first sight of the world as God sees it, as a single sphere reflecting light in the darkness.

As the Apollo astronauts flew over the moon's gray surface on Christmas Eve, they spoke to us of the beauty of earth—and in that voice so clear across the lunar distance, we heard them invoke God's blessing on its goodness.

In that moment, their view from the moon moved poet Archibald MacLeish to write:

"To see the earth as it truly is, small and blue and beautiful in that eternal silence where it floats, is to see ourselves as riders on the earth together, brothers on that bright loveliness in the eternal cold—brothers who know now they are truly brothers."

In that moment of surpassing technological triumph, men turned their thoughts toward home and humanity—seeing in that far perspective that man's destiny on earth is not divisible; telling us that however far we reach into the cosmos, our destiny lies not in the stars but on Earth itself, in our own hands, in our own hearts.

We have endured a long night of the American spirit. But as our eyes catch the dimness of the first rays of dawn, let us not curse the remaining dark. Let us gather the light.

Our destiny offers, not the cup of despair, but the chalice of opportunity. So let us seize it, not in fear, but in gladness—and, "riders on the earth together," let us go forward, firm in our faith, steadfast in our purpose, cautious of the dangers; but sustained by our confidence in the will of God and the promise of man.

Richard Nixon

Saturday, January 20, 1973

$\mathcal{T}$he United States still was fighting its longest war when Richard Milhous Nixon took the oath of office for a second time. But an end to the hostilities was imminent, allowing Nixon to talk not of war, but of peace. "As we meet here today," he said, "we stand on the threshold of a new era of peace in the world." A treaty ending American involvement in the war was signed days later in Paris.

In winning a second term in 1972, Richard Nixon threw a final spade of earth on the grave of the New Deal coalition. He swept the electoral map save for Massachusetts and the District of Columbia. The significance of the victory went beyond mere numbers, although they were impressive. Nixon won 520 votes in the Electoral College, and he defeated Democrat George McGovern by an astonishing 23.3 percentage points. Beyond the numbers was the larger story: for the first time ever a Republican presidential candidate captured every southern state. Nixon had achieved more than a personal victory. He had realigned national politics.

Nixon's smashing victory was all the more noteworthy because he had not brought an end to the conflict that had driven his predecessor, Lyndon Johnson, out of office. In fact, Nixon actually had widened the conflict, ordering a secret invasion of Cambodia, where enemy troops took refuge, and the mining of North Vietnamese harbors. Demonstrations against the war were as intense as any Johnson faced, and confrontations between National Guard troops and protestors led to the deaths of students at Kent State University in Ohio and Jackson State University in Mississippi in 1970. To help build political support for his war policy, Nixon delivered a televised appeal to a group he called the "great silent majority," asking for its help in achieving peace with honor.

Domestically, Nixon governed as a moderate, to the chagrin of conservative Republicans who supported the war but who associated activist government with the Democrats. Overseas, Nixon stunned the nation, and the world, by traveling to Communist China in February 1972 in an effort to reduce tensions between the two nations. Nixon continued to ease the Cold War with a visit to the Soviet Union later in the year.

Amid the triumphs abroad and policy victories at home, however, there was the war in Vietnam. More than twenty-seven thousand U.S. troops and hundreds of thousands of Vietnamese died during Nixon's first term. Before his reelection campaign began in earnest, Nixon started reducing U.S. troop levels in Vietnam, ended the draft, embarked on a policy of relying more on the South Vietnamese army to fight the Communists, and entered into preliminary peace talks with the North Vietnamese.

Still, dissatisfaction with the war remained widespread, inspiring a host of well-known Democrats to enter the party's presidential primaries in 1972. The primary system, a series of presidential preference contests spread out from late January to early June in an election year, had been slowly eroding the power of party leaders in choosing presidential candidates since the 1960s. By 1972, thanks to the proliferation of primaries and new rules governing the Democratic Party's nomination process, the primary system had become the decisive factor in selecting presidential candidates.

Richard Nixon and his top political aides paid close attention to the Democratic Party's nominating process. That attention turned to obsession when members of Nixon's reelection committee, the Committee to Re-elect the President, broke into the office of Democratic National Chairman Lawrence O'Brien in the Watergate Hotel in June.

During the bitter Democratic primary season, antiwar Democrats rallied behind the candidacy of South Dakota senator George McGovern, a World War II bomber pilot and avowed opponent of the Vietnam conflict. McGovern won the party's nomination and named Senator Thomas Eagleton of Missouri as his running mate. Not long after the convention broke up, however, Eagleton revealed that he suffered from depression and had been treated with electroshock therapy. McGovern stood by Eagleton at first but eventually withdrew his support, forcing Eagleton to withdraw. McGovern replaced him with Sargent Shriver, who served as director of the Peace Corps during the administration of his late brother-in-law, John Kennedy.

McGovern's campaign never gained traction, even among voters between the ages of eighteen and twenty who were voting in a national election for the first time in 1972. (The Twenty-sixth Amendment, ratified in 1971, lowered the voting age from twenty-one to eighteen.) And when Nixon's national security adviser, Henry Kissinger, announced in late October that peace in Vietnam was "at hand," so was a presidential landslide.

With McGovern soundly defeated, Richard Nixon took the oath of office a second time with an air of personal vindication. His inaugural address, while centered on his vague outlines for a post-Vietnam foreign policy, contained elements of a chip-on-the-shoulder defiance and a bit of maudlin self-pity that could only be described as Nixonian. "Our children have been taught to be ashamed of their country, ashamed of their parents, ashamed of America's record at home and of its role in the world," he said. "At every turn, we have been beset by those who find everything wrong with America and little that is right. But I am confident that this will not be the judgment of history on these remarkable times in which we are privileged to live."

The centerpiece of the speech was an outline of Nixon's vision of foreign policy conducted in the shadow of Vietnam. Nixon wisely made no claims to victory in Vietnam. Instead, he conceded that mistakes had been made, that policies had failed, and that new thinking was required. "The time has passed when America will make every other nation's conflict our own, or make every other nation's future our responsibility, or presume to tell the people of other nations how to manage their own affairs," he said.

Standing near Nixon as he spoke was his vice president, Spiro Agnew of Maryland. Even as the two men celebrated their victory, investigations were under way that would spiral out of their formidable powers of control. Agnew was caught up on an old-fashioned kickback scheme dating to his time as governor of Maryland. He resigned as vice president on October 10, 1973. Agnew's unscheduled departure allowed Nixon to nominate a new vice president, Representative Gerald Ford of Michigan.

Less than a year later, Ford became President when Nixon became the first President to resign his office. Nixon quit after the House Judiciary Committee recommended that the President be impeached on Watergate-related charges, and days after a secret tape recording revealed that Nixon had ordered a cover-up of White House involvement in the break-in.

Nixon's Shakespearean fall led to the inauguration, on August 9, 1974, of the nation's first truly unelected President, Gerald Ford.

Mr. Vice President, Mr. Speaker, Mr. Chief Justice,
Senator Cook, Mrs. Eisenhower, and my fellow citizens of
this great and good country we share together:

When we met here four years ago, America was bleak in spirit, depressed by the prospect of seemingly endless war abroad and of destructive conflict at home.

As we meet here today, we stand on the threshold of a new era of peace in the world.

The central question before us is: how shall we use that peace? Let us resolve that this era we are about to enter will not be what other postwar periods have so often been: a time of retreat and isolation that leads to stagnation at home and invites new danger abroad.

Let us resolve that this will be what it can become: a time of great responsibilities greatly borne, in which we renew the spirit and the promise of America as we enter our third century as a nation.

This past year saw far-reaching results from our new policies for peace. By continuing to revitalize our traditional friendships, and by our missions to Peking and to Moscow, we were able to establish the base for a new and more durable pattern of relationships among the nations of the world. Because of America's bold initiatives, 1972 will be long remembered as the year of the greatest progress since the end of World War II toward a lasting peace in the world.

The peace we seek in the world is not the flimsy peace which is merely an interlude between wars, but a peace which can endure for generations to come.

It is important that we understand both the necessity and the limitations of America's role in maintaining that peace.

Unless we in America work to preserve the peace, there will be no peace.

Unless we in America work to preserve freedom, there will be no freedom.

But let us clearly understand the new nature of America's role, as a result of the new policies we have adopted over these past four years.

We shall respect our treaty commitments.

We shall support vigorously the principle that no country has the right to impose its will or rule on another by force.

We shall continue, in this era of negotiation, to work for the limitation of nuclear arms, and to reduce the danger of confrontation between the great powers.

We shall do our share in defending peace and freedom in the world. But we shall expect others to do their share.

The time has passed when America will make every other nation's conflict our own, or make every other nation's future our responsibility, or presume to tell the people of other nations how to manage their own affairs.

Just as we respect the right of each nation to determine its own future, we also recognize the responsibility of each nation to secure its own future.

Just as America's role is indispensable in preserving the world's peace, so is each nation's role indispensable in preserving its own peace.

Together with the rest of the world, let us resolve to move forward from the beginnings we have made. Let us continue to bring down the walls of hostility which have divided the world for too long, and to build in their place bridges of understanding—so that despite profound differences between systems of government, the people of the world can be friends.

Let us build a structure of peace in the world in which the weak are as safe as the strong—in which each respects the right of the other to live by a different system—in which those who would influence others will do so by the strength of their ideas, and not by the force of their arms.

Let us accept that high responsibility not as a burden, but gladly—gladly because the chance to build such a peace is the noblest endeavor in which a nation can engage; gladly, also, because only if we act greatly in meeting our responsibilities abroad will we remain a great nation, and only if we remain a great nation will we act greatly in meeting our challenges at home.

We have the chance today to do more than ever before in our history to make life better in America—to ensure better education, better health, better housing, better transportation, a cleaner environment—to restore respect for law, to make our communities more livable, and to ensure the God-given right of every American to full and equal opportunity.

Because the range of our needs is so great—because the reach of our opportunities is so great—let us be bold in our determination to meet those needs in new ways.

Just as building a structure of peace abroad has required turning away from old policies that failed, so building a new era of progress at home requires turning away from old policies that have failed.

Abroad, the shift from old policies to new has not been a retreat from our responsibilities, but a better way to peace.

And at home, the shift from old policies to new will not be a retreat from our responsibilities, but a better way to progress.

Abroad and at home, the key to those new responsibilities lies in the placing and the division of responsibility. We have lived too long with the consequences of attempting to gather all power and responsibility in Washington.

Abroad and at home, the time has come to turn away from the condescending policies of paternalism—of "Washington knows best."

A person can be expected to act responsibly only if he has responsibility. This is human nature. So let us encourage individuals at home and nations abroad to do more for themselves, to decide more for themselves. Let us locate responsibility in more places. Let us measure what we will do for others by what they will do for themselves.

That is why today I offer no promise of a purely governmental solution for every problem. We have lived too long with that false promise. In trusting too much in government, we have asked of it more than it can deliver. This leads only to inflated expectations, to reduced individual effort, and to a disappointment and frustration that erode confidence both in what government can do and in what people can do.

Government must learn to take less from people so that people can do more for themselves.

Let us remember that America was built not by government, but by people—not by welfare, but by work—not by shirking responsibility, but by seeking responsibility.

In our own lives, let each of us ask—not just what will government do for me, but what can I do for myself?

In the challenges we face together, let each of us ask—not just how can government help, but how can I help?

Your national government has a great and vital role to play. And I pledge to you that where this government should act, we will act boldly and we will lead boldly. But just as important is the role that each and every one of us must play, as an individual and as a member of his own community.

From this day forward, let each of us make a solemn commitment in his own heart: to bear his responsibility, to do his part, to live his ideals—so that together, we can see the dawn of a new age of progress for America, and together, as we celebrate our two hundredth anniversary as a nation, we can do so proud in the fulfillment of our promise to ourselves and to the world.

As America's longest and most difficult war comes to an end, let us again learn to debate our differences with civility and decency. And let each of us reach out for that one precious quality government cannot provide—a new level of respect for the rights and feelings of one another, a new level of respect for the individual human dignity which is the cherished birthright of every American.

Above all else, the time has come for us to renew our faith in ourselves and in America.

In recent years, that faith has been challenged.

Our children have been taught to be ashamed of their country, ashamed of their parents, ashamed of America's record at home and of its role in the world.

At every turn, we have been beset by those who find everything wrong with America and little that is right. But I am confident that this will not be the judgment of history on these remarkable times in which we are privileged to live.

America's record in this century has been unparalleled in the world's history for its responsibility, for its generosity, for its creativity, and for its progress.

Let us be proud that our system has produced and provided more freedom and more abundance, more widely shared, than any other system in the history of the world.

Let us be proud that in each of the four wars in which we have been engaged in this century, including the one we are now bringing to an end, we have fought not for our selfish advantage, but to help others resist aggression.

Let us be proud that by our bold, new initiatives, and by our steadfastness for peace with honor, we have made a breakthrough toward creating in the world what the world has not known before—a structure of peace that can last, not merely for our time, but for generations to come.

We are embarking here today on an era that presents challenges great as those any nation, or any generation, has ever faced.

We shall answer to God, to history, and to our conscience for the way in which we use these years.

As I stand in this place, so hallowed by history, I think of others who have stood here before me. I think of the dreams they had for America, and I think of how each recognized that he needed help far beyond himself in order to make those dreams come true.

Today, I ask your prayers that in the years ahead I may have God's help in making decisions that are right for America, and I pray for your help so that together we may be worthy of our challenge.

Let us pledge together to make these next four years the best four years in America's history, so that on its two hundredth birthday America will be as young and as vital as when it began, and as bright a beacon of hope for all the world.

Let us go forward from here confident in hope, strong in our faith in one another, sustained by our faith in God who created us, and striving always to serve His purpose.

Gerald R. Ford

· SUCCESSION SPEECH ·

Friday, August 9, 1974

Never before had an accidental President's swearing-in been such a public occasion. Gerald R. Ford's inauguration as Richard Nixon's successor had none of the pomp and pageantry of a regularly scheduled inauguration, but it was very much a public event and, in that way, very unlike emergency successions of the past.

The drama of Nixon's fall, the negotiations leading to his resignation, and his speech on August 8 announcing his decision to resign effective at noon the following day set the stage for public interest in Ford's inauguration. Unlike successions brought about by a President's sudden death, this one was planned as carefully as could be, given the circumstances. Ford himself was involved in drawing up a list of about 275 guests to be invited to the ceremony, and he called old friends in Congress for last-minute advice. House majority leader Thomas "Tip" O'Neill, a Democrat from Massachusetts, wished Ford well before adding, "Christ, Jerry, isn't this a wonderful country? Here we can talk like this and you and I can be friends, and eighteen months from now I'll be going around the country kicking your ass in."[1]

Thoughts of the 1976 election, however, were far from the minds of most citizens and politicians on that Friday morning in Washington as Richard Nixon bade farewell to his staff in a rambling, sad speech recalling the gentle presence of his mother, who, he said, was a saint. As television cameras brought the scene to living rooms across the nation, Ford escorted Nixon and his family to a waiting helicopter, then returned to the White House to take the oath of office from Chief Justice Warren Burger,

---

1. Gerald R. Ford, *A Time to Heal* (New York: Harper and Row, 1979), 32.

who flew from the Netherlands to Washington overnight in order to administer the oath.

Because the occasion was so public, because the nation had been through a political calamity unlike any other in its history, because so many people were watching on television, the occasion demanded a speech of some sort. Ford, a modest man, said he felt it was his "duty" to speak to the nation. It was, he said, "not an inaugural address, not a fireside chat, not a campaign speech—just a little straight talk among friends. And I intend it to be the first of many."

The phrase "a little straight talk among friends" spoke to the public's disgust with the lies and deceptions of Watergate and revealed to voters the informal, ingenuous side of their new President. He developed the theme further in his speech, saying that "I believe that truth is the glue that holds government together, not only our government but civilization itself. That bond, though badly strained, is unbroken at home and abroad."

Ford tried to reassure a suspicious public that the long, painful, and disillusioning scandal known as Watergate was now at an end. "My fellow Americans, our long national nightmare is over," he said. "Our Constitution works; our great Republic is a government of laws and not of men. Here the people rule."

The new President's plainspoken style, reminiscent in a way of Harry Truman (whose picture Ford installed in the White House shortly after taking office), won him a brief honeymoon, during which he appointed former New York governor Nelson Rockefeller as his vice president. For the first time in U.S. history, both the President and vice president were appointed, not elected.

Ford's assertion that the nation's nightmare was over proved to be too optimistic. While a discredited President had been peacefully forced out of office, the nation's malaise continued. The economy was wracked by an annual inflation rate of about 12 percent and had still not recovered from an embargo of Arab-supplied oil in 1973 and the subsequent huge increases in the price of gasoline.

A month after taking office, Gerald Ford pardoned Richard Nixon for any offenses he might have committed as President. A fledgling criminal prosecution of the former President came to an abrupt end. And enraged public reaction suggested that the nightmare was not, in fact, over.

Mr. Chief Justice, my dear friends, my fellow Americans:

The oath that I have taken is the same oath that was taken by George Washington and by every President under the Constitution. But I assume the presidency under extraordinary circumstances never before experienced by Americans. This is an hour of history that troubles our minds and hurts our hearts.

Therefore, I feel it is my first duty to make an unprecedented compact with my countrymen. Not an inaugural address, not a fireside chat, not a campaign speech—just a little straight talk among friends. And I intend it to be the first of many.

I am acutely aware that you have not elected me as your President by your ballots, and so I ask you to confirm me as your President with your prayers. And I hope that such prayers will also be the first of many.

If you have not chosen me by secret ballot, neither have I gained office by any secret promises. I have not campaigned either for the presidency or the vice presidency. I have not subscribed to any partisan platform. I am indebted to no man, and only to one woman—my dear wife—as I begin this very difficult job.

I have not sought this enormous responsibility, but I will not shirk it. Those who nominated and confirmed me as vice president were my friends and are my friends. They were of both parties, elected by all the people and acting under the Constitution in their name. It is only fitting then that I should pledge to them and to you that I will be the President of all the people.

Thomas Jefferson said the people are the only sure reliance for the preservation of our liberty. And down the years, Abraham Lincoln renewed this American article of faith asking, "Is there any better way or equal hope in the world?"

I intend, on Monday next, to request of the Speaker of the House of Representatives and the President pro tempore of the Senate the privilege of appearing before the Congress to share with my former colleagues and with you, the American people, my views on the priority business of the nation and to solicit your views and their views. And may I say to the Speaker and the others, if I could meet with you right after these remarks, I would appreciate it.

Even though this is late in an election year, there is no way we can go forward except together and no way anybody can win except by serving the people's urgent needs. We cannot stand still or slip backwards. We must go forward now together.

To the peoples and the governments of all friendly nations, and I hope that could encompass the whole world, I pledge an uninterrupted and sincere search for peace. America will remain strong and united, but its strength will remain dedicated to the safety and sanity of the entire family of man, as well as to our own precious freedom.

I believe that truth is the glue that holds government together, not only our government but civilization itself. That bond, though strained, is unbroken at home and abroad.

In all my public and private acts as your President, I expect to follow my instincts of openness and candor with full confidence that honesty is always the best policy in the end.

My fellow Americans, our long national nightmare is over.

Our Constitution works; our great Republic is a government of laws and not of men. Here the people rule. But there is a higher Power, by whatever name we honor Him, who ordains not only righteousness but love, not only justice but mercy.

As we bind up the internal wounds of Watergate, more painful and more poisonous than those of foreign wars, let us restore the Golden Rule to our political process, and let brotherly love purge our hearts of suspicion and of hate.

In the beginning, I asked you to pray for me. Before closing, I ask again your prayers, for Richard Nixon and for his family. May our former President, who brought peace to millions, find it for himself. May God bless and comfort his wonderful wife and daughters, whose love and loyalty will forever be a shining legacy to all who bear the lonely burdens of the White House.

I can only guess at those burdens, although I have witnessed at close hand the tragedies that befell three Presidents and the lesser trials of others.

With all the strength and all the good sense I have gained from life, with all the confidence my family, my friends, and my dedicated staff impart to me, and with the good will of countless Americans I have encountered in recent visits to forty states, I now solemnly reaffirm my promise I made to you last December 6: to uphold the Constitution, to do what is right as God gives me to see the right, and to do the very best I can for America.

God helping me, I will not let you down.

Thank you.

Jimmy Carter

· INAUGURAL ADDRESS ·

Thursday, January 20, 1977

*N*ot since Herbert Hoover acknowledged the services of Calvin Coolidge on Inauguration Day 1929 had a new President paid tribute to his predecessor in his inaugural address. But on a bitterly cold Thursday afternoon in 1977, Jimmy Carter began his first speech as President with a tribute to the defeated incumbent, Gerald Ford. "For myself and for our nation," Carter said, "I want to thank my predecessor for all he has done to heal our land."

When the crowd burst into sustained applause, the former President stood up, acknowledged the ovation, and then shook his successor's hand. It was a symbolic act of unity after a tumultuous era in American politics.

Carter's rise to the nation's highest political office stunned his fellow Democrats, the press, and most political professionals. A one-term governor of Georgia, he declared his candidacy for president in December 1974, which was considered extremely early—at least by those who even noticed. Carter was utterly unknown outside his state, destined, it seemed, to be overshadowed by better-known candidates. But he understood that the plethora of presidential primaries had taken power out of the hands of the party establishment. Earnest and soft-spoken, blessed with a pleasant smile, Carter outworked his better-known rivals, finishing ahead of all of them in the Iowa caucus and winning the New Hampshire primary. In what seemed like an instant, he jumped from curiosity to front-runner to nominee.

Carter capitalized on a singular moment in American political history when voters were looking for a fresh face after the trauma of Watergate. The nightmare was over, but the nation's spirits remained low. Abroad, American prestige suffered a blow when the South Vietnamese government

collapsed in the face of renewed war with the North Vietnamese and Viet-
cong in 1975. At home, the unemployment rate was 8 percent, while inflation was at 6 percent. Ford had tried to tackle the cycle of rising prices and
wages, even urging citizens to wear pin-on buttons reading WIN, for
"whip inflation now." But neither the buttons nor a tightening of credit
seemed to help.

Republicans, already embattled, declined to rally around their incumbent President. Instead, many conservatives supported an insurgency led
by former California governor Ronald Reagan, who challenged Ford in the
party's primaries. The two men fought bitterly through the spring, and
when the party met for its convention in Kansas City, neither one had the
necessary delegates to win a first-ballot nomination. After frantic, closed-
door meetings, Ford won a first-ballot nomination, finishing with 1,187
delegates, about 57 more than he needed.

The general election campaign featured the first presidential debates
between a sitting President, Ford, and his challenger. (The famous Kennedy-
Nixon debates featured two challengers, not an incumbent President.)
Ford committed a well-publicized gaffe when he denied that Communist-
ruled Poland was dominated by the Soviet Union. Nevertheless, the incumbent sliced into Carter's lead in the polls, and by Election Day the race
was a dead heat.

Carter finished with 40.8 million votes and 297 electoral votes to Ford's
39.1 million and 240 electoral votes. It was the closest electoral vote in
more than half a century.

Although he proved to be a dogged campaigner whose soft drawl belied his shrewd political instincts, Carter saw his inaugural address as a
chance to heal the wounds that had been festering since Watergate and
even before, when the nation was divided by Vietnam and civil rights. After reading his predecessors' speeches, he was taken with Woodrow Wilson's sermonlike first inaugural, which spoke of a government that had
been "too often debauched and made an instrument of evil" but which
would now return to its high ideals. A devout southern Baptist, Carter
chose to make his point with a verse from the Book of Micah, implicitly
calling on the nation to "do justly" and to "love mercy."

Carter's speech was relatively short, at slightly more than twelve hundred words, and took just about eight minutes to deliver. While the phrases
are unadorned and plain, in keeping with the image he fostered as a candidate, Carter later wrote that he labored over the speech. During his re-
hearsals, he forced himself to speak slower.

Carter could hardly have ignored the tumult of the previous three years. He did so gently—"I believe America can be better"—and with a Wilsonian recommitment to the country's better angels. "Let us create together a new national spirit of unity and trust," he said. "Your strength can compensate for my weaknesses, and your wisdom can help to minimize my mistakes." Those words were among the most humble ever uttered at a presidential inaugural.

Humility found its way into another passage, in which Carter foresaw a new era of limits for Americans and for American power. "We have learned that 'more' is not necessarily 'better,' that even our great nation has its recognized limits, and that we can neither answer all questions nor solve all problems," Carter said. "We cannot afford to do everything, nor can we afford to lack boldness as we meet the future. So, together, in a spirit of individual sacrifice for the common good, we must simply do our best." The notion of limits rarely made its way into American political rhetoric, particularly in inaugural addresses, when new Presidents prefer to speak about expansion—of influence, of economic growth, of ambition.

Carter, however, was the first President elected after the Vietnam War, after the shock of the Arab oil embargo of 1973, after the nation learned that it was not, in fact, omnipotent. His invocation of limits won applause on Inauguration Day. But such sentiments would serve Carter poorly when he was confronted, four years later, by a man who spoke of sunny horizons without limits.

After the new President reviewed the traditional parade, he and his wife chose to walk, rather than ride in a car, from the Capitol to the White House along Pennsylvania Avenue. The sight of the President and First Lady waving to crowds during the short walk seemed to symbolize a break from the imperial presidency of Richard Nixon and suggested the beginning of a new era for the presidency and for the nation.

The feeling of renewal, however, was to be short-lived.

Bibliographic Note

Jimmy Carter, *Keeping Faith* (New York: Bantam Books, 1982).

Peter Bourne, *Jimmy Carter* (New York: Scribner, 1997).

FOR MYSELF AND FOR OUR NATION, I want to thank my predecessor for all
he has done to heal our land.

In this outward and physical ceremony we attest once again to the in-
ner and spiritual strength of our nation. As my high school teacher, Miss
Julia Coleman, used to say: "We must adjust to changing times and still
hold to unchanging principles."

Here before me is the Bible used in the inauguration of our first President, in 1789, and I have just taken the oath of office on the Bible my
mother gave me a few years ago, opened to a timeless admonition from the
ancient prophet Micah:

"He hath showed thee, O man, what is good; and what doth the Lord
require of thee, but to do justly, and to love mercy, and to walk humbly
with thy God." (Micah 6:8)

This inauguration ceremony marks a new beginning, a new dedication
within our government, and a new spirit among us all. A President may
sense and proclaim that new spirit, but only a people can provide it.

Two centuries ago our nation's birth was a milestone in the long quest
for freedom, but the bold and brilliant dream which excited the founders
of this nation still awaits its consummation. I have no new dream to set
forth today, but rather urge a fresh faith in the old dream.

Ours was the first society openly to define itself in terms of both spiri-
tuality and of human liberty. It is that unique self-definition which has
given us an exceptional appeal, but it also imposes on us a special obligation, to take on those moral duties which, when assumed, seem invariably
to be in our own best interests.

You have given me a great responsibility—to stay close to you, to be
worthy of you, and to exemplify what you are. Let us create together a new
national spirit of unity and trust. Your strength can compensate for my
weakness, and your wisdom can help to minimize my mistakes.

Let us learn together and laugh together and work together and pray
together, confident that in the end we will triumph together in the right.

The American Dream endures. We must once again have full faith in
our country—and in one another. I believe America can be better. We can
be even stronger than before.

Let our recent mistakes bring a resurgent commitment to the basic principles of our nation, for we know that if we despise our own government we have no future. We recall in special times when we have stood briefly, but magnificently, united. In those times no prize was beyond our grasp.

But we cannot dwell upon remembered glory. We cannot afford to drift. We reject the prospect of failure or mediocrity or an inferior quality of life for any person. Our government must at the same time be both competent and compassionate.

We have already found a high degree of personal liberty, and we are now struggling to enhance equality of opportunity. Our commitment to human rights must be absolute, our laws fair, our natural beauty preserved; the powerful must not persecute the weak, and human dignity must be enhanced.

We have learned that "more" is not necessarily "better," that even our great nation has its recognized limits, and that we can neither answer all questions nor solve all problems. We cannot afford to do everything, nor can we afford to lack boldness as we meet the future. So, together, in a spirit of individual sacrifice for the common good, we must simply do our best.

Our nation can be strong abroad only if it is strong at home. And we know that the best way to enhance freedom in other lands is to demonstrate here that our democratic system is worthy of emulation.

To be true to ourselves, we must be true to others. We will not behave in foreign places so as to violate our rules and standards here at home, for we know that the trust which our nation earns is essential to our strength.

The world itself is now dominated by a new spirit. Peoples more numerous and more politically aware are craving and now demanding their place in the sun—not just for the benefit of their own physical condition, but for basic human rights.

The passion for freedom is on the rise. Tapping this new spirit, there can be no nobler nor more ambitious task for America to undertake on this day of a new beginning than to help shape a just and peaceful world that is truly humane.

We are a strong nation, and we will maintain strength so sufficient that it need not be proven in combat—a quiet strength based not merely on the size of an arsenal, but on the nobility of ideas.

We will be ever vigilant and never vulnerable, and we will fight our wars against poverty, ignorance, and injustice—for those are the enemies against which our forces can be honorably marshaled.

We are a purely idealistic nation, but let no one confuse our idealism with weakness.

Because we are free we can never be indifferent to the fate of freedom elsewhere. Our moral sense dictates a clear-cut preference for these societies which share with us an abiding respect for individual human rights. We do not seek to intimidate, but it is clear that a world which others can dominate with impunity would be inhospitable to decency and a threat to the well-being of all people.

The world is still engaged in a massive armaments race designed to ensure continuing equivalent strength among potential adversaries. We pledge perseverance and wisdom in our efforts to limit the world's armaments to those necessary for each nation's own domestic safety. And we will move this year a step toward our ultimate goal—the elimination of all nuclear weapons from this earth. We urge all other people to join us, for success can mean life instead of death.

Within us, the people of the United States, there is evident a serious and purposeful rekindling of confidence. And I join in the hope that when my time as your President has ended, people might say this about our nation:

—that we had remembered the words of Micah and renewed our search for humility, mercy, and justice;

—that we had torn down the barriers that separated those of different race and region and religion, and where there had been mistrust, built unity, with a respect for diversity;

—that we had found productive work for those able to perform it;

—that we had strengthened the American family, which is the basis of our society;

—that we had ensured respect for the law, and equal treatment under the law, for the weak and the powerful, for the rich and the poor;

—and that we had enabled our people to be proud of their own government once again.

I would hope that the nations of the world might say that we had built a lasting peace, built not on weapons of war but on international policies which reflect our own most precious values.

These are not just my goals, and they will not be my accomplishments, but the affirmation of our nation's continuing moral strength and our belief in an undiminished, ever-expanding American Dream.

Ronald Reagan

· FIRST INAUGURAL ADDRESS ·

Tuesday, January 20, 1981

In the waning hours of his presidency, through his last night in the White House, Jimmy Carter tried to bring an end to a crisis that had helped doom his chances for reelection. Fifty-two Americans were being held as hostages in Iran, seized by young adherents of Iran's Islamic revolutionary government who stormed the United States embassy in Tehran. The U.S. had been powerless to end the crisis, which had dragged on for more than four hundred days.

The hostage crisis was but one of many burdens for Carter in 1980. The American economy was in a profound slump, bedeviled by the very same problems that had haunted Gerald Ford in 1976. Unemployment remained stuck at around 7 percent through Carter's term, while inflation skyrocketed from 6.5 percent in 1977 to more than 13 percent by 1980. The Federal Reserve attempted to check inflation by raising interest rates, making it harder for consumers and businesses to borrow and spend. In December 1980, the prime rate hit 21.5 percent.

With the nation in turmoil, Carter became the second consecutive incumbent President to face a challenge to his nomination. Senator Edward Kennedy of Massachusetts—younger brother of John and Robert Kennedy—attempted to wrest the nomination from Carter, but Carter prevailed. Kennedy didn't help himself when he fumbled in trying to answer a question from CBS reporter Roger Mudd, who asked the Senator why he wanted to be President.

The Republicans chose the man who had nearly denied Gerald Ford the party's nomination in 1976, Ronald Reagan. The former actor's victory in the primaries was a milestone for the Republican Party, for it represented the triumph of Sunbelt conservatism over East Coast moderation. Reagan, like Barry Goldwater in 1964, was an unapologetic conservative who preached a tougher line against the Soviet Union and insisted that big government was to blame for many of the country's ills.

In a face-to-face debate between the two major-party candidates, Reagan's unwarlike demeanor and rhetoric deflected Carter's charges that he was intent on dangerous confrontations with the Soviets. The debate proved to be Carter's undoing and Reagan's salvation. Carter never recovered from the derision that greeted an anecdote he told about his discussions with his twelve-year-old daughter, Amy, about nuclear weapons policy, and from the well-delivered power of Reagan's closing statement. Speaking directly to viewers, Reagan asked: "Are you better off than you were four years ago?" Almost overnight, momentum shifted to Reagan. On Election Day, he won in an a landslide, taking forty-three million popular votes and 489 electoral votes to Carter's thirty-five million and 49, respectively. Third-party candidate John Anderson finished with more than five million popular votes and no electoral votes. Reagan, at the age of sixty-nine (he would turn seventy on February 6), became the oldest person ever elected to the presidency.

The hostage crisis continued to dominate the news during the transition from Carter to Reagan, culminating in the outgoing President's frantic attempts to negotiate the captives' release before leaving office. As he prepared for his inauguration on the morning of January 20, Reagan received word from Carter that a deal had been struck and that the hostages might soon leave Iranian airspace. But the Americans were still in Iran when Reagan picked up Carter for the traditional ride of Presidents from the White House to the Capitol, and, according to Richard Reeves in his book *President Reagan,* as Reagan rose to recite the oath of office, he glanced at Carter to see if there were any developments he could announce in his inaugural address. Carter, according to Reeves, looked at Reagan, shook his head, and whispered, "Not yet."[1] The hostages would not be released until later that afternoon.

The outgoing and incoming Presidents, along with the thousands who witnessed the ceremony, were gathered together on the west side of the Capitol, instead of the traditional spot on the Capitol's East Portico. The move was loaded with symbolism: Reagan was from the West, a place associated with new beginnings and fresh starts.

In the second paragraph of his speech, Reagan pointedly thanked his

---

1. Richard Reeves, *President Reagan: Triumph of the Imagination* (New York: Simon & Schuster, 2005), 2.

vanquished foe for his cooperation during the transition—and, implicitly, for his last-minute work on behalf of the hostages.

The rest of Reagan's speech was a firm declaration of principle, filled with rhetorical devices that would become familiar during his presidency. He at times spoke directly to his audience. He used anecdotes to illustrate his belief in America's special mission in the world. And he cast himself as a populist outsider, chosen by the people to do battle with "an elite group" in government. "In this present crisis," he said, "government is not the solution to our problem, government is the problem." Later, he asserted that, "We are a nation that has a government—not the other way around. And this makes us special among the nations of the earth." Reagan hailed average Americans as heroes. "Those who say that we are in a time when there are no heroes just don't know where to look. You can see heroes every day going in and out of factory gates."

Although a devout anti-Communist, Reagan broke with many Cold War Presidents by devoting large chunks of his address to domestic issues. "These United States are confronted with an economic affliction of great proportions," he said. "We suffer from the longest and one of the worst sustained inflations in our national history. It distorts our economic decisions, penalizes thrift, and crushes the struggling young and the fixed-income elderly alike."

Reagan promised to "get government back within its means, and to lighten our punitive tax burden. And these will be our first priorities, and on these principles, there will be no compromise."

Given the size of Reagan's victory, many Democrats in the House of Representatives—the lone bastion of Democratic Party control after Republicans took the Senate in November—were not in a mood to demand compromises from Reagan, especially after he survived an assassination attempt on March 30. Public sympathy for the President made it that much easier for Reagan to get his programs through Congress.

Reagan promised change in his inaugural address. When he ran for re-election in 1984, nobody accused him of failing to deliver on that promise.

To a few of us here today, this is a solemn and most momentous occasion; and yet, in the history of our nation, it is a commonplace occurrence. The orderly transfer of authority as called for in the Constitution routinely takes place as it has for almost two centuries and few of us stop to think how unique we really are. In the eyes of many in the world, this every-four-year ceremony we accept as normal is nothing less than a miracle.

Mr. President, I want our fellow citizens to know how much you did to carry on this tradition. By your gracious cooperation in the transition process, you have shown a watching world that we are a united people pledged to maintaining a political system which guarantees individual liberty to a greater degree than any other, and I thank you and your people for all your help in maintaining the continuity which is the bulwark of our Republic.

The business of our nation goes forward. These United States are confronted with an economic affliction of great proportions. We suffer from the longest and one of the worst sustained inflations in our national history. It distorts our economic decisions, penalizes thrift, and crushes the struggling young and the fixed-income elderly alike. It threatens to shatter the lives of millions of our people.

Idle industries have cast workers into unemployment, causing human misery and personal indignity. Those who do work are denied a fair return for their labor by a tax system which penalizes successful achievement and keeps us from maintaining full productivity.

But great as our tax burden is, it has not kept pace with public spending. For decades, we have piled deficit upon deficit, mortgaging our future and our children's future for the temporary convenience of the present. To continue this long trend is to guarantee tremendous social, cultural, political, and economic upheavals.

You and I, as individuals, can, by borrowing, live beyond our means, but for only a limited period of time. Why, then, should we think that collectively, as a nation, we are not bound by that same limitation?

We must act today in order to preserve tomorrow. And let there be no misunderstanding—we are going to begin to act, beginning today.

The economic ills we suffer have come upon us over several decades. They will not go away in days, weeks, or months, but they will go away. They will go away because we, as Americans, have the capacity now, as we have had in the past, to do whatever needs to be done to preserve this last and greatest bastion of freedom.

In this present crisis, government is not the solution to our problem, government is the problem.

From time to time, we have been tempted to believe that society has become too complex to be managed by self-rule, that government by an elite group is superior to government for, by, and of the people. But if no one among us is capable of governing himself, then who among us has the capacity to govern someone else? All of us together, in and out of government, must bear the burden. The solutions we seek must be equitable, with no one group singled out to pay a higher price.

We hear much of special interest groups. Our concern must be for a special interest group that has been too long neglected. It knows no sectional boundaries or ethnic and racial divisions, and it crosses political party lines. It is made up of men and women who raise our food, patrol our streets, man our mines and our factories, teach our children, keep our homes, and heal us when we are sick—professionals, industrialists, shopkeepers, clerks, cabbies, and truckdrivers. They are, in short, "We the people," this breed called Americans.

Well, this administration's objective will be a healthy, vigorous, growing economy that provides equal opportunity for all Americans, with no barriers born of bigotry or discrimination. Putting America back to work means putting all Americans back to work. Ending inflation means freeing all Americans from the terror of runaway living costs. All must share in the productive work of this "new beginning" and all must share in the bounty of a revived economy. With the idealism and fair play which are the core of our system and our strength, we can have a strong and prosperous America at peace with itself and the world.

So, as we begin, let us take inventory. We are a nation that has a government—not the other way around. And this makes us special among the nations of the earth. Our government has no power except that granted it by the people. It is time to check and reverse the growth of government which shows signs of having grown beyond the consent of the governed.

It is my intention to curb the size and influence of the federal establishment and to demand recognition of the distinction between the powers granted to the federal government and those reserved to the states or to the people. All of us need to be reminded that the federal government did not create the states; the states created the federal government.

Now, so there will be no misunderstanding, it is not my intention to do away with government. It is, rather, to make it work—work with us, not over us; to stand by our side, not ride on our back. Government can and must provide opportunity, not smother it; foster productivity, not stifle it.

If we look to the answer as to why, for so many years, we achieved so much, prospered as no other people on earth, it was because here, in this land, we unleashed the energy and individual genius of man to a greater extent than has ever been done before. Freedom and the dignity of the individual have been more available and assured here than in any other place on earth. The price for this freedom at times has been high, but we have never been unwilling to pay that price.

It is no coincidence that our present troubles parallel and are proportionate to the intervention and intrusion in our lives that result from unnecessary and excessive growth of government. It is time for us to realize that we are too great a nation to limit ourselves to small dreams. We are not, as some would have us believe, doomed to an inevitable decline. I do not believe in a fate that will fall on us no matter what we do. I do believe in a fate that will fall on us if we do nothing. So, with all the creative energy at our command, let us begin an era of national renewal. Let us renew our determination, our courage, and our strength. And let us renew our faith and our hope.

We have every right to dream heroic dreams. Those who say that we are in a time when there are no heroes just don't know where to look. You can see heroes every day going in and out of factory gates. Others, a handful in number, produce enough food to feed all of us and then the world beyond. You meet heroes across a counter—and they are on both sides of that counter. There are entrepreneurs with faith in themselves and faith in an idea who create new jobs, new wealth and opportunity. They are individuals and families whose taxes support the government and whose voluntary gifts support church, charity, culture, art, and education. Their patriotism is quiet but deep. Their values sustain our national life.

I have used the words "they" and "their" in speaking of these heroes. I could say "you" and "your" because I am addressing the heroes of whom I speak—you, the citizens of this blessed land. Your dreams, your hopes, your goals are going to be the dreams, the hopes, and the goals of this administration, so help me God.

We shall reflect the compassion that is so much a part of your makeup. How can we love our country and not love our countrymen, and loving them, reach out a hand when they fall, heal them when they are sick, and provide opportunities to make them self-sufficient so they will be equal in fact and not just in theory?

Can we solve the problems confronting us? Well, the answer is an unequivocal and emphatic "yes." To paraphrase Winston Churchill, I did not take the oath I have just taken with the intention of presiding over the dissolution of the world's strongest economy.

In the days ahead I will propose removing the roadblocks that have slowed our economy and reduced productivity. Steps will be taken aimed at restoring the balance between the various levels of government. Progress may be slow—measured in inches and feet, not miles—but we will progress. Is it time to reawaken this industrial giant, to get government back within its means, and to lighten our punitive tax burden. And these will be our first priorities, and on these principles, there will be no compromise.

On the eve of our struggle for independence a man who might have been one of the greatest among the Founding Fathers, Dr. Joseph Warren, President of the Massachusetts Congress, said to his fellow Americans, "Our country is in danger, but not to be despaired of . . . On you depend the fortunes of America. You are to decide the important questions upon which rests the happiness and the liberty of millions yet unborn. Act worthy of yourselves."

Well, I believe we, the Americans of today, are ready to act worthy of ourselves, ready to do what must be done to ensure happiness and liberty for ourselves, our children, and our children's children.

And as we renew ourselves here in our own land, we will be seen as having greater strength throughout the world. We will again be the exemplar of freedom and a beacon of hope for those who do not now have freedom.

To those neighbors and allies who share our freedom, we will strengthen our historic ties and assure them of our support and firm commitment. We will match loyalty with loyalty. We will strive for mutually beneficial relations. We will not use our friendship to impose on their sovereignty, for our own sovereignty is not for sale.

As for the enemies of freedom, those who are potential adversaries, they will be reminded that peace is the highest aspiration of the American people. We will negotiate for it, sacrifice for it; we will not surrender for it—now or ever.

Our forbearance should never be misunderstood. Our reluctance for conflict should not be misjudged as a failure of will. When action is required to preserve our national security, we will act. We will maintain sufficient strength to prevail if need be, knowing that if we do so we have the best chance of never having to use that strength.

Above all, we must realize that no arsenal, or no weapon in the arsenals of the world, is so formidable as the will and moral courage of free men and women. It is a weapon our adversaries in today's world do not have. It is a weapon that we as Americans do have. Let that be understood by those who practice terrorism and prey upon their neighbors.

I am told that tens of thousands of prayer meetings are being held on this day, and for that I am deeply grateful. We are a nation under God, and I believe God intended for us to be free. It would be fitting and good, I think, if on each Inauguration Day in future years it should be declared a day of prayer.

This is the first time in history that this ceremony has been held, as you have been told, on this West Front of the Capitol. Standing here, one faces a magnificent vista, opening up on this city's special beauty and history. At the end of this open mall are those shrines to the giants on whose shoulders we stand.

Directly in front of me, the monument to a monumental man: George Washington, father of our country. A man of humility who came to greatness reluctantly. He led America out of revolutionary victory into infant nationhood. Off to one side, the stately memorial to Thomas Jefferson. The Declaration of Independence flames with his eloquence.

And then beyond the Reflecting Pool the dignified columns of the Lincoln Memorial. Whoever would understand in his heart the meaning of America will find it in the life of Abraham Lincoln.

Beyond those monuments to heroism is the Potomac River, and on the far shore the sloping hills of Arlington National Cemetery with its row on row of simple white markers bearing crosses or Stars of David. They add up to only a tiny fraction of the price that has been paid for our freedom.

Each one of those markers is a monument to the kinds of hero I spoke of earlier. Their lives ended in places called Belleau Wood, the Argonne, Omaha Beach, Salerno, and halfway around the world on Guadalcanal, Tarawa, Pork Chop Hill, the Chosin Reservoir, and in a hundred rice paddies and jungles of a place called Vietnam.

Under one such marker lies a young man—Martin Treptow—who left his job in a small town barbershop in 1917 to go to France with the famed

Rainbow Division. There, on the western front, he was killed trying to carry a message between battalions under heavy artillery fire.

We are told that on his body was found a diary. On the flyleaf under the heading, "My Pledge," he had written these words: "America must win this war. Therefore, I will work, I will save, I will sacrifice, I will endure, I will fight cheerfully and do my utmost, as if the issue of the whole struggle depended on me alone."

The crisis we are facing today does not require of us the kind of sacrifice that Martin Treptow and so many thousands of others were called upon to make. It does require, however, our best effort, and our willingness to believe in ourselves and to believe in our capacity to perform great deeds; to believe that together, with God's help, we can and will resolve the problems which now confront us.

And, after all, why shouldn't we believe that? We are Americans. God bless you, and thank you.

Ronald Reagan

Monday, January 21, 1985

*R*onald Reagan's reelection in 1984 was a landslide of historic pro-portions: he won forty-nine states—his opponent, former vice president Walter Mondale, won only his home state of Minnesota and the District of Columbia—and captured nearly 59 percent of the popular vote.

But the true measure of Reagan's accomplishment went beyond num-bers. He had governed as he had promised—as an unabashed conservative ready to confront the Soviet Union after years of détente, to question the bipartisan consensus on social welfare spending and government regula-tion, and to cut taxes for wealthy earners to stimulate spending and in-vestment. Twenty years after the nation soundly rejected Barry Goldwater's candidacy, which made similar promises, it embraced the Reagan Revolu-tion and so moved the center of American politics to the right.

Reagan's smashing victory was a milestone for the conservative move-ment, which had labored on the fringes of American politics from the 1950s to Reagan's election in 1980. It also was a milestone for the office of the presidency itself. Tragedy and scandal had left deep scars on the of-fice. So had the sense of powerlessness that engulfed the Carter years. In his first term as the most ideological President since Franklin Roosevelt, Ronald Reagan restored the presidency's image and its importance in pub-lic life. With his charisma, communication skills, and larger-than-life presence, Reagan dominated domestic and global politics in ways that in-vited comparisons to FDR, whose legacy Reagan sought to undermine.

There was a moment during the 1984 campaign when Reagan, the old-est person to occupy the nation's highest office, seemed vulnerable, in ev-ery sense of the word. During a nationally televised debate with Mondale, Reagan seemed flustered, and even admitted, "I'm all confused now." Mondale clearly won the debate, and commentators wondered if Reagan was too old for another term. When the two candidates met again, Reagan was asked about the age issue. "I am not going to exploit, for political purposes, my opponent's youth and inexperience," the President said, prompting laughter. The age issue disappeared—and so did Mondale's chances. On Election Day, Reagan won fifty-four million votes to Mondale's thirty-seven million, and 525 electoral votes to Mondale's astonishingly small 13.

Inauguration Day 1985 was one of the coldest in history. With temperatures hovering in the single digits, the traditional outdoor ceremony and parades were canceled. Ronald Reagan delivered his inaugural address in the warmth of the Capitol's Rotunda.

He opened his speech with a personal touch. Among those gathered to hear the speech was Senator John Stennis of Mississippi, a Democrat who was recovering from cancer surgery that required amputation of his left leg. Missing from the audience was Congressman Gillis Long, a Democrat from Virginia. He had died only hours earlier of a heart ailment. Reagan opened his speech by acknowledging Stennis and by asking for a "moment of silent prayer" for Long. The gracious gesture, unprecedented for an inaugural address, perhaps explains part of the Reagan mystique: he rhetorically stepped aside to allow the spotlight to shine on a colleague, however briefly.

Always quick to seize on the power of symbols and popular history, Reagan noted that his inauguration was the fiftieth in history—it was, not counting midterm successions caused by death or resignation. More than any President since John Kennedy, Reagan deployed historic references to help explain his critique of the present and his vision of the future. The fiftieth inaugural celebration allowed him to invoke folk memories of George Washington and the very first inaugural. "When the first President, George Washington, placed his hand upon the Bible," he said, "he stood less than a single day's journey by horseback from raw, untamed wilderness. There were four million Americans in a union of thirteen States. Today we are sixty times as many in a union of fifty states. We have lighted the world with our inventions, gone to the aid of mankind wherever in the world there was a cry for help, journeyed to the moon and safely returned. So much has changed. And yet we stand together as we did two centuries ago."

Reagan's unambiguous celebration of American achievement had served him well during his first term and during his reelection campaign. "In this blessed land," he continued, "there is always a better tomorrow." Those sentiments connected Reagan to his supporters and softened the edges of his ideological agenda.

Emboldened by one of the greatest landslides in American history, Reagan set out an ambitious program for his second term. Now that "freedom and incentives" had unleashed the forces of prosperity in America, Washington ought to keep cutting taxes and simplify the tax system, "tear down economic barriers" that stood in the way of investment and growth, and freeze government spending. Returning to an old theme, he emphasized the importance of balancing the federal budget, although the size of the deficit skyrocketed during his first four years, from about $60 billion in 1980 to $200 billion in 1985.

While Reagan emphasized domestic affairs, he did not neglect the Cold War and the Soviet Union. He made a pitch for his Strategic Defense Initiative, which envisioned a weapon that could destroy nuclear missiles from space—earning the not-always-affectionate nickname of Star Wars. But missing from this speech is the fiery anti-Communist rhetoric that had helped propel him to national attention, and that had been a staple of past inaugurals since Harry Truman. A new round of arms reduction talks with Moscow was scheduled for March, but before they convened, Soviet leader Konstantin Chernenko died after just over a year in office. His predecessor, Yuri Andropov, had served just fifteen months after succeeding longtime leader Leonid Brezhnev.

The Soviet leadership chose Mikhail Gorbachev to replace Chernenko. He was fifty-four years old, a generation younger than Reagan, and far more charismatic than his predecessors. He and Reagan would form one of the century's most unlikely political and strategic alliances during the President's second term, leading to a thaw in superpower relations by the time Reagan left office. In January 1989, with his term about to expire, Reagan visited Gorbachev in Russia. The Cold War, which Reagan framed as a struggle between good and evil, was about to come to an end. Reagan, who many feared would provoke nuclear war, instead seized the chance to bring about peace. Not for the first time, he had confounded skeptics and antagonists.

This day has been made brighter with the presence here of one who, for a time, has been absent—Senator John Stennis.

God bless you and welcome back.

There is, however, one who is not with us today: Representative Gillis Long of Louisiana left us last night. I wonder if we could all join in a moment of silent prayer. [Moment of silent prayer.] Amen.

There are no words adequate to express my thanks for the great honor that you have bestowed on me. I will do my utmost to be deserving of your trust.

This is, as Senator Mathias told us, the fiftieth time that we the people have celebrated this historic occasion. When the first President, George Washington, placed his hand upon the Bible, he stood less than a single day's journey by horseback from raw, untamed wilderness. There were four million Americans in a union of thirteen States. Today we are sixty times as many in a union of fifty states. We have lighted the world with our inventions, gone to the aid of mankind wherever in the world there was a cry for help, journeyed to the moon and safely returned. So much has changed. And yet we stand together as we did two centuries ago.

When I took this oath four years ago, I did so in a time of economic stress. Voices were raised saying we had to look to our past for the greatness and glory. But we, the present-day Americans, are not given to looking backward. In this blessed land, there is always a better tomorrow.

Four years ago, I spoke to you of a new beginning and we have accomplished that. But in another sense, our new beginning is a continuation of that beginning created two centuries ago when, for the first time in history, government, the people said, was not our master, it is our servant; its only power that which we the people allow it to have.

That system has never failed us, but, for a time, we failed the system. We asked things of government that government was not equipped to give. We yielded authority to the national government that properly belonged to states or to local governments or to the people themselves. We allowed taxes and inflation to rob us of our earnings and savings and watched the great industrial machine that had made us the most productive people on earth slow down and the number of unemployed increase.

By 1980, we knew it was time to renew our faith, to strive with all our strength toward the ultimate in individual freedom consistent with an orderly society.

We believed then and now there are no limits to growth and human progress when men and women are free to follow their dreams.

And we were right to believe that. Tax rates have been reduced, inflation cut dramatically, and more people are employed than ever before in our history.

We are creating a nation once again vibrant, robust, and alive. But there are many mountains yet to climb. We will not rest until every American enjoys the fullness of freedom, dignity, and opportunity as our birthright. It is our birthright as citizens of this great Republic, and we'll meet this challenge.

These will be years when Americans have restored their confidence and tradition of progress; when our values of faith, family, work, and neighborhood were restated for a modern age; when our economy was finally freed from government's grip; when we made sincere efforts at meaningful arms reduction, rebuilding our defenses, our economy, and developing new technologies, and helped preserve peace in a troubled world; when Americans courageously supported the struggle for liberty, self-government, and free enterprise throughout the world, and turned the tide of history away from totalitarian darkness and into the warm sunlight of human freedom.

My fellow citizens, our nation is poised for greatness. We must do what we know is right and do it with all our might. Let history say of us, "These were golden years—when the American Revolution was reborn, when freedom gained new life, when America reached for her best."

Our two-party system has served us well over the years, but never better than in those times of great challenge when we came together not as Democrats or Republicans, but as Americans united in a common cause.

Two of our Founding Fathers, a Boston lawyer named Adams and a Virginia planter named Jefferson, members of that remarkable group who met in Independence Hall and dared to think they could start the world over again, left us an important lesson. They had become political rivals in the presidential election of 1800. Then years later, when both were retired, and age had softened their anger, they began to speak to each other again through letters. A bond was reestablished between those two who had helped create this government of ours.

In 1826, the fiftieth anniversary of the Declaration of Independence, they both died. They died on the same day, within a few hours of each other, and that day was the Fourth of July.

In one of those letters exchanged in the sunset of their lives, Jefferson wrote: "It carries me back to the times when, beset with difficulties and dangers, we were fellow laborers in the same cause, struggling for what is most valuable to man, his right to self-government. Laboring always at the same oar, with some wave ever ahead threatening to overwhelm us, and yet passing harmless . . . we rode through the storm with heart and hand."

Well, with heart and hand, let us stand as one today: one people under God determined that our future shall be worthy of our past. As we do, we must not repeat the well-intentioned errors of our past. We must never again abuse the trust of working men and women, by sending their earnings on a futile chase after the spiraling demands of a bloated federal establishment. You elected us in 1980 to end this prescription for disaster, and I don't believe you reelected us in 1984 to reverse course.

At the heart of our efforts is one idea vindicated by twenty-five straight months of economic growth: freedom and incentives unleash the drive and entrepreneurial genius that are the core of human progress. We have begun to increase the rewards for work, savings, and investment; reduce the increase in the cost and size of government and its interference in people's lives.

We must simplify our tax system, make it more fair, and bring the rates down for all who work and earn. We must think anew and move with a new boldness, so every American who seeks work can find work; so the least among us shall have an equal chance to achieve the greatest things—to be heroes who heal our sick, feed the hungry, protect peace among nations, and leave this world a better place.

The time has come for a new American emancipation—a great national drive to tear down economic barriers and liberate the spirit of enterprise in the most distressed areas of our country. My friends, together we can do this, and do it we must, so help me God.

From new freedom will spring new opportunities for growth, a more productive, fulfilled, and united people, and a stronger America—an

America that will lead the technological revolution, and also open its mind and heart and soul to the treasures of literature, music, and poetry, and the values of faith, courage, and love.

A dynamic economy, with more citizens working and paying taxes, will be our strongest tool to bring down budget deficits. But an almost unbroken fifty years of deficit spending has finally brought us to a time of reckoning. We have come to a turning point, a moment for hard decisions. I have asked the cabinet and my staff a question, and now I put the same question to all of you: If not us, who? And if not now, when? It must be done by all of us going forward with a program aimed at reaching a balanced budget. We can then begin reducing the national debt.

I will shortly submit a budget to the Congress aimed at freezing government program spending for the next year. Beyond that, we must take further steps to permanently control government's power to tax and spend. We must act now to protect future generations from government's desire to spend its citizens' money and tax them into servitude when the bills come due. Let us make it unconstitutional for the federal government to spend more than the federal government takes in.

We have already started returning to the people and to state and local governments responsibilities better handled by them. Now, there is a place for the federal government in matters of social compassion. But our fundamental goals must be to reduce dependency and upgrade the dignity of those who are infirm or disadvantaged. And here a growing economy and support from family and community offer our best chance for a society where compassion is a way of life, where the old and infirm are cared for, the young and, yes, the unborn protected, and the unfortunate looked after and made self-sufficient.

And there is another area where the federal government can play a part. As an older American, I remember a time when people of different race, creed, or ethnic origin in our land found hatred and prejudice installed in social custom and, yes, in law. There is no story more heartening in our history than the progress that we have made toward the "brotherhood of man" that God intended for us. Let us resolve there will be no turning back or hesitation on the road to an America rich in dignity and abundant with opportunity for all our citizens.

Let us resolve that we the people will build an American opportunity society in which all of us—white and black, rich and poor, young and old—will go forward together arm in arm. Again, let us remember that though our heritage is one of bloodlines from every corner of the earth, we are all Americans pledged to carry on this last, best hope of man on earth.

I have spoken of our domestic goals and the limitations which we should put on our national government. Now let me turn to a task which is the primary responsibility of national government—the safety and security of our people.

Today, we utter no prayer more fervently than the ancient prayer for peace on earth. Yet history has shown that peace will not come, nor will our freedom be preserved, by good will alone. There are those in the world who scorn our vision of human dignity and freedom. One nation, the Soviet Union, has conducted the greatest military buildup in the history of man, building arsenals of awesome offensive weapons.

We have made progress in restoring our defense capability. But much remains to be done. There must be no wavering by us, nor any doubts by others, that America will meet her responsibilities to remain free, secure, and at peace.

There is only one way safely and legitimately to reduce the cost of national security, and that is to reduce the need for it. And this we are trying to do in negotiations with the Soviet Union. We are not just discussing limits on a further increase of nuclear weapons. We seek, instead, to reduce their number. We seek the total elimination one day of nuclear weapons from the face of the earth.

Now, for decades, we and the Soviets have lived under the threat of mutual assured destruction; if either resorted to the use of nuclear weapons, the other could retaliate and destroy the one who had started it. Is there either logic or morality in believing that if one side threatens to kill tens of millions of our people, our only recourse is to threaten killing tens of millions of theirs?

I have approved a research program to find, if we can, a security shield that would destroy nuclear missiles before they reach their target. It wouldn't kill people, it would destroy weapons. It wouldn't militarize space, it would help demilitarize the arsenals of earth. It would render nuclear weapons obsolete. We will meet with the Soviets, hoping that we can agree on a way to rid the world of the threat of nuclear destruction.

We strive for peace and security, heartened by the changes all around us. Since the turn of the century, the number of democracies in the world has grown fourfold. Human freedom is on the march, and nowhere more so than our own hemisphere. Freedom is one of the deepest and noblest aspirations of the human spirit. People, worldwide, hunger for the right of self-determination, for those inalienable rights that make for human dignity and progress.

America must remain freedom's staunchest friend, for freedom is our best ally.

And it is the world's only hope, to conquer poverty and preserve peace. Every blow we inflict against poverty will be a blow against its dark allies of oppression and war. Every victory for human freedom will be a victory for world peace.

So we go forward today, a nation still mighty in its youth and powerful in its purpose. With our alliances strengthened, with our economy leading the world to a new age of economic expansion, we look forward to a world rich in possibilities. And all this because we have worked and acted together, not as members of political parties, but as Americans.

My friends, we live in a world that is lit by lightning. So much is changing and will change, but so much endures, and transcends time.

History is a ribbon, always unfurling; history is a journey. And as we continue our journey, we think of those who traveled before us. We stand together again at the steps of this symbol of our democracy—or we would have been standing at the steps if it hadn't gotten so cold. Now we are standing inside this symbol of our democracy. Now we hear again the echoes of our past: a general falls to his knees in the hard snow of Valley Forge; a lonely President paces the darkened halls, and ponders his struggle to preserve the Union; the men of the Alamo call out encouragement to each other; a settler pushes west and sings a song, and the song echoes out forever and fills the unknowing air.

It is the American sound. It is hopeful, bighearted, idealistic, daring, decent, and fair. That's our heritage; that is our song. We sing it still. For all our problems, our differences, we are together as of old, as we raise our voices to the God who is the Author of this most tender music. And may He continue to hold us close as we fill the world with our sound—in unity, affection, and love—one people under God, dedicated to the dream of freedom that He has placed in the human heart, called upon now to pass that dream on to a waiting and hopeful world.

God bless you and may God bless America.

George H. W. Bush

· INAUGURAL ADDRESS ·

Friday, January 20, 1989

*H*istory was not on George H. W. Bush's side as he campaigned for the presidency in 1988. No sitting vice president, as Bush was that year, had been elected president since Martin Van Buren succeeded Andrew Jackson in 1836. What's more, Bush seemed to represent the aging, moderate wing of the Republican Party, a group that Ronald Reagan had pushed to the margins with his conservative ideology and middle-class, Sunbelt constituency.

Although he moved from Connecticut to Texas as a young adult, Bush was the product of centrist, eastern Republicanism. His father, Prescott Bush, was a three-term senator from Connecticut in the 1950s and '60s. George Bush had served a term as a congressman from Texas before becoming chairman of the Republican National Committee in the early 1970s and director of the Central Intelligence Agency from 1976 to 1977. He was an establishment figure, running for President at a time when the Republican Party's old guard seemed out of favor with the new Reagan coalition of white ethnics and religious conservatives.

Making matters bleaker still for Bush were scandals that tarnished the Reagan administration midway through its second term. The public learned that White House staff had been selling arms illegally to Iran and using the profits to help fund a guerilla group in Nicaragua, the contras. Democrats tried to link Bush, as Reagan's vice president, to the scandal.

But Bush's association with Reagan helped him more than it hurt him. Reagan remained personally popular despite the Iran-contra scandal and general second-term malaise. He became the first President since Eisenhower to serve two full terms and was widely credited with restoring American pride in the presidency. His partnership with Soviet leader Gorbachev eased Cold War tensions dramatically. And his articulation of

America's special mission in the world remained enormously appealing at home.

Reagan gave Bush his blessing in a highly contested race for the Republican nomination. Bush was one of eight serious candidates, a field that included former secretary of state Al Haig, Senator Bob Dole of Kansas, Congressman Jack Kemp of New York, former (and future) defense secretary Donald Rumsfeld, and, in a sign of the growing power of religious conservatives, televangelist Pat Robertson. Bush had a good deal of party support, helping him to defeat his strongest foe, Dole, in New Hampshire after the vice president portrayed the senator as a tax hiker.

The Democrats also produced a crowded field, from which the governor of Massachusetts, Michael Dukakis, emerged after scoring unexpected victories in the early primaries. Dukakis triumphed after better-known Democrats like senators Al Gore of Tennessee, Joseph Biden of Delaware, and former senator Gary Hart of Colorado either imploded or failed to catch on with primary voters. One of Dukakis's most persistent challengers was the Reverend Jesse Jackson, a civil rights leader and a charismatic orator.

Dukakis, the son of Greek immigrants, officially claimed the nomination at a highly successful convention in Atlanta. For a moment in midsummer, the Democrat enjoyed a double-digit lead in head-to-head polls against Bush.

The Bush campaign, however, was quick to launch an offensive against Dukakis, who found himself portrayed as an out-of-touch liberal who opposed the recitation of the Pledge of Allegiance in public schools and who allowed dangerous criminals to walk the streets. Dukakis, who once admitted to reading a book about land-use issues in Sweden while on vacation, seemed incapable of responding to the Republican assault. Bush successfully framed the campaign as a referendum on Dukakis's tenure as governor of Massachusetts rather than focusing on the issues that faced the United States as the long Cold War neared an end. Bush also gained in the polls by promising that he would impose "no new taxes."

Dukakis's image as a technocrat strong on policy details but weak in the language and symbols of politics was solidified in the second of two debates with Bush, when he responded to a hypothetical question about the rape and murder of his wife with a lawyer's brief against the death penalty.

Bush overcame Dukakis's early lead and easily won election as Ronald Reagan's heir. He took more than forty-eight million popular votes and

426 electoral votes to Dukakis's forty-one million and 112, respectively. With Bush's victory, the Republicans won their fifth presidential election in six campaigns, starting with 1968.

Bush opened his inaugural as Jimmy Carter did in 1977, by acknowledging his predecessor. And, like Eisenhower at his first swearing-in, Bush recited a prayer before launching into the substance of his address.

Although the Cold War was not over, and the Soviet Union remained a symbolic if not real antagonist, Bush's speech proclaimed an end to the era of East-West tensions. "The totalitarian era is passing, its old ideas blown away like leaves from an ancient, lifeless tree," he said. "A new breeze is blowing, and a nation refreshed by freedom stands ready to push on."

The new breeze could not help but put wind in the rhetorical sails of the free world's new leader. "We know what's right," he said. "Freedom is right. We know how to secure a more just and prosperous life for man on earth: through free markets, free speech, free elections, and the exercise of free will unhampered by the state."

There was a triumphant tone to the speech—America's triumph in the Cold War, Bush suggested, settled some of humankind's greatest and most persistent questions. "For the first time in this century, for the first time in perhaps all history, man does not have to invent a system by which to live," he said. "We don't have to talk late into the night about which form of government is better."

The impending end of the Cold War, then, meant the beginning of a worldwide political consensus. Americans listening that day had little reason to question Bush's bold assertions.

MR. CHIEF JUSTICE, MR. PRESIDENT, VICE PRESIDENT QUAYLE, SENATOR MITCHELL, SPEAKER WRIGHT, SENATOR DOLE, CONGRESSMAN MICHEL, AND FELLOW CITIZENS, NEIGHBORS, AND FRIENDS:

There is a man here who has earned a lasting place in our hearts and in our history. President Reagan, on behalf of our nation, I thank you for the wonderful things that you have done for America.

I have just repeated word for word the oath taken by George Washington two hundred years ago, and the Bible on which I placed my hand is the

Bible on which he placed his. It is right that the memory of Washington be with us today, not only because this is our bicentennial inauguration, but because Washington remains the Father of Our Country. And he would, I think, be gladdened by this day; for today is the concrete expression of a stunning fact: our continuity these two hundred years since our government began.

We meet on democracy's front porch, a good place to talk as neighbors and as friends. For this is a day when our nation is made whole, when our differences, for a moment, are suspended.

And my first act as President is a prayer. I ask you to bow your heads:

Heavenly Father, we bow our heads and thank You for Your love. Accept our thanks for the peace that yields this day and the shared faith that makes its continuance likely. Make us strong to do Your work, willing to heed and hear Your will, and write on our hearts these words: "Use power to help people." For we are given power not to advance our own purposes, nor to make a great show in the world, nor a name. There is but one just use of power, and it is to serve people. Help us to remember it, Lord. Amen.

I come before you and assume the presidency at a moment rich with promise. We live in a peaceful, prosperous time, but we can make it better. For a new breeze is blowing, and a world refreshed by freedom seems reborn; for in man's heart, if not in fact, the day of the dictator is over. The totalitarian era is passing, its old ideas blown away like leaves from an ancient, lifeless tree. A new breeze is blowing, and a nation refreshed by freedom stands ready to push on. There is new ground to be broken, and new action to be taken. There are times when the future seems thick as a fog; you sit and wait, hoping the mists will lift and reveal the right path. But this is a time when the future seems a door you can walk right through into a room called tomorrow.

Great nations of the world are moving toward democracy through the door to freedom. Men and women of the world move toward free markets through the door to prosperity. The people of the world agitate for free expression and free thought through the door to the moral and intellectual satisfactions that only liberty allows.

We know what works: freedom works. We know what's right: freedom is right. We know how to secure a more just and prosperous life for man on earth: through free markets, free speech, free elections, and the exercise of free will unhampered by the state.

For the first time in this century, for the first time in perhaps all history, man does not have to invent a system by which to live. We don't have to talk late into the night about which form of government is better. We don't have to wrest justice from the kings. We only have to summon it from within ourselves. We must act on what we know. I take as my guide the hope of a saint: in crucial things, unity; in important things, diversity; in all things, generosity.

America today is a proud, free nation, decent and civil, a place we cannot help but love. We know in our hearts, not loudly and proudly, but as a simple fact, that this country has meaning beyond what we see, and that our strength is a force for good. But have we changed as a nation even in our time? Are we enthralled with material things, less appreciative of the nobility of work and sacrifice?

My friends, we are not the sum of our possessions. They are not the measure of our lives. In our hearts we know what matters. We cannot hope only to leave our children a bigger car, a bigger bank account. We must hope to give them a sense of what it means to be a loyal friend, a loving parent, a citizen who leaves his home, his neighborhood and town, better than he found it. What do we want the men and women who work with us to say when we are no longer there? That we were more driven to succeed than anyone around us? Or that we stopped to ask if a sick child had gotten better, and stayed a moment there to trade a word of friendship?

No President, no government, can teach us to remember what is best in what we are. But if the man you have chosen to lead this government can help make a difference; if he can celebrate the quieter, deeper successes that are made not of gold and silk, but of better hearts and finer souls; if he can do these things, then he must.

America is never wholly herself unless she is engaged in high moral principle. We as a people have such a purpose today. It is to make kinder the face of the nation and gentler the face of the world. My friends, we have work to do. There are the homeless, lost and roaming. There are the children who have nothing, no love, no normalcy. There are those who cannot free themselves of enslavement to whatever addiction—drugs, welfare, the demoralization that rules the slums. There is crime to be conquered, the rough crime of the streets. There are young women to be helped who are about to become mothers of children they can't care for and might not love. They need our care, our guidance, and our education, though we bless them for choosing life.

The old solution, the old way, was to think that public money alone could end these problems. But we have learned that is not so. And in any case, our funds are low. We have a deficit to bring down. We have more will than wallet; but will is what we need. We will make the hard choices, looking at what we have and perhaps allocating it differently, making our decisions based on honest need and prudent safety. And then we will do the wisest thing of all: we will turn to the only resource we have that in times of need always grows—the goodness and the courage of the American people.

I am speaking of a new engagement in the lives of others, a new activism, hands-on and involved, that gets the job done. We must bring in the generations, harnessing the unused talent of the elderly and the unfocused energy of the young. For not only leadership is passed from generation to generation, but so is stewardship. And the generation born after the Second World War has come of age.

I have spoken of a thousand points of light, of all the community organizations that are spread like stars throughout the nation, doing good. We will work hand in hand, encouraging, sometimes leading, sometimes being led, rewarding. We will work on this in the White House, in the cabinet agencies. I will go to the people and the programs that are the brighter points of light, and I will ask every member of my government to become involved. The old ideas are new again because they are not old, they are timeless: duty, sacrifice, commitment, and a patriotism that finds its expression in taking part and pitching in.

We need a new engagement, too, between the executive and the Congress. The challenges before us will be thrashed out with the House and the Senate. We must bring the federal budget into balance. And we must ensure that America stands before the world united, strong, at peace, and fiscally sound. But, of course, things may be difficult. We need compromise; we have had dissension. We need harmony; we have had a chorus of discordant voices.

For Congress, too, has changed in our time. There has grown a certain divisiveness. We have seen the hard looks and heard the statements in which not each other's ideas are challenged, but each other's motives. And our great parties have too often been far apart and untrusting of each other. It has been this way since Vietnam. That war cleaves us still. But, friends, that war began in earnest a quarter of a century ago; and surely the statute of limitations has been reached. This is a fact: the final lesson of Vietnam is that no great nation can long afford to be sundered by a memory. A new breeze is blowing, and the old bipartisanship must be made new again.

To my friends—and yes, I do mean friends—in the loyal opposition—

and yes, I mean loyal: I put out my hand. I am putting out my hand to you, Mr. Speaker. I am putting out my hand to you Mr. Majority Leader. For this is the thing: this is the age of the offered hand. We can't turn back clocks, and I don't want to. But when our fathers were young, Mr. Speaker, our differences ended at the water's edge. And we don't wish to turn back time, but when our mothers were young, Mr. Majority Leader, the Congress and the executive were capable of working together to produce a budget on which this nation could live. Let us negotiate soon and hard. But in the end, let us produce. The American people await action. They didn't send us here to bicker. They ask us to rise above the merely partisan. "In crucial things, unity"—and this, my friends, is crucial.

To the world, too, we offer new engagement and a renewed vow: we will stay strong to protect the peace. The "offered hand" is a reluctant fist; but once made, strong, and can be used with great effect. There are today Americans who are held against their will in foreign lands, and Americans who are unaccounted for. Assistance can be shown here, and will be long remembered. Good will begets good will. Good faith can be a spiral that endlessly moves on.

Great nations like great men must keep their word. When America says something, America means it, whether a treaty or an agreement or a vow made on marble steps. We will always try to speak clearly, for candor is a compliment, but subtlety, too, is good and has its place. While keeping our alliances and friendships around the world strong, ever strong, we will continue the new closeness with the Soviet Union, consistent both with our security and with progress. One might say that our new relationship in part reflects the triumph of hope and strength over experience. But hope is good, and so are strength and vigilance.

Here today are tens of thousands of our citizens who feel the understandable satisfaction of those who have taken part in democracy and seen their hopes fulfilled. But my thoughts have been turning the past few days to those who would be watching at home, to an older fellow who will throw a salute by himself when the flag goes by, and the woman who will tell her sons the words of the battle hymns. I don't mean this to be sentimental. I mean that on days like this, we remember that we are all part of a continuum, inescapably connected by the ties that bind.

Our children are watching in schools throughout our great land. And to them I say, thank you for watching democracy's big day. For democracy belongs to us all, and freedom is like a beautiful kite that can go higher and higher with the breeze. And to all I say: no matter what your circumstances or where you are, you are part of this day, you are part of the life of our great nation.

A President is neither prince nor pope, and I don't seek a window on men's souls. In fact, I yearn for a greater tolerance, an easygoingness about each other's attitudes and way of life.

There are few clear areas in which we as a society must rise up united and express our intolerance. The most obvious now is drugs. And when that first cocaine was smuggled in on a ship, it may as well have been a deadly bacteria, so much has it hurt the body, the soul of our country. And there is much to be done and to be said, but take my word for it: this scourge will stop.

And so, there is much to do; and tomorrow the work begins. I do not mistrust the future; I do not fear what is ahead. For our problems are large, but our heart is larger. Our challenges are great, but our will is greater. And if our flaws are endless, God's love is truly boundless.

Some see leadership as high drama, and the sound of trumpets calling, and sometimes it is that. But I see history as a book with many pages, and each day we fill a page with acts of hopefulness and meaning. The new breeze blows, a page turns, the story unfolds. And so today a chapter begins, a small and stately story of unity, diversity, and generosity—shared, and written, together.

Thank you. God bless you and God bless the United States of America.

Bill Clinton

Wednesday, January 20, 1993

*I*n 1963, the year he turned seventeen, William J. Clinton shook hands with John F. Kennedy at the White House during a reception for young men sponsored by the American Legion.

On a day not unlike that on which Kennedy took office—cold and sunny—Bill Clinton became, at the age of forty-six, the youngest President elected since Kennedy. And like JFK, Clinton saw his election as not just a change in leadership but a change of generations. His vanquished opponent, George H. W. Bush, was one of the millions of Americans who fought in World War II, but who were now in their sixties and seventies. John Kennedy was the first of the GI generation to become President; Bush would be the last.

In defeating Bush, Clinton became the first baby boomer—one of the seventy-six million children born between 1946 and 1964—to become President. He was not a veteran of the armed services; in fact, he was among those who protested against the Vietnam War. During the 1992 campaign, he admitted to smoking marijuana, although he insisted that he had never inhaled, and he was confronted with allegations of womanizing after his marriage to attorney Hillary Rodham in 1975.

His less-than-tidy youth and personal life might have doomed a politician of lesser gifts, but Clinton, who was elected governor of Arkansas in 1978 at the age of thirty-two, emerged from relative obscurity to dominate the Democratic primaries of 1992. He brought to the campaign Kennedy-like charisma, youthful energy, and a mastery of public policy. Still, he faced an uphill battle in trying to unseat the incumbent, Bush, who had successfully led a multinational war against Iraq after its dictator, Saddam Hussein, ordered an invasion of the oil-rich Persian Gulf nation of Kuwait in the late summer of 1990.

Bush was enormously popular after U.S.-led forces routed Hussein's troops in a short, decisive, and relatively pain-free war. Many leading Democrats, including Senator Al Gore of Tennessee, Senator Bill Bradley of New Jersey, and New York governor Mario Cuomo, chose not to challenge Bush, leaving the field open to lesser-known candidates like Clinton, Senator Bob Kerrey of Nebraska, former Massachusetts senator Paul Tsongas, and former California governor Jerry Brown. After fending off charges of marital infidelity, Clinton finished a surprising second to Tsongas in the New Hampshire primary, in which Tsongas was the overwhelming favorite. He went on to win the Democratic nomination with ease.

The general election campaign of 1992 was notable, not only because of the generational conflict between Bush and Clinton, but also because it was the first presidential campaign of the post–Cold War era. That conflict, which dominated U.S. politics since the administration of Harry Truman, was over. The Soviet Union collapsed in 1991, two years after the Berlin Wall fell, and along with it Soviet domination of Eastern Europe.

But at this moment of triumph, with the United States as the lone superpower in the world, the nation plunged into a steep recession that robbed Bush of the popularity he gained after the Gulf War. Unemployment neared 8 percent just as the general election campaign of 1992 got underway in earnest. Bush's foreign policy successes seemed less important than his management of the economy, giving Clinton, who had no foreign policy experience, an opportunity to focus his attention on domestic issues.

Texas businessman H. Ross Perot was the wild card in the race. Running as an independent, with the resources to finance a credible national campaign, Perot hammered away at the nation's annual budget deficit, which was approaching $300 billion. Although he offered an alternative to both Clinton and Bush, his criticism of federal spending and budgeting seemed directed more at the incumbent than at the young governor of Arkansas.

On Election Day, Clinton won easily, with 44.9 million popular votes, although he captured only 43 percent of the three-way vote. Bush had 39 million votes, just 37 percent, the worst performance by an incumbent since William Howard Taft in 1912. Perot won 19.7 million votes, just shy of 19 percent. Clinton took 370 electoral votes to Bush's 168. Perot won no electoral votes.

Inauguration Day 1993, became a generational milestone for baby boomers as one of their own replaced a man who represented the authority and power of their parents' generation. The new President played on the theme of generational politics, but not before continuing a new trend in inaugural addresses, that of thanking his predecessor for his service. He also thanked "the millions of men and women whose steadfastness and sacrifice triumphed over Depression, fascism, and communism."

Clinton's speech self-consciously played on the changes that his victory and the end of the Cold War represented. He spoke of the need to "renew America," and asserted that the "American people have summoned the change we celebrate today."

"Today," he said, "as an old order passes, the new world is more free but less stable. Communism's collapse has called forth old animosities and new dangers. Clearly America must continue to lead the world we did so much to make." With the Cold War over and the Soviet Union out of business, the United States could not afford to turn inward, Clinton said. America had done so once before, after World War I. It could not do so again.

There are traces of Kennedy's inaugural throughout Clinton's speech—in the generational theme, in his call to young Americans to a "season of service," and even in his citation from Scripture: "And let us not be weary in well-doing, for in due season, we shall reap, if we faint not." But Clinton's agenda for change, his own attempts to get the country moving again, soon faced an obstacle: the landslide election of a Republican Congress in 1994, inspired in part by public disenchantment with Clinton's plan to reform health insurance.

Bibliographic Note
Colin Campbell and Bert A. Rockman, eds., *The Clinton Legacy*
(New York: Seven Bridges Press, 2000).

My fellow citizens:

Today we celebrate the mystery of American renewal.

This ceremony is held in the depth of winter. But, by the words we speak and the faces we show the world, we force the spring. A spring reborn in the world's oldest democracy, that brings forth the vision and courage to reinvent America.

When our founders boldly declared America's independence to the world and our purposes to the Almighty, they knew that America, to endure, would have to change. Not change for change's sake, but change to preserve America's ideals: life, liberty, the pursuit of happiness. Though we march to the music of our time, our mission is timeless.

Each generation of Americans must define what it means to be an American.

On behalf of our nation, I salute my predecessor, President Bush, for his half century of service to America. And I thank the millions of men and women whose steadfastness and sacrifice triumphed over Depression, fascism, and communism.

Today, a generation raised in the shadows of the Cold War assumes new responsibilities in a world warmed by the sunshine of freedom but threatened still by ancient hatreds and new plagues.

Raised in unrivaled prosperity, we inherit an economy that is still the world's strongest, but is weakened by business failures, stagnant wages, increasing inequality, and deep divisions among our people.

When George Washington first took the oath I have just sworn to uphold, news traveled slowly across the land by horseback and across the ocean by boat. Now, the sights and sounds of this ceremony are broadcast instantaneously to billions around the world.

Communications and commerce are global; investment is mobile; technology is almost magical; and ambition for a better life is now universal. We earn our livelihood in peaceful competition with people all across the earth.

Profound and powerful forces are shaking and remaking our world, and the urgent question of our time is whether we can make change our friend and not our enemy.

This new world has already enriched the lives of millions of Americans who are able to compete and win in it. But when most people are working harder for less; when others cannot work at all; when the cost of health care devastates families and threatens to bankrupt many of our enterprises, great and small; when fear of crime robs law-abiding citizens of their freedom; and when millions of poor children cannot even imagine the lives we are calling them to lead, we have not made change our friend.

We know we have to face hard truths and take strong steps. But we have not done so. Instead, we have drifted, and that drifting has eroded our resources, fractured our economy, and shaken our confidence.

Though our challenges are fearsome, so are our strengths. And Americans have ever been a restless, questing, hopeful people. We must bring to our task today the vision and will of those who came before us.

From our Revolution to the Civil War, to the Great Depression to the civil rights movement, our people have always mustered the determination to construct from these crises the pillars of our history.

Thomas Jefferson believed that to preserve the very foundations of our nation, we would need dramatic change from time to time. Well, my fellow citizens, this is our time. Let us embrace it.

Our democracy must be not only the envy of the world but the engine of our own renewal. There is nothing wrong with America that cannot be cured by what is right with America.

And so today, we pledge an end to the era of deadlock and drift; a new season of American renewal has begun. To renew America, we must be bold. We must do what no generation has had to do before. We must invest more in our own people, in their jobs, in their future, and at the same time cut our massive debt. And we must do so in a world in which we must compete for every opportunity. It will not be easy; it will require sacrifice. But it can be done, and done fairly, not choosing sacrifice for its own sake, but for our own sake. We must provide for our nation the way a family provides for its children.

Our founders saw themselves in the light of posterity. We can do no less. Anyone who has ever watched a child's eyes wander into sleep knows what posterity is. Posterity is the world to come; the world for whom we hold our ideals, from whom we have borrowed our planet, and to whom we bear sacred responsibility. We must do what America does best: offer more opportunity to all and demand responsibility from all.

It is time to break the bad habit of expecting something for nothing, from our government or from each other. Let us all take more responsibility, not only for ourselves and our families, but for our communities and our country. To renew America, we must revitalize our democracy.

This beautiful capital, like every capital since the dawn of civilization, is often a place of intrigue and calculation. Powerful people maneuver for position and worry endlessly about who is in and who is out, who is up and who is down, forgetting those people whose toil and sweat sends us here and pays our way.

Americans deserve better, and in this city today there are people who want to do better. And so I say to all of us here, let us resolve to reform our politics, so that power and privilege no longer shout down the voice of the people. Let us put aside personal advantage so that we can feel the pain and see the promise of America. Let us resolve to make our government a place for what Franklin Roosevelt called "bold, persistent experimentation," a government for our tomorrows, not our yesterdays. Let us give this capital back to the people to whom it belongs.

To renew America, we must meet challenges abroad as well at home. There is no longer division between what is foreign and what is domestic; the world economy, the world environment, the world AIDS crisis, the world arms race: they affect us all.

Today, as an old order passes, the new world is more free but less stable. Communism's collapse has called forth old animosities and new dangers. Clearly America must continue to lead the world we did so much to make.

While America rebuilds at home, we will not shrink from the challenges, nor fail to seize the opportunities, of this new world. Together with our friends and allies, we will work to shape change, lest it engulf us.

When our vital interests are challenged, or the will and conscience of the international community is defied, we will act—with peaceful diplomacy whenever possible, with force when necessary. The brave Americans serving our nation today in the Persian Gulf, in Somalia, and wherever else they stand are testament to our resolve.

But our greatest strength is the power of our ideas, which are still new in many lands. Across the world, we see them embraced, and we rejoice. Our hopes, our hearts, our hands are with those on every continent who are building democracy and freedom. Their cause is America's cause.

The American people have summoned the change we celebrate today. You have raised your voices in an unmistakable chorus. You have cast your votes in historic numbers. And you have changed the face of Congress, the presidency, and the political process itself. Yes, you, my fellow Americans have forced the spring. Now, we must do the work the season demands.

To that work I now turn, with all the authority of my office. I ask the Congress to join with me. But no President, no Congress, no government, can undertake this mission alone. My fellow Americans, you, too, must play your part in our renewal. I challenge a new generation of young Americans to a season of service; to act on your idealism by helping troubled children, keeping company with those in need, reconnecting our torn communities. There is so much to be done; enough indeed for millions of others who are still young in spirit to give of themselves in service, too.

In serving, we recognize a simple but powerful truth, we need each other. And we must care for one another. Today, we do more than celebrate America; we rededicate ourselves to the very idea of America.

An idea born in revolution and renewed through two centuries of challenge. An idea tempered by the knowledge that, but for fate, we, the fortunate and the unfortunate, might have been each other. An idea ennobled by the faith that our nation can summon from its myriad diversity the deepest measure of unity. An idea infused with the conviction that America's long heroic journey must go forever upward.

And so, my fellow Americans, at the edge of the twenty-first century, let us begin with energy and hope, with faith and discipline, and let us work until our work is done. The scripture says, "And let us not be weary in well-doing, for in due season, we shall reap, if we faint not."

From this joyful mountaintop of celebration, we hear a call to service in the valley. We have heard the trumpets. We have changed the guard. And now, each in our way, and with God's help, we must answer the call.

Thank you, and God bless you all.

Bill Clinton

· SECOND INAUGURAL ADDRESS ·

Monday, January 20, 1997

The last inauguration of the twentieth century also was the first ever to be held on a national holiday—Martin Luther King Jr. Day. The holiday, mandated by an act of Congress in 1983 and signed by President Reagan, falls on the third Monday of January, near King's birthday, January 15. In 1997, the third Monday in January also happened to be January 20, the day Bill Clinton took the oath of office for a second time.

The holiday and the looming end of the twentieth century provided Clinton with themes for his second inaugural. He looked back at an eventful century and forward to the possibilities of the information age, but rather than hold up traditional heroes of the American narrative as examples and guides—past Presidents or other national icons—Clinton focused on the slain civil rights leader as a representative of American aspiration and ideals. "His quest is our quest: the ceaseless striving to live out our true creed," he said of King.

Clinton, coatless despite another chilly Inauguration Day, used the metaphor of a bridge to address both the divisions of race and the nation's impending journey from the twentieth to the twenty-first century. "Yes, let us build our bridge," he said. "A bridge wide enough and strong enough for every American to cross over to a blessed land of new promise."

The bridge theme was more than mere words. In the days and hours leading up to Clinton's speech, some of the quarter million people who gathered on the Washington mall for the ceremonies crossed a small-scale suspension bridge to an exhibit of new technology. There, they toured a specially constructed pavilion featuring a dozen computers that allowed visitors to send messages to Clinton via a new technology, e-mail.

The last President of the twentieth century had won reelection over the last member of the World War II generation to run for the presidency—

former senator Bob Dole of Kansas—after a tumultuous first term. Clinton's legislative agenda collapsed when he failed to persuade Congress to approve a complicated health insurance program that his wife, Hillary Rodham Clinton, oversaw. Republicans seized on the miscalculation and won the Senate and, for the first time in a half century, the House of Representatives. Newt Gingrich of Georgia, a sharp critic of the President, became Speaker of the House.

A feud between Gingrich and Clinton led to a budget impasse in late 1995, and it in turn led to a shutdown of government services in late '95 and early 1996. The embarrassing spectacle wound up damaging Gingrich more than it did Clinton, at least at first. But it was during the shutdown that the President began an extramarital affair with a White House intern named Monica Lewinsky. A second-term investigation of that affair led to Clinton's impeachment on charges that he lied to a grand jury about his relationship with the aide.

Despite his missteps, and his declaration in early 1996 that the "era of big government is over"—a phrase that did not endear him to his party's core supporters—Clinton rallied in 1996, in part by co-opting Republican issues like welfare reform. He had little trouble beating Dole, who finally won the Republican Party presidential nomination in his third try. After an unmemorable campaign, Clinton won reelection with forty-seven million popular votes, about 49 percent, while Dole won thirty-nine million. Third-party candidate Ross Perot could not repeat the success he enjoyed in 1992, taking eight million votes, or about 8 percent. The incumbent won 379 electoral votes to Dole's 159 to become the first Democratic President to win reelection since Franklin Roosevelt in 1944.

His address contrasted with his hero Kennedy's, and his own in 1993, in its humility. It contained no ambitious initiatives: rather, it envisioned a "government that is smaller, lives within its means, and does more with less." Those were the words of a chastened Democratic President who even in victory faced the prospect of a hostile Republican Congress.

Although Clinton pleaded for the nation to rally behind "one common destiny," the next four years would bring more acrimony and bitterness to Washington. Even as Clinton spoke, one of the men seated near the presidential podium, Newt Gingrich, had just admitted that he improperly used taxpayer money for political purposes. Republicans would continue to investigate alleged ethical abuses by the President, one of them centered on a land deal in Arkansas. Charges and countercharges would culminate in Clinton's impeachment in 1998 for lying about the Lewinsky matter, while Gingrich resigned his seat that same year and would later admit to having an extramarital affair while he pressed for Clinton's impeachment.

Even as Washington wallowed in tawdry scandal during Clinton's second term, overseas, Americans came under attack from a shadowy terrorist group. U.S. embassies in Kenya and Tanzania were bombed in August 1998, prompting Clinton to authorize missile attacks on terrorist training bases in Afghanistan and the Sudan. The missiles missed their presumed target, a terrorist leader whose name few Americans had ever heard, Osama bin Laden.

Bibliographic Note
Joe Klein, *The Natural: The Misunderstood Presidency of Bill Clinton*
(Broadway Books, 2002).

My fellow citizens:

At this last presidential inauguration of the twentieth century, let us lift our eyes toward the challenges that await us in the next century. It is our great good fortune that time and chance have put us not only at the edge of a new century, in a new millennium, but on the edge of a bright new prospect in human affairs, a moment that will define our course, and our character, for decades to come. We must keep our old democracy forever young. Guided by the ancient vision of a promised land, let us set our sights upon a land of new promise.

The promise of America was born in the eighteenth century out of the bold conviction that we are all created equal. It was extended and preserved in the nineteenth century, when our nation spread across the continent, saved the Union, and abolished the awful scourge of slavery.

Then, in turmoil and triumph, that promise exploded onto the world stage to make this the American century.

And what a century it has been. America became the world's mightiest industrial power; saved the world from tyranny in two world wars and a long Cold War; and time and again, reached out across the globe to millions who, like us, longed for the blessings of liberty.

Along the way, Americans produced a great middle class and security in old age; built unrivaled centers of learning and opened public schools to all; split the atom and explored the heavens; invented the computer and the microchip; and deepened the wellspring of justice by making a revolution in civil rights for African Americans and all minorities, and extending the circle of citizenship, opportunity, and dignity to women.

Now, for the third time, a new century is upon us, and another time to choose. We began the nineteenth century with a choice, to spread our nation from coast to coast. We began the twentieth century with a choice, to harness the industrial revolution to our values of free enterprise, conservation, and human decency. Those choices made all the difference.

At the dawn of the twenty-first century a free people must now choose to shape the forces of the information age and the global society, to unleash the limitless potential of all our people, and, yes, to form a more perfect union.

When last we gathered, our march to this new future seemed less certain than it does today. We vowed then to set a clear course to renew our nation.

In these four years, we have been touched by tragedy, exhilarated by challenge, strengthened by achievement. America stands alone as the world's indispensable nation. Once again, our economy is the strongest on earth. Once again, we are building stronger families, thriving communities, better educational opportunities, a cleaner environment. Problems that once seemed destined to deepen now bend to our efforts: our streets are safer and record numbers of our fellow citizens have moved from welfare to work.

And once again, we have resolved for our time a great debate over the role of government. Today we can declare: government is not the problem, and government is not the solution. We, the American people, we are the solution. Our founders understood that well and gave us a democracy strong enough to endure for centuries, flexible enough to face our common challenges and advance our common dreams in each new day.

As times change, so government must change. We need a new government for a new century—humble enough not to try to solve all our problems for us, but strong enough to give us the tools to solve our problems for ourselves; a government that is smaller, lives within its means, and does more with less. Yet where it can stand up for our values and interests in the world, and where it can give Americans the power to make a real difference in their everyday lives, government should do more, not less. The preeminent mission of our new government is to give all Americans an opportunity, not a guarantee, but a real opportunity to build better lives.

Beyond that, my fellow citizens, the future is up to us. Our founders taught us that the preservation of our liberty and our Union depends upon responsible citizenship. And we need a new sense of responsibility for a new century. There is work to do, work that government alone cannot do: teaching children to read; hiring people off welfare rolls; coming out from behind locked doors and shuttered windows to help reclaim our streets from drugs and gangs and crime; taking time out of our own lives to serve others.

Each and every one of us, in our own way, must assume personal responsibility, not only for ourselves and our families, but for our neighbors and our nation. Our greatest responsibility is to embrace a new spirit of community for a new century. For any one of us to succeed, we must succeed as one America.

The challenge of our past remains the challenge of our future, will we be one nation, one people, with one common destiny, or not? Will we all come together, or come apart?

The divide of race has been America's constant curse. And each new wave of immigrants gives new targets to old prejudices. Prejudice and contempt, cloaked in the pretense of religious or political conviction, are no different. These forces have nearly destroyed our nation in the past. They plague us still. They fuel the fanaticism of terror. And they torment the lives of millions in fractured nations all around the world.

These obsessions cripple both those who hate and, of course, those who are hated, robbing both of what they might become. We cannot, we will not, succumb to the dark impulses that lurk in the far regions of the soul everywhere. We shall overcome them. And we shall replace them with the generous spirit of a people who feel at home with one another.

Our rich texture of racial, religious, and political diversity will be a godsend in the twenty-first century. Great rewards will come to those who can live together, learn together, work together, forge new ties that bind together.

As this new era approaches, we can already see its broad outlines. Ten years ago, the Internet was the mystical province of physicists; today, it is a commonplace encyclopedia for millions of schoolchildren. Scientists now are decoding the blueprint of human life. Cures for our most feared illnesses seem close at hand.

The world is no longer divided into two hostile camps. Instead, now we are building bonds with nations that once were our adversaries. Growing connections of commerce and culture give us a chance to lift the fortunes and spirits of people the world over. And for the very first time in all of history, more people on this planet live under democracy than dictatorship.

My fellow Americans, as we look back at this remarkable century, we may ask, can we hope not just to follow, but even to surpass the achievements of the twentieth century in America and to avoid the awful bloodshed that stained its legacy? To that question, every American here and every American in our land today must answer a resounding "yes."

This is the heart of our task. With a new vision of government, a new sense of responsibility, a new spirit of community, we will sustain America's journey. The promise we sought in a new land we will find again in a land of new promise.

In this new land, education will be every citizen's most prized possession. Our schools will have the highest standards in the world, igniting the spark of possibility in the eyes of every girl and every boy. And the doors of higher education will be open to all. The knowledge and power of the information age will be within reach not just of the few, but of every classroom, every library, every child. Parents and children will have time not only to work, but to read and play together. And the plans they make at their kitchen table will be those of a better home, a better job, the certain chance to go to college.

Our streets will echo again with the laughter of our children, because no one will try to shoot them or sell them drugs anymore. Everyone who can work, will work, with today's permanent underclass part of tomorrow's growing middle class. New miracles of medicine at last will reach not only those who can claim care now, but the children and hardworking families too long denied.

We will stand mighty for peace and freedom, and maintain a strong defense against terror and destruction. Our children will sleep free from the threat of nuclear, chemical, or biological weapons. Ports and airports, farms and factories will thrive with trade and innovation and ideas. And the world's greatest democracy will lead a whole world of democracies.

Our land of new promise will be a nation that meets its obligations, a nation that balances its budget but never loses the balance of its values. A nation where our grandparents have secure retirement and health care, and their grandchildren know we have made the reforms necessary to sustain those benefits for their time. A nation that fortifies the world's most productive economy even as it protects the great natural bounty of our water, air, and majestic land.

And in this land of new promise, we will have reformed our politics so that the voice of the people will always speak louder than the din of narrow interests, regaining the participation and deserving the trust of all Americans.

Fellow citizens, let us build that America, a nation ever moving forward toward realizing the full potential of all its citizens. Prosperity and power, yes, they are important, and we must maintain them. But let us never forget: the greatest progress we have made, and the greatest progress we have yet to make, is in the human heart. In the end, all the world's wealth and a thousand armies are no match for the strength and decency of the human spirit.

Thirty-four years ago, the man whose life we celebrate today spoke to us down there, at the other end of this mall, in words that moved the conscience of a nation. Like a prophet of old, he told of his dream that one day America would rise up and treat all its citizens as equals before the law and in the heart. Martin Luther King's dream was the American Dream. His quest is our quest: the ceaseless striving to live out our true creed. Our history has been built on such dreams and labors. And by our dreams and labors we will redeem the promise of America in the twenty-first century.

To that effort I pledge all my strength and every power of my office. I ask the members of Congress here to join in that pledge. The American people returned to office a President of one party and a Congress of another. Surely, they did not do this to advance the politics of petty bickering and extreme partisanship they plainly deplore. No, they call on us instead to be repairers of the breach, and to move on with America's mission.

America demands and deserves big things from us, and nothing big ever came from being small. Let us remember the timeless wisdom of Cardinal Bernardin, when facing the end of his own life. He said, "It is wrong to waste the precious gift of time on acrimony and division."

Fellow citizens, we must not waste the precious gift of this time. For all of us are on that same journey of our lives, and our journey, too, will come to an end. But the journey of our America must go on.

And so, my fellow Americans, we must be strong, for there is much to dare. The demands of our time are great and they are different. Let us meet them with faith and courage, with patience and a grateful and happy heart. Let us shape the hope of this day into the noblest chapter in our history. Yes, let us build our bridge. A bridge wide enough and strong enough for every American to cross over to a blessed land of new promise.

May those generations whose faces we cannot yet see, whose names we may never know, say of us here that we led our beloved land into a new century with the American Dream alive for all her children; with the American promise of a more perfect union a reality for all her people; with America's bright flame of freedom spreading throughout all the world.

From the height of this place and the summit of this century, let us go forth. May God strengthen our hands for the good work ahead, and always, always bless our America.

George W. Bush

Saturday, January 20, 2001

$\mathcal{T}$he first presidential election of the twenty-first century was settled not on Election Day but weeks later, and not by voters but by members of the U.S. Supreme Court.

Beyond dispute was the popular vote: Democrat Al Gore, the sitting vice president, won 51 million votes, while Republican George W. Bush, son of former President George H. W. Bush, won 50.4 million. But the election hinged on the electoral votes in a single state, Florida, whose governor happened to be Jeb Bush, the Republican candidate's brother. The Florida result was so close, showing Bush with a lead of a few hundred votes, that a recount was necessary. It was a messy, controversial process, extending into mid-December and inspiring a flurry of lawsuits. Finally, the Supreme Court ordered an end to the recount. Bush was awarded the state's twenty-five electoral votes, and thus the presidency, by a margin of 537 popular votes.

The decision ended a campaign season that emphasized personality and leadership style more than substance and policy. Bush, a two-time governor of Texas, campaigned as a conservative who sought compassionate solutions to social ills. Gore was in the difficult position of being the heir to a President, Bill Clinton, who had been impeached, but who also remained popular. That popularity, at least among core Democrats, was due in part to the impeachment process—many Democrats believed the proceeding was an act of partisanship by a Republican-controlled Congress. By the same token, many Republicans sincerely believed that Clinton's lies under oath about his extramarital affair constituted an impeachable offense.

With the Cold War nothing more than a memory, the economy still sound despite signs of a slowdown, and few apparent crises around the world, the contest centered on issues like Bush's service, or lack thereof, in the Texas Air National Guard during the Vietnam War and perceptions that Gore was a wooden, inauthentic product of Washington politics. When the all-important electoral votes finally were tallied, Bush had 271, one more than the absolute minimum required, while Gore had 266. Gore was hurt in Florida by the presence of third-party candidate Ralph Nader, who won more than ninety-seven thousand votes in a state Gore lost by fewer than six hundred votes. In theory, many of Nader's voters would have preferred Gore to Bush. Without the chance to cast a protest vote, Nader voters might have supported Gore, and so given him the state's electoral votes and ultimate victory.

Instead, Florida was Gore's downfall. So George W. Bush became the fourth President to take the oath of office despite losing the popular vote. Among those seated near the new President as he recited the oath was his father, former President George H. W. Bush.

In an unprecedented gesture, Bush thanked not only his predecessor, Clinton, but his vanquished opponent, Gore. Both men, representing the outgoing administration, were seated near Bush for the inaugural ceremony.

Bush's acknowledgement of Gore's "spirit" during the campaign and his "grace" upon its conclusion was the only reference to the bitterness that preceded the inauguration. Bush chose not to dwell on the controversy of the Florida recount and his own status as a winner despite having lost the popular vote. Instead, he built on his campaign theme of compassionate conservatism. Poverty, he said, "is unworthy of our nation's promise. Whatever our views of its cause, we can agree that children at risk are not at fault . . . Where there is suffering, there is duty."

Eight years earlier, in his inaugural address, Bush's father had emphasized volunteerism and charity as alternatives to a bureaucratic welfare state. The younger Bush returned to that theme, emphasizing the ways in which religious communities could play a role in building a more compassionate society. "Some needs and hurts are so deep that they only respond to a mentor's touch or a pastor's prayer," he said. "Church and charity, synagogue and mosque lend our communities their humanity, and they will have an honored place in our plans and in our laws."

Unlike his father, the younger Bush had little personal experience in international relations, and the speech reflected the country's post–Cold War disinterest in events elsewhere. Bush, in fact, seemed to acknowledge the nation's outlook by warning other nations not to misjudge America's preoccupation with prosperity at home. "The enemies of liberty and our country should make no mistake," he said. "America remains engaged in the world by history and by choice, shaping a balance of power that favors freedom. We will defend our allies and our interests; we will show purpose without arrogance; we will meet aggression and bad faith with resolve and strength; and to all nations, we will speak for the values that gave our nation birth."

Bush's reference to the country's enemies came at a time when the nation's most recent antagonist, the Soviet Union, was defunct. International communism was a spent force, and the United States was the world's lone superpower. Who, then, were "the enemies of liberty and our country?"

That question was answered nine months later, on September 11, 2001.

CHIEF JUSTICE REHNQUIST, PRESIDENT CARTER, PRESIDENT BUSH,
PRESIDENT CLINTON, DISTINGUISHED GUESTS AND
MY FELLOW CITIZENS:

The peaceful transfer of authority is rare in history, yet common in our country. With a simple oath, we affirm old traditions and make new beginnings.

As I begin, I thank President Clinton for his service to our nation, and I thank Vice President Gore for a contest conducted with spirit and ended with grace.

I am honored and humbled to stand here, where so many of America's leaders have come before me, and so many will follow.

We have a place, all of us, in a long story. A story we continue, but whose end we will not see. It is the story of a new world that became a friend and liberator of the old, a story of a slaveholding society that became a servant of freedom, the story of a power that went into the world to protect but not possess, to defend but not to conquer. It is the American story. A story of flawed and fallible people, united across the generations by grand and enduring ideals. The grandest of these ideals is an unfolding American promise that everyone belongs, that everyone deserves a chance, that no insignificant person was ever born. Americans are called upon to enact this promise in our lives and in our laws; and though our nation has sometimes halted, and sometimes delayed, we must follow no other course.

Through much of the last century, America's faith in freedom and democracy was a rock in a raging sea. Now it is a seed upon the wind, taking root in many nations. Our democratic faith is more than the creed of our country, it is the inborn hope of our humanity, an ideal we carry but do not own, a trust we bear and pass along; and even after nearly 225 years, we have a long way yet to travel.

While many of our citizens prosper, others doubt the promise, even the justice, of our own country. The ambitions of some Americans are limited by failing schools and hidden prejudice and the circumstances of their birth; and sometimes our differences run so deep, it seems we share a continent, but not a country. We do not accept this, and we will not allow it. Our unity, our Union, is the serious work of leaders and citizens in every generation; and this is my solemn pledge: I will work to build a single nation of justice and opportunity. I know this is in our reach because we are guided by a power larger than ourselves who creates us equal in His image, and we are confident in principles that unite and lead us onward.

America has never been united by blood or birth or soil. We are bound by ideals that move us beyond our backgrounds, lift us above our interests, and teach us what it means to be citizens. Every child must be taught these principles. Every citizen must uphold them; and every immigrant, by embracing these ideals, makes our country more, not less, American.

Today, we affirm a new commitment to live out our nation's promise through civility, courage, compassion, and character. America, at its best, matches a commitment to principle with a concern for civility. A civil society demands from each of us good will and respect, fair dealing and forgiveness. Some seem to believe that our politics can afford to be petty because, in a time of peace, the stakes of our debates appear small. But the stakes for America are never small. If our country does not lead the cause of freedom, it will not be led. If we do not turn the hearts of children toward knowledge and character, we will lose their gifts and undermine their idealism. If we permit our economy to drift and decline, the vulnerable will suffer most. We must live up to the calling we share. Civility is not a tactic or a sentiment. It is the determined choice of trust over cynicism, of community over chaos. This commitment, if we keep it, is a way to shared accomplishment.

America, at its best, is also courageous. Our national courage has been clear in times of depression and war, when defending common dangers defined our common good. Now we must choose if the example of our fathers and mothers will inspire us or condemn us. We must show courage in a time of blessing by confronting problems instead of passing them on to future generations.

Together, we will reclaim America's schools, before ignorance and apathy claim more young lives; we will reform Social Security and Medicare, sparing our children from struggles we have the power to prevent; we will reduce taxes, to recover the momentum of our economy and reward the effort and enterprise of working Americans; we will build our defenses beyond challenge, lest weakness invite challenge; and we will confront weapons of mass destruction, so that a new century is spared new horrors.

The enemies of liberty and our country should make no mistake, America remains engaged in the world by history and by choice, shaping a balance of power that favors freedom. We will defend our allies and our interests; we will show purpose without arrogance; we will meet aggression and bad faith with resolve and strength; and to all nations we will speak for the values that gave our nation birth.

America, at its best, is compassionate. In the quiet of American conscience, we know that deep, persistent poverty is unworthy of our nation's promise. Whatever our views of its cause, we can agree that children at risk are not at fault. Abandonment and abuse are not acts of God, they are failures of love. The proliferation of prisons, however necessary, is no substitute for hope and order in our souls. Where there is suffering, there is duty. Americans in need are not strangers, they are citizens, not problems, but priorities, and all of us are diminished when any are hopeless. Government has great responsibilities for public safety and public health, for civil rights and common schools. Yet compassion is the work of a nation, not just a government. Some needs and hurts are so deep they will only respond to a mentor's touch or a pastor's prayer. Church and charity, synagogue and mosque lend our communities their humanity, and they will have an honored place in our plans and in our laws. Many in our country do not know the pain of poverty, but we can listen to those who do. I can pledge our nation to a goal: When we see that wounded traveler on the road to Jericho, we will not pass to the other side.

America, at its best, is a place where personal responsibility is valued and expected. Encouraging responsibility is not a search for scapegoats, it is a call to conscience. Though it requires sacrifice, it brings a deeper fulfillment. We find the fullness of life not only in options but in commitments. We find that children and community are the commitments that set us free. Our public interest depends on private character, on civic duty and family bonds and basic fairness, on uncounted, unhonored acts of decency which give direction to our freedom. Sometimes in life we are called to do great things. But as a saint of our times has said, every day we are called to do small things with great love. The most important tasks of a democracy are done by everyone. I will live and lead by these principles, to advance my convictions with civility, to pursue the public interest with courage, to speak for greater justice and compassion, to call for responsibility and try to live it as well. In all of these ways, I will bring the values of our history to the care of our times.

What you do is as important as anything government does. I ask you to seek a common good beyond your comfort; to defend needed reforms against easy attacks; to serve your nation, beginning with your neighbor. I ask you to be citizens. Citizens, not spectators; citizens, not subjects; responsible citizens, building communities of service and a nation of character.

Americans are generous and strong and decent, not because we believe in ourselves, but because we hold beliefs beyond ourselves. When this spirit of citizenship is missing, no government program can replace it. When this spirit is present, no wrong can stand against it.

After the Declaration of Independence was signed, Virginia statesman John Page wrote to Thomas Jefferson, "We know the race is not to the swift nor the battle to the strong. Do you not think an angel rides in the whirlwind and directs this storm?" Much time has passed since Jefferson arrived for his inauguration. The years and changes accumulate, but the themes of this day he would know, "our nation's grand story of courage and its simple dream of dignity."

We are not this story's author, who fills time and eternity with His purpose. Yet His purpose is achieved in our duty, and our duty is fulfilled in service to one another. Never tiring, never yielding, never finishing, we renew that purpose today: to make our country more just and generous; to affirm the dignity of our lives and every life.

This work continues. This story goes on. And an angel still rides in the whirlwind and directs this storm.

God bless you all, and God bless America.

George W. Bush

· SECOND INAUGURAL ADDRESS ·

Thursday, January 20, 2005

$\mathscr{I}$mages of an embattled America returned to inaugural rhetoric on January 20, 2005, as George W. Bush took the oath of office for the second time.

Terrorist attacks in September 2001 killed three thousand people, brought down the World Trade Center in downtown Manhattan, and damaged a portion of the Pentagon. The catastrophe brought an end to the post–Cold War era of reduced international tensions and the illusion that great ideological conflicts belonged to the past. Having faced and defeated fascism and communism, the United States now confronted a global insurrection led by Islamic militants who sought to purge the Muslim world of Western influences.

The attacks of September 11 rallied the country around President Bush, whose first several months in office had been unsteady and tentative. As smoke filled the skies over Manhattan, turning New York City into a war zone, American diplomats built an international case for invading Afghanistan, which harbored the terrorist group, al Qaeda, that launched the attacks on American soil. A U.S.-led international coalition forcibly removed the Afghan regime from power and sent al Qaeda's leadership into hiding. But the well-coordinated campaign failed to capture the group's leader, Osama bin Laden.

Victory in Afghanistan boosted Bush's popularity and emboldened members of his administration to push for a U.S. invasion of Iraq to remove Saddam Hussein from power. The administration argued that Iraq possessed weapons of mass destruction that might easily be passed to al Qaeda and so presented a danger to American national security. In March 2003, the Bush administration ordered an invasion of Iraq as part of its larger war on terrorism. Hussein was driven from power and was later captured and executed by a new Iraqi government.

But by the campaign season of 2004, the triumphant victory over Hussein had given way to the bitterness of a bloody occupation. American and coalition forces were challenged by private militias and loosely organized guerillas. Rival sects and tribes engaged in a virtual civil war. The President's poll ratings began to decline. Hussein's weapons of mass destruction were never found, although their existence was the putative reason for the U.S.-led invasion.

Sensing Bush's vulnerability, the Democrats nominated John Kerry, a senator from Massachusetts, to challenge the incumbent in 2004. Kerry was a Vietnam War veteran who turned against the war upon his return home. Ironically, though, as a presidential candidate, Kerry highlighted his service in Vietnam, implicitly contrasting his record with Bush, who served in the Air National Guard at home during the war.

While Kerry focused on the quagmire in Iraq, Bush asked the nation for patience in a larger, global conflict he said would last for many years. Despite his drop in popularity since late 2001, Bush still had the aura of a wartime President. He won reelection with sixty-two million votes and 286 electoral votes to Kerry's fifty-nine million and 251, respectively.

Bush's second inaugural reflected the nation's changed outlook since his inward-looking first inaugural in January, 2001. "At this second gathering, our duties are defined not by the words I use, but by the history we have seen together," he said. "We have seen our vulnerability—and we have seen its deepest source. For as long as whole regions of the world simmer in resentment and tyranny—prone to ideologies that feed hatred and excuse murder—violence will gather, and multiply in destructive power, and cross the most defended borders, and raise a mortal threat."

Bush announced that his "most solemn duty" would be to "protect this nation and its people against further attacks and emerging threats."

So, after an absence of perhaps a dozen years, conflict and ideological struggle returned as important themes in an inaugural address. Bush framed the new conflict as his predecessors had framed the old one: as a challenge to liberty and freedom, and a choice between good and evil.

George W. Bush's successor may, or may not, differ on matters of policy and approach. But it seems certain that regardless of who takes the oath of office in 2009, the war on terrorism will define his or her inaugural address, and perhaps the inaugural speeches of an entire generation.

VICE PRESIDENT CHENEY, MR. CHIEF JUSTICE,
PRESIDENT CARTER, PRESIDENT BUSH, PRESIDENT CLINTON,
REVEREND CLERGY, DISTINGUISHED GUESTS,
FELLOW CITIZENS:

On this day, prescribed by law and marked by ceremony, we celebrate the durable wisdom of our Constitution and recall the deep commitments that unite our country. I am grateful for the honor of this hour, mindful of the consequential times in which we live, and determined to fulfill the oath that I have sworn and you have witnessed.

At this second gathering, our duties are defined not by the words I use, but by the history we have seen together. For a half century, America defended our own freedom by standing watch on distant borders. After the shipwreck of communism came years of relative quiet, years of repose, years of sabbatical—and then there came a day of fire.

We have seen our vulnerability—and we have seen its deepest source. For as long as whole regions of the world simmer in resentment and tyranny—prone to ideologies that feed hatred and excuse murder—violence will gather, and multiply in destructive power, and cross the most defended borders, and raise a mortal threat. There is only one force of history that can break the reign of hatred and resentment, and expose the pretensions of tyrants, and reward the hopes of the decent and tolerant, and that is the force of human freedom.

We are led, by events and common sense, to one conclusion: the survival of liberty in our land increasingly depends on the success of liberty in other lands. The best hope for peace in our world is the expansion of freedom in all the world.

America's vital interests and our deepest beliefs are now one. From the day of our founding, we have proclaimed that every man and woman on this earth has rights, and dignity, and matchless value because they bear the image of the Maker of heaven and earth. Across the generations we have proclaimed the imperative of self-government, because no one is fit to be a master, and no one deserves to be a slave. Advancing these ideals is the mission that created our nation. It is the honorable achievement of our fathers. Now it is the urgent requirement of our nation's security, and the calling of our time.

So it is the policy of the United States to seek and support the growth of democratic movements and institutions in every nation and culture, with the ultimate goal of ending tyranny in our world.

This is not primarily the task of arms, though we will defend ourselves and our friends by force of arms when necessary. Freedom, by its nature, must be chosen, and defended by citizens, and sustained by the rule of law and the protection of minorities. And when the soul of a nation finally speaks, the institutions that arise may reflect customs and traditions very different from our own. America will not impose our own style of government on the unwilling. Our goal instead is to help others find their own voice, attain their own freedom, and make their own way.

The great objective of ending tyranny is the concentrated work of generations. The difficulty of the task is no excuse for avoiding it. America's influence is not unlimited, but fortunately for the oppressed, America's influence is considerable, and we will use it confidently in freedom's cause.

My most solemn duty is to protect this nation and its people against further attacks and emerging threats. Some have unwisely chosen to test America's resolve and have found it firm.

We will persistently clarify the choice before every ruler and every nation: the moral choice between oppression, which is always wrong, and freedom, which is eternally right. America will not pretend that jailed dissidents prefer their chains, or that women welcome humiliation and servitude, or that any human being aspires to live at the mercy of bullies.

We will encourage reform in other governments by making clear that success in our relations will require the decent treatment of their own people. America's belief in human dignity will guide our policies, yet rights must be more than the grudging concessions of dictators; they are secured by free dissent and the participation of the governed. In the long run, there is no justice without freedom, and there can be no human rights without human liberty.

Some, I know, have questioned the global appeal of liberty—though this time in history, four decades defined by the swiftest advance of freedom ever seen, is an odd time for doubt. Americans, of all people, should never be surprised by the power of our ideals. Eventually, the call of freedom comes to every mind and every soul. We do not accept the existence of permanent tyranny because we do not accept the possibility of permanent slavery. Liberty will come to those who love it.

Today, America speaks anew to the peoples of the world:

All who live in tyranny and hopelessness can know: the United States will not ignore your oppression or excuse your oppressors. When you stand for your liberty, we will stand with you.

Democratic reformers facing repression, prison, or exile can know: America sees you for who you are, the future leaders of your free country.

The rulers of outlaw regimes can know that we still believe as Abraham Lincoln did: "Those who deny freedom to others deserve it not for themselves; and, under the rule of a just God, cannot long retain it."

The leaders of governments with long habits of control need to know: to serve your people you must learn to trust them. Start on this journey of progress and justice, and America will walk at your side.

And all the allies of the United States can know: we honor your friendship, we rely on your counsel, and we depend on your help. Division among free nations is a primary goal of freedom's enemies. The concerted effort of free nations to promote democracy is a prelude to our enemies' defeat.

Today, I also speak anew to my fellow citizens:

From all of you, I have asked patience in the hard task of securing America, which you have granted in good measure. Our country has accepted obligations that are difficult to fulfill and would be dishonorable to abandon. Yet because we have acted in the great liberating tradition of this nation, tens of millions have achieved their freedom. And as hope kindles hope, millions more will find it. By our efforts, we have lit a fire as well—a fire in the minds of men. It warms those who feel its power, it burns those who fight its progress, and one day this untamed fire of freedom will reach the darkest corners of our world.

A few Americans have accepted the hardest duties in this cause: in the quiet work of intelligence and diplomacy, the idealistic work of helping raise up free governments, the dangerous and necessary work of fighting our enemies. Some have shown their devotion to our country in deaths that honored their whole lives—and we will always honor their names and their sacrifice.

All Americans have witnessed this idealism, and some for the first time. I ask our youngest citizens to believe the evidence of your eyes. You have seen duty and allegiance in the determined faces of our soldiers. You have seen that life is fragile, and evil is real, and courage triumphs. Make the choice to serve in a cause larger than your wants, larger than yourself—

and in your days you will add not just to the wealth of our country but to its character.

America has need of idealism and courage because we have essential work at home—the unfinished work of American freedom. In a world moving toward liberty, we are determined to show the meaning and promise of liberty.

In America's ideal of freedom, citizens find the dignity and security of economic independence instead of laboring on the edge of subsistence. This is the broader definition of liberty that motivated the Homestead Act, the Social Security Act, and the GI Bill of Rights. And now we will extend this vision by reforming great institutions to serve the needs of our time. To give every American a stake in the promise and future of our country, we will bring the highest standards to our schools and build an ownership society. We will widen the ownership of homes and businesses, retirement savings, and health insurance—preparing our people for the challenges of life in a free society. By making every citizen an agent of his or her own destiny, we will give our fellow Americans greater freedom from want and fear, and make our society more prosperous and just and equal.

In America's ideal of freedom, the public interest depends on private character—on integrity, and tolerance toward others, and the rule of conscience in our own lives. Self-government relies, in the end, on the governing of the self. That edifice of character is built in families, supported by communities with standards, and sustained in our national life by the truths of Sinai, the Sermon on the Mount, the words of the Koran, and the varied faiths of our people. Americans move forward in every generation by reaffirming all that is good and true that came before—ideals of justice and conduct that are the same yesterday, today, and forever.

In America's ideal of freedom, the exercise of rights is ennobled by service, and mercy, and a heart for the weak. Liberty for all does not mean independence from one another. Our nation relies on men and women who look after a neighbor and surround the lost with love. Americans, at our best, value the life we see in one another and must always remember that even the unwanted have worth. And our country must abandon all the habits of racism, because we cannot carry the message of freedom and the baggage of bigotry at the same time.

From the perspective of a single day, including this day of dedication, the issues and questions before our country are many. From the viewpoint of centuries, the questions that come to us are narrowed and few. Did our generation advance the cause of freedom? And did our character bring credit to that cause?

These questions that judge us also unite us, because Americans of every party and background, Americans by choice and by birth, are bound to one another in the cause of freedom. We have known divisions, which must be healed to move forward in great purposes, and I will strive in good faith to heal them. Yet those divisions do not define America. We felt the unity and fellowship of our nation when freedom came under attack, and our response came like a single hand over a single heart. And we can feel that same unity and pride whenever America acts for good, and the victims of disaster are given hope, and the unjust encounter justice, and the captives are set free.

We go forward with complete confidence in the eventual triumph of freedom. Not because history runs on the wheels of inevitability; it is human choices that move events. Not because we consider ourselves a chosen nation; God moves and chooses as He wills. We have confidence because freedom is the permanent hope of mankind, the hunger in dark places, the longing of the soul. When our founders declared a new order of the ages; when soldiers died in wave upon wave for a union based on liberty; when citizens marched in peaceful outrage under the banner Freedom Now—they were acting on an ancient hope that is meant to be fulfilled. History has an ebb and flow of justice, but history also has a visible direction, set by liberty and the Author of Liberty.

When the Declaration of Independence was first read in public and the Liberty Bell was sounded in celebration, a witness said, "It rang as if it meant something." In our time it means something still. America, in this young century, proclaims liberty throughout all the world and to all the inhabitants thereof. Renewed in our strength—tested, but not weary—we are ready for the greatest achievements in the history of freedom.

May God bless you, and may He watch over the United States of America.